Non-Connected Convexities and Applications

Applied Optimization

Volume 68

The titles published in this series are listed at the end of this volume.

Non-Connected Convexities and Applications

by

Gabriela Cristescu

Aurel Vlaicu University of Arad,
Arad, România

and

Liana Lupşa

Babeş-Bolyai University of Cluj-Napoca,
Cluj-Napoca, România

SPRINGER-SCIENCE+BUSINESS MEDIA, B.V.

A C.I.P. Catalogue record for this book is available from the Library of Congress.

ISBN 978-1-4613-4881-8 ISBN 978-1-4615-0003-2 (eBook)
DOI 10.1007/978-1-4615-0003-2

Printed on acid-free paper

This book is dedicated to our families

CLEOPATRA and MIHAI CRISTESCU
and
NICOLAE, LUCIANA and RADU LUPŞA

Table of Contents

Preface

Lectori salutem! The kind reader opens the book that its authors would have liked to read it themselves, but it was not written yet. Then, their only choice was to write this book, to fill a gap in the mathematical literature.

The idea of convexity has appeared in the human mind since the antiquity and its fertility has led to a huge diversity of notions and of applications. A student intending a thoroughgoing study of convexity has the sensation of swimming into an ocean. It is due to two reasons: the first one is the great number of properties and applications of the classical convexity and second one is the great number of generalisations for various purposes. As a consequence, a tendency of writing huge books guiding the reader in convexity appeared during the last twenty years (for example, the books of P.M. Gruber and J.M. Willis (1993) and R.J. Webster (1994)). Another last years' tendency is to order, from some point of view, as many convexity notions as possible (for example, the book of I. Singer (1997)). These approaches to the domain of convexity follow the previous point of view of axiomatizing it (A. Ghika (1955), W. Prenowitz (1961), D. Voiculescu (1967), V.W. Bryant and R.J. Webster (1969)).

Following this last tendency, our book proposes to the reader two classifications of convexity properties for sets, both of them starting from the internal mechanism of defining them. The objects of our classification study are in our collection, which contains 100 distinct notions of convexity for sets. Obviously, one of their common qualities is that all of them include, as a particular case, the classical convexity. The first classification is presented in chapter one. The classifier of this chapter is the method of obtaining new convexity concepts, which, eventually, do not preserve the connectivity, starting from the classical convexity and its properties. The second classification presented in Chapter 8, starts with the identification of the common key intimate elements of all 100 convexities included in our collection. All the possible manners of using the key elements to build a definition are discussed, resulting in a general method of defining the convexity for sets. The general definitions obtained here are both classifiers and new directions to introduce concepts of convexity. The clustering process is described in Chapter 8, having as a result 11 classes of convexities for sets. But, a careful analyse of the general notions lead to the remark that there are definitions, which do not represent a class of existing convexities (in our collection), proving that our study opens new ways to obtain new and useful convexity properties for sets.

Chapters 1 and 8 have as purpose to close the process of ordering the enormously extended domain of convexity properties for sets and to transform the ocean in which each student or researcher usually swims in the stage of learning the convexity into a mild river. But, in fact, our classifications might be just the beginning or an intermediate stage. Of course, they are perfectible because they are based only on our collection, which is not exhaustive. But we are convinced that they contain enough resources to include the majority (perhaps all) of the remained convexities.

The most properties and applications of classically convex sets due to their behaviour of imitating at a reduced scale (between the members of each pair of its points) the structure of the real interval [0, 1]. As a consequence, the connectivity of a convex set is one of its most important sources of applications. But the generalisations of the classical convexity easily break the connectivity. The concepts of convexity that do not imply the connectivity are called non-connected convexities. The possible ways to obtain a definition of a non-connected convexity are discussed in Chapter 1.

Chapters 2 –7 present our own non-connected convexity concepts. The framework for most of them is a space, in which a notion of straight-line segment is defined. The convexity notions, which take into account that sometimes it is necessary to look into a set through another one (a network, a grid, etc.) are studied in chapters 2, 4, 5, 6. These situations first occured in integer and in rational programming and later in image analysis. The convexities assuming that another condition besides that of belonging to the set is supposed to be satisfied by the straight-line segment are studied in chapters 3 and 5. In order to define the convexities from chapters 3, 4 and 5, points outside the straight-line segment are considered in relation to the given set. Chapter 6 presents the manner in which it is possible to induce a convexity in a new space starting from an existing convexity in a known space. Chapter 7 discusses about the convexities obtained through the modification of the convex combination condition from the definition of the straight-line segment by placing its end points in another context, using a transformation of the space.

The second part of the book deals with applications. Chapter 9 emphasises the way in which the convexities from Chapter 5 are involved in applications in image analysis. Chapter 10 establishes alternative theorems. They refer to the situation when an integer convex polytope exists and another is empty, or there are two integer convex polytopes with a relationship between them. The last paragraph of Chapter 10 presents some applications of alternative theorems in duality theory for integer programming problems.

Almost all concepts of convexity for sets have related notions of convexity for functions. The behaviour of the graph of such a function or of a region of the space related to this graph identifies the class of convex functions. Even if the aim of this book is not to study convexities for functions, we dedicate Chapter 11 to some classes of convex functions related to some above discussed convexities for sets. They prepare the results contained in the chapters dealing with applications in mathematical programming. Also, we give some interesting models of construction of some generalisations of convex functions using divided differences.

The notion of d-base, discussed in Chapter 12, permits the spotlight of necessary conditions or sufficient conditions or necessary and sufficient conditions for a feasible solution of mathematical programming problem to be an optimal solution. Duality theorems for discrete programming problem, where the feasible solution is an integer convex discrete polyhedral set are proved.

Chapter 13 contains applications of discrete programming in various problems arising in pharmaco-economics, but they are also useful in sociology. They are based on the results of Luciana Lupşa, obtained and experienced within the Research Department of “I. Chiricuţă” Oncological Institute of Cluj-Napoca. More approaches of multiple-criteria optimisation problems in pharmaco-economics are presented: the choice of the min-efficient elements from an admissible set, the construction of an effectiveness index to characterise a decision, the construction of a preference relation on a set of points. Each element (point, decision) means a drug (treatment protocol, vaccine) or social policy. Under this variety of techniques providing the user (doctor, hospital, and pharmacy) with information on the effectiveness of their treatment methods or policies, an ethical debate is hidden. It refers both to researcher’s approach to such a problem and to user’s attitude concerning the practical application of the information provided by a study of efficiency. The methods determining the min-efficient points are dictatorial, offering only the solution, which is the best according to a certain set of criteria. The methods of preference relations or of effectiveness indexes allow an interactive approach by taking into account more new criteria related to patient’s particularities. The same mathematical methods might be used in sociology, to solve problems related to efficiency evaluation of a social policy within a population.

We are happy to write, in the context of the twentieth century research in convexity for sets, about the results of the members of the Convexity School developed at Babeş- Bolyai University of Cluj-Napoca (România) under the leadership of Academician Dr. Tiberiu Popoviciu and Prof. Dr. Doc. Elena Popoviciu. We grew up in this school, here we started to learn about convexity and we published our first papers. I. Singer, W. Breckner, A. Aleman, I. Kolumbán, D. Duca, E. Duca, L. Blaga, G. Toader grew up in the environment provided by the same research school. Their names are related to generalised convexity notions for sets and applications.

Our intention was to write not only a self-contained book, but also a book with self-contained chapters. Our achievement is a self-contained book with enough autonomous chapters to allow both a partial reading according to certain purposes and an integral reading.

This book is useful for good students and young researchers that begin the study of convexity. It is of help to all researchers working in convexity and its applications. The non-connected convexity extends the domain of applications of convexity techniques to integer, rational, discrete programming, or to more types of non-convex optimisation. Therefore, it is useful for decision-makers or researchers in the fields of decision aid, wherever it appears in economics, in ecology, in social sciences, in management, in engineering, a.s.o.

The Authors

Acknowledgements

All our gratitude is due to God for making us meet, becoming friends, cooperating in our research work and opening our eyes and our minds, enabling us to understand a drop of the secrets and the beauties of our World.

Our greatest thanks are due to our mentor, distinguished Prof. dr. doc. Elena Popoviciu, who discovered our special skills for scientific research since we were students. She introduced us, during the research seminar in Mathematical Analysis for students that she organized and supervised, in the rich field of convexity. We are grateful to our colleagues Lucia Blaga, Wolfgang Breckner, Ştefan Cobzaş, Dorel Duca, Eugenia Duca, Gheorghe Toader for the helpful discussions concerning their results in convexity and their connections.

We are delighted to place on record our gratitude to Marilyn Breen, Jean Marc Chassery, Luciana Lupşa and Azriel Rosenfeld who kindly allowed us to quote, in extenso, their results in convexity and their applications in our book. Also, we direct our gratitude to PAMM – Centre of the Technical University of Budapest for allowing us to quote some of our results in the fields of convexity that appeared first at PAMM – Centre.

We are enormously grateful to our good colleague and friend Otilia Huţiu for carefully reading the manuscript and for her precious remarks.

Our gratitude is directed to our mathematics teachers who encouraged our efforts to understand and to learn mathematics during secondary school and high school: Ştefan Dârţu, Atanasie Florescu, Aurel Feldiorean and Vasile Paşcu.

The elaboration of this book would not have been possible without the continuous support of our families, Cleopatra and Mihai Cristescu and Nicolae, Luciana and Radu Lupşa. We are enormously grateful for their commitment and for their effective daily help.

We are grateful to Kluwer Academic Publishers for publishing this book and especially to John Martindale and Angela Quilici for their prompt, useful and friendly advice during the preparation of the manuscript and for the moral support.

Counting on the interest of kind readers, we say usually: *Lectori salutem!*

The Authors

Main notations

$\mathbf{C}$	set of complex numbers
$\mathbf{R}$	set of real numbers
$\mathbf{Q}$	set of rational numbers
$\mathbf{Z}$	set of integer numbers
$\mathbf{N}$	set of natural numbers
$\mathbf{R}_+$ ($\mathbf{Q}_+$, $\mathbf{Z}_+$)	set of non-negative real (rational, integer) numbers
$\mathbf{R}_-$ ($\mathbf{Q}_-$, $\mathbf{Z}_-$)	set of non-positive real (rational, integer) numbers
$R_+^*(Q_+^*, Z_+^*)$	set of positive real (rational, integer) numbers
$R_-^*(Q_-^*, Z_-^*)$	set of negative real (rational, integer) numbers
A, B, X_1, X_2, …	sets
a, b, …	elements of a set
{a, b, c, …}	set of distinct elements a, b, c, …
2^X	set of all subsets of X
$\varnothing$	empty set
$\overline{A}$	closure of set A
Int (A)	interior of set A
Ext (A)	exterior of set A
Fr (A)	boundary of set A
sp (A)	linear hull of A
aff (A)	affine hull of A

conv (A)	convex hull of A
$\cup$	union of sets
$\cap$	intersection of sets
$\neq$	not equal
$\backslash$	difference between two sets
[x]	integer part of x
]x[	fractional part of x
$\lvert x \rvert$	absolute value of x
$\lVert x \rVert$	norm or length of x
$\mathbf{R}^n$	n-dimensional Euclidean space
$M_{m \times n}(X)$	set of m×n matrices with elements from X
A^T	transpose of matrix (or vector) A
dim (X)	dimension of set X
d (a, b)	distance between elements a and b
c (a, b)	Chebyshev distance between a and b
$<x, y>$	straight-line segment between x and y
$<x, y$	half-line with origin in x and containing y
[a, b]	straight-line segment in R (closed interval)
(x | y)	scalar product of elements x and y
$x < y, x > y$	strict inequalities between x and y
$x \leq y, x \geq y$	non-strict inequalities between x and y (coordinatewise defined in $\mathbf{R}^n$)
$x \leqq y, x \geqq y$	non-strict inequalities between x and y, coordinatewise defined in $\mathbf{R}^n$, such that there is a coordinate for which the inequality is strict
$pr_j(X)$	projection of set X on the j-th axis

Part 1. Non-connected convexity properties

Chapter 1

The fields of non-connected convexity properties

The historical investigation on the convexity leads to contributions from the antiquity, even if the systematical study of the domain started by the end of the XIX-th century only. The great number of applications that are consequences of the presence of a convexity property resulted in an amazing diversification of the domain (see the monographs and the collections of papers of S. Schaible, W.T. Ziemba (1981), J. M. Chassery (1984 [43]), V.P. Soltan (1984), M. Avriel, W.E. Diewert, S. Schaible, I. Zang (1988), P. M. Gruber and J. M. Willis (1993), I. Singer (1997), G. Cristescu (2000 [71])). Many directions of generalisation of the convexity appeared, some of them allowing an axiomatic approach, as it can be seen in the above mentioned books. Others apply to sets with less geometrical properties, for example not connected sets from the necessity of extending the tools provided by the convexity to more general frameworks.

This chapter contains an investigation of the property of convexity for sets in order to identify the ways in which it is possible to obtain new convexities that do not imply the connectivity. Our study follows both the historical approach and thematically presents the directions of developing various non-connected convexity theories.

The starting point of our research is in the definition of the classical convexity for sets, as it appeared in the papers of H. Brunn (1887), H. Minkowski (1911). We investigate the origin of the connectivity of the convex sets, in the classical sense, in order to identify the intimate mechanism that allows extending the tools of convexity to sets that are not connected. In fact, we point out the main directions of generalisation of the convexity, which avoid at least one of the elements that lead to connectivity. For each direction of generalisation, we discuss shape properties of the new type of convex sets in the direction of connectivity or lack of connectivity.

1.1 Classical convexity for sets and the connectivity

The roots of the notion of convex set may be found in antiquity. Many results of Euclid (330-275 BC) and Archimedes (287?-212 BC) refer to properties and remarks on various particular convex sets: straight-line segments, half-lines, straight-lines, half-planes, planes, disks, ellipses, and polygons. The study of sets having some properties found at these particular sets have been enriched with the approach of classes of sets characterised in a precise manner (for example sets of points having tangent at each point of their frontier, sets of points that may be separated by hyperplanes) until the end of the XIXth century. H. Brunn (1887) has formulated for the first time the definition of the property of convexity for sets (we shall refer to it as classical convexity). Using an algebraic definition, H. Minkowski (1911) characterised the classical convexity in the framework provided by the linear structure of the space in which is placed the study.

The classical framework for the convexity is a nonempty set X in which the notion of straight-line segment is defined. The following definition will introduce the notion, which we shall refer to as the classical convexity.

Definition 1.1.1. *A subset* A *of* X *is* ***convex*** *if whenever it contains two points, it also contains the straight-line segment joining them.*

Therefore, in a convex set, for each two points $x \in A$ and $y \in A$, the straight-line segment joining them, denoted by $<x, y>$, is also contained in A. In a real linear space, $(X, +, .\mathbf{R})$, the straight-line segment joining the points x and y is defined by $<x, y> = \{z \in X \mid z = t\,x + (1-t)\,y,\ 0 \leq t \leq 1\}$. From this definition we see that the straight-line segment is a set isomorphic to interval [0, 1], the isomorphism being $s: [0, 1] \to <x, y>$ defined by $s(t) = t\,x + (1-t)\,y$. It means that the straight-line segment imitates the real interval [0, 1], implying the connectivity of a classically convex set.

Before discussing on the connectivity of convex sets let us recall the concept of connectivity.

In a topological space (X, T), two sets $A \subset X$ and $B \subset X$ are separated by open sets if there are two disjoint open sets M and N such that $A \subset M$ and $B \subset N$. The sets $A \subset X$ and $B \subset X$ are separated by a continuous function if there is a continuous function $f: X \to \mathbf{R}$ such that $f(a) = 0$ for $a \in A$ and $f(a) = 1$ for $a \in B$. A set $G \subseteq X$ is *connected* if there are not two separated sets $A \subset G$ and $B \subset G$ such that $G = A \cup B$.

In a nonempty set X, in which a notion of segment (or arc) determined by a pair of points is defined for certain pair of points of X, for each two points $a, b \in X$, the segment (arc) $<a, b>$ is the standard connected element of the space. A chain in X is a succession of segments (or arcs) $<a_1, a_2>$, $<a_2, a_3>$,

..., $<a_{n-1}, a_n>$ of X. A set $A \subset X$ is said to be connected by arcs if for each pair of points $a, b \in A$ there is a chain such that $<a, a_1> \subset A$, $<a_1, a_2> \subset A$, ..., $<a_k, a_{k+1}> \subset A$, ..., $<a_n, b> \subset A$.

In fact, the classical convexity is defined by the condition that a standard convex element, namely the straight-line segment joining two points, belongs to the set under investigation, in all its possible appearances in this set. Therefore, the connectivity of a convex set depends on the connectivity of the straight-line segment involved in its definition.

The connectivity of the straight-line segment depends, first of all, on the definition of the connectivity in the space X. If the connectivity is defined by means of the straight-line segment (the usual arcwise connectivity) then the straight-line segment itself is connected and, therefore, each convex set is connected. If the connectivity is defined by means of a topological structure of the space, then the straight-line segment might be not connected. In this situation it is possible to obtain a convexity notion that does not imply the connectivity. The connectivity and the convexity of the straight-line segment were thoroughly discussed by L.M. Blumenthal (1953).

1.2 Axiomatic convexity

The first attempt to do an axiomatic approach to the convexity for sets seems to belong to G. Birkhoff (1948). The most general notion of convexity was defined by D.C. Kay and E.W. Womble (1971), within the most general framework called a convexity space.

Definition 1.2.1. *A convexity space is a pair* (X, C) *consisting of a nonempty set* X *and a family* C *of subsets of* X, *which:*

1. *contains* X *and the empty set* $\varnothing$;
2. *is closed under arbitrary intersection.*

Members of C are called *convex*. The convex hull operator on 2^X generated by C is defined by $\mathsf{C}(A) = \cap \{C \in \mathsf{C} \mid A \subset C\}$, for each $A \subset X$. The possibility of linearization of the spaces having this structure have been studied by R. Precup (1980).

A similar approach is due to K. Keimel and A. Wieczorek (1988), and A. Wieczorek (1992), in the framework offered by a topological space X. A family of closed sets $\mathsf{B} \subset 2^X$ that is stable under arbitrary intersections is called *a convexity on* X. The sets $B \in \mathsf{B}$ are *closed convex sets.*

The general idea is to define a space in which there is a notion of joining two points, called join space.

Generally, a join space is a nonempty set X, in which two binary operations are defined:

$$\bullet: X^2 \to 2^X \text{ and } /: X^2 \to 2^X,$$

subjected to some axioms. The straight-line segment is defined by the first operation, $<x, y> = x \bullet y$, for $x \in X$ and $y \in X$. Both operations are extended to operations between subsets of X as it follows: if $A, B \subseteq X$ then

$$A \bullet B = \cup\{a \bullet b \mid a \in A, b \in B\},$$

$$A / B = \cup\{a / b \mid a \in A, b \in B\}.$$

Then, a set $A \subseteq X$ is said to be convex if $A \bullet A \subseteq A$. This definition is equivalent to the following statement: a set $A \subseteq X$ is said to be *convex* if for each pair of points $x \in A$ and $y \in A$, the subset $<x, y>$ is included in A.

Various authors tried to elaborate axiomatic systems for these operations in order to obtain spaces in which more classical geometric and topologic properties are retrieved. We shall quote some of them in this paragraph and other in chapter 8 of this book.

A. Ghika (1955) defined the *lineat non-vectorial space* that includes more sets in which it is not possible to define a linear structure as, for example, the hyperbolic space of Lobacevski and the space of Riemann. But in these spaces it is possible to extend the notion of segment and of convexity and also to have separation properties for convex sets. Their topology, of a very general type, allows the extension of some non-linear functionals. The lineat non-vectorial space is a join space having the operations described as it follows.

Let E be a nonempty set and C (E) a set of parts of E, containing both $\varnothing$ and E, ordered by inclusion, satisfying the conditions: the supremum of the sets $X, Y \in$ C (E) is denoted by $X \vee Y$ (and if $x, y \in E$ the notation $x \vee y$ will simplify $\{x\} \vee \{y\}$) and satisfies

(L1) $\forall x \in E$, $\{x\} \in$ C (E);

(L2) $\forall x, y \in E$, $\exists\, x \vee y$;

(L3) If $x, y \in X \subseteq E$ implies $x \vee y \subseteq X$, then $X \in$ C (E);

(L4) $\forall X \in$ C (E) and $x, y \in X$, $x \vee y \subseteq X$;

(L5) If $x \vee y$ and $x' \vee y'$ have two distinct common points then there are two points $u, v \in E$ such that $(x \vee y) \vee (x' \vee y') \subseteq u \vee v$;

(L6) $\forall$ z $\in$ x $\vee$ y, (x $\vee$ z) $\cap$ (z $\vee$ y) = {z};

(L7) $\forall$ y' $\in$ x $\vee$ y and z' $\in$ x $\vee$ z, (y' $\vee$ z) $\cap$ (y $\vee$ z') $\neq \varnothing$;

(L8) (x $\vee$ y) $\vee$ (x' $\vee$ y') = $\cup$ {z $\vee$ z' | z $\in$ x $\vee$ y, z' $\in$ x' $\vee$ y'}

for each x, y, x', y' $\in$ E.

The conditions (L1) – (L8) are not contradicting, but they are not independent: for example, (L3) and (L4) depend on (L2).

Definition 1.2.2. *A set* E *satisfying* (L1) – (L8) *is called a lineat space and the elements of* C (E) *are called convex sets. The smallest convex set* x$\vee$y *containing two points* x, y $\in$ E *is called the segment joining the points* x *and* y.

Definition 1.2.3. *For each two distinct points* a, b $\in$ E, *the union*

$$L(a, b) = \cup \{x \vee y \mid a \vee b \subseteq x \vee y\}$$

of all segments including a $\vee$ b *is called the line defined by* a *and* b.

Definition 1.2.4. *A subset* A *of* E *is called a lineat set if for each pair of points* x, y $\in$ E *the line* L (x, y) *lies in* A.

It follows that each lineat subset of E is convex. Now, it is obvious that the lineat spaces are join spaces. Axioms (L3) and (L4) imply the closure for intersection of the set C (E).

W. Prenowitz (1961) and later V.W. Bryant and R.J. Webster (1969) defined and studied another type of convexity space.

Definition 1.2.5. *A convexity space is a nonempty set* X, *in which two binary operations are defined:*

$$\cdot: X^2 \to 2^X \text{ and } /: X^2 \to 2^X,$$

subjected to following axioms:

(i) *Existence law.* a b $\neq \varnothing$, a / b $\neq \varnothing$;

(ii) *Idempotent laws.* a a = a = a / a;

(iii) *Commutative law.* a b = b a;

(iv) *Associative law.* a (b c) = (a b) c;

(v) *Transposition law.* a / b $\cap$ c / d $\neq \varnothing \Rightarrow$ a d $\cap$ b c $\neq \varnothing$,

for all a, b, c, d $\in$ X, *and* A B = $\cup$ {a b | a$\in$A, b$\in$B}, A / B = $\cup${a / b | a$\in$A, b$\in$B} *for all* A, B $\subseteq$ X.

Obviously, the convexity space is a join space. The straight-line segment is defined, as usual in a join space, by $\langle x, y \rangle = x\,y$, for $x \in X$ and $y \in X$. A set $A \subseteq X$ is *convex* if $A\,A \subseteq A$.

V. W. Bryant and R. J. Webster (1972, 1973, 1977) studied a particular type of convexity spaces, defined by the axioms (i) – (v) and

(vi) $a\,b \cap a\,c \neq \varnothing \Rightarrow b = c$ *or* $b \in a\,c$ *or* $c \in a\,b$.

The axioms are not independent because the commutative law is a consequence of laws (i), (ii), (iv) and (v). Also, according to R. Precup (1980), $a\,b \neq \varnothing$ is a consequence of $a\,/\,b \neq \varnothing$ and (v). R. Precup (1980) proved that each geometry organised within a space satisfying the incidence and order axioms according to the Hilbert axiomatic system and the axiom of Pasch is a convexity space satisfying (i) –(vi).

D. Voiculescu (1967 [290]) defines a geometric space by a nonempty set E, consisting from elements called points. To each pair of points (a, b) from E corresponds the set E (a, b), called *segment*, that satisfies the axioms:

A1) $E(a, b) = E(b, a)$;

A2) $E(a, a) = \{a\}$;

A3) $a \in E(a, b)$;

A4) $x \in E(a, b) \Rightarrow E(a, b) = E(a, x) \cup E(x, b)$;

A5) $x \in E(a, b) \Rightarrow E(a, x) \cap E(x, b) = \{x\}$;

A6) If $E(x_1, x_2) \cap E(x_3, x_4) \supset \{x, y\}$ $(x \neq y)$, then there are two indices k, l $(k \neq l)$ such that $E(x_k, x_l) \supset \{x_1, x_2, x_3, x_4\}$.

Definition 1.2.6. *A set* $M \subseteq E$ *is convex if for each two points* $x, y \in M$, *the inclusion* $E(x, y) \subseteq M$ *holds.*

For a set $P \subseteq E$, the intersection of all convex sets containing P is called *the convex hull* of P and is denoted by E (P). D. Voiculescu (1967 [291]) completes the definition of the geometric space with the following two axioms:

A7) $E(a, b) \setminus \{a, b\} \neq \varnothing$ $(a \neq b)$;

A8) If $y \in E(M) \setminus E(M \setminus \{x\})$ $(x \in M)$, then there is $z \in E(M \setminus \{x\})$ such that $y \in E(x, z)$.

A set E, which satisfies the axioms A1) – A8) is called *set with convexity.* D. Voiculescu proves that in a set with convexity we can find the same separation properties of the space as the separation properties of the triangle in the Euclidean plane. Also, the topological properties of a set with convexity are similar to the Euclidean plane.

These axiomatic approaches of the convexity lead to convex sets that are not necessarily connected. The following example presents a convexity space according to V. W. Bryant and R. J. Webster (1972), definition 1.2.5 (i) – (vi). Let (X, ≤) a totally ordered set having the property: for each pair of points a, b ∈ X, a < b (meaning a ≤ b and a ≠ b), there are c, d, e ∈ X such that c < a < d < b < e. The operation •: $X^2 \to 2^X$ is defined by

$$a \bullet b = \begin{cases} a, & a = b \\ \{c \mid a < c < b\}, & a < b, \\ \{c \mid b < c < a\}, & b < a \end{cases}$$

for each a, b ∈ X. The pair (X, •) is a convexity space. The connectivity is not implied. Indeed, for X = **Q**, the set of all rational numbers, the required property fulfils. But **Q** is not a connected subset of **R**. We notice that each dense subset of **R** might be taken instead of **Q**.

1.3 Convexities defined by segmential methods

In this paragraph we intend to discuss the methods of defining various concepts of convex set by modifying the straight-line segment. We shall use the term of *segmential methods* to call all these techniques.

The segmential methods may be grouped in several classes, according to the way in which each of them is elaborated.

1. Replacing the linear structure of the space by another type of structure and defining a notion of straight-line segment.
2. In a linear space, choosing a particular subset of the straight-line segment and using it instead of the straight-line segment.
3. In a linear space, choosing a certain type of straight-line segments from the set of all the straight-line segments.
4. In a space, in which a notion of straight-line segment is defined, another condition besides that of belonging to the set is supposed to be satisfied by the straight-line segment.
5. In a space, in which a notion of straight-line segment is defined, points outside the straight-line segment are considered in relation with the given set.
6. Weak segmential case, when either the straight-line segment determined by two points is not unique or it must be situated in a particular context related to the set under discussion.

We shall present each class of segmential methods historically, according to the bibliography quoted at the end of the book.

1.3.1 Convexity in non-linear structures

K. Menger (1928) introduced the convexity in a metric space, calling it **metric convexity**, by defining the metric segment. Let (X, d) be a metric space and for each $x \in X$ and $y \in X$ the metric segment is defined by

$$<x, y> = \{z \in X \mid d(x, z) + d(z, y) = d(x, y)\}.$$

A set $A \subseteq X$ is said to be *metric convex* if for all two points $x, y \in A$, the whole metric segment $<x, y>$ lies in A.

In an ordered space there are more approaches of convexity for sets.

G. Birkhoff (1948) first defined the **order convexity** as it follows: let us consider a partially ordered set $X = (X, \leq)$ and two elements $x \in X$ and $y \in X$ such that $x \leq y$. The order segment determined by x and y is defined by $<x,y> = \{z \in X \mid x \leq z \leq y\}$. The condition $x \leq y$ is required to ensure that $<x,y> \neq \emptyset$ and that $x \in <x, y>$ and $y \in <x, y>$. A set $A \subseteq X$ is said to be *order convex* if for each two points $x \in A$ and $y \in A$ with $x \leq y$ the condition $<x, y> \subseteq A$ occurs.

S.V. Ovchinnikov (1980) works in a distributive lattice, defining the **convexity in subsets of lattices.** Let $(L, \leq)$ be a lattice and $M \subset L$. A ternary relation on L is called a betweenness relation and is denoted by $B(x_1, x_2, x_3)$. As L.M. Blumenthal (1953), the relation $B(x, x_1, x_2)$ is said to be true if and only if $x_1 \wedge x_2 \leq x \leq x_1 \vee x_2$. An element $x \in L$ lies between x_1 and x_2 if the relation $B(x, x_1, x_2)$ is true. Let the points $a, b \in M$. An M-interval I [a, b] is the set of all the elements of M which lie between a and b,

$$I[a, b] = \{x \in M \mid a \wedge b \leq x \leq a \vee b\} = I[a \wedge b, a \vee b].$$

This ternary relation is extended to a (k+1)-ary relation B on L, called also betweenness. The relation $B(x, x_1, x_2, \ldots, x_k)$ is true if and only if

$$\bigwedge_{i=1}^{k} x_i \leq x \leq \bigvee_{i=1}^{k} x_i,$$

meaning that the point x lies between $x_1, x_2, \ldots, x_k$. The straight-line segment is replaced by a M-polyhedron $I[x_1, x_2, \ldots, x_k]$, which is the set of all the elements of M which lie between $x_1, x_2, \ldots, x_k$,

$$I[x_1, x_2, \ldots, x_k] = \{x \in M \mid \bigwedge_{i=1}^{k} x_i \leq x \leq \bigvee_{i=1}^{k} x_i\} = I[\bigwedge_{i=1}^{k} x_i, \bigvee_{i=1}^{k} x_i].$$

A set $A \subseteq M$ is r-*convex with respect to* M $(r \geq 2)$ if and only if the truth of B $(x, x_1, x_2, \ldots, x_r)$ implies $x \in A$ for all $x \in M$ and all r-tuples of points $x_1, x_2, \ldots, x_r$ belonging to A.

J.E. Martínez-Legaz and I. Singer (1990) extend the order convexity to the situation in which more order relations are defined in a set X. They call the new notion **segmential multi-order convexity.** A multi-ordered set is an ordered pair (X, O), where X is a set and O is a nonempty family of partial orders on X. Let (X, O) be a multi-ordered set. Given $x, y \in X$, we define the O-segment $<x, y> = <x, y>_O = \cap \{<x, y>_\rho \mid \rho \in O\}$. Here, $<x, y>_\rho = \{z \in X \mid \text{either } x \rho z \rho y \text{ or } y \rho z \rho x\}$ for each partial order $\rho \in O$. A set $A \subseteq X$ is said to be O – *segmentially convex* (or, briefly, O_{seg} – *convex*) if $<x, y> \subseteq A$ for each $x, y \in A$.

J.R. Calder (1971) generalised the notion of straight-line segment in a non-organised set of points, X, by defining the general interval. As consequence, he obtains the notion of **interval convexity.** Suppose that X is a point set. The set of the nonempty subsets of X is denoted by $P(X) = 2^X \setminus \varnothing$. A transformation $T: X^2 \to P(X)$ is said to be an interval convexity on X. A set $A \subseteq X$ is said to be *intervally convex* if $T(x, y) \subseteq A$ for each two points $x \in A$ and $y \in A$.

In the paper of V. Soltan (1984) the interval transformation T is replaced by the convex hull operator g, leading to the concept of **g-convexity.** In a nonempty set X, a mapping $g: 2^X \to 2^X$ is called a convex hull operator on X if:

(a) $A \subseteq g(A)$,

(b) $g(g(A)) = g(A)$,

(c) $g(B) \subseteq g(A)$,

for any $A, B \in 2^X$, $B \subseteq A$. If *g* is a convex hull operator defined on X, and

$$G = \{A \subseteq X \mid g(A) = A\},$$

then the pair (X, G) is a convexity space on X, in the sense of definition 1.2.1, called the convexity space generated by g. Conversely, if (X, G) is a convexity space on X, then a convex hull operator on X is defined by the mapping $g: 2^X \to 2^X$ given by $g(A) = \cap \{C \in G \mid A \subseteq C\}$, for all $A \in 2^X$. This function g is called the convex hull operator generated by G. If g is a convex hull operator defined on X then the straight-line segment determined by two points $x \in X$ and $y \in X$ is the set

$$<x, y> = g(\{x, y\}).$$

As usual, a set $A \subseteq X$ is said to be g-*convex* if for each two points $x \in A$ and $y \in A$, the subset $<x, y>$ is included in A.

A. Wieczorek (1989) used, instead the interval transformation T, another type of function, called spot operator. He obtained the concept of **u – convexity.** Given a nonempty set X, a *spot operator* in X is any operator

$$u: \text{Dom}(u) \to 2^X,$$

where the domain of u, denoted by Dom (u) is a family of subsets of X. The set u (C) replaces the straight-line segment corresponding to the set $C \subseteq X$. A set $A \subseteq X$ is u-*convex* if $u(F) \subseteq A$ for each $F \in \text{Dom}(u)$ with $F \subseteq A$.

V. Soltan (1984) also works in a general nonempty set in order to generalise the convex combination and obtains a series of **convexity properties defined by generalised convex combinations**. He calls them **Φ-convexity** and, in a more general context, **Ψ-convexity**. Consider X a nonempty set and the family Φ of functions,

$$\Phi = \{\varphi_{x,y} \mid \varphi_{x,y}: [0, 1] \to X, \varphi_{x,y}(0) = y, \varphi_{x,y}(1) = x, x \in X, y \in X\}.$$

The value $\varphi_{x,y}(t)$ is called an abstract convex combination of the points x and y. The straight-line segment determined by two points $x, y \in X$ is defined by $<x, y> = \{\varphi_{x,y}(t) \mid t \in [0, 1]\}$. A set $A \subseteq X$ is said to be Φ - *convex* if for each two points $x \in A$ and $y \in A$, the subset $<x, y>$ is included in A. A generalisation of the abstract convex combination $\varphi_{x,y}(t)$, together with its related notions of straight-line segment and convex set is obtained by considering, for each points $x, y \in X$, the multifunction $\psi_{x,y} : [0, 1] \to 2^X$, with the properties $y \in \psi_{x,y}(0)$ and $x \in \psi_{x,y}(1)$. The set $\psi_{x,y}(t)$ is also called an abstract convex combination and, for each pair of elements $x, y \in X$, the straight-line segment determined by these points is defined by

$$<x, y> = \bigcup_{0 \le t \le 1} \psi_{x,y}(t).$$

As usually, a set $A \subseteq X$ is called Ψ - *convex* if for each pair of points $x \in A$ and $y \in A$, the subset $<x, y>$ is included in A.

In the same year, I. Singer (1984) considers another type of generalised convex combination in a locally convex Hausdorff topological vector space and obtains the notion of **quasi-convex sets.** This convexity is discussed in chapter 8.

L. Blaga and L. Lupşa (1998) studied a special method of generating a notion of segment in a set X transporting it from a set in which we have either an interval operator, or a hull operator, or another type of straight-line segment. They obtained the **induced seg-convexity with respect to p, f and M**. Let X, T be two arbitrary sets, M be a nonempty subset of T and let us consider the functions $f: X \to T$ and $p: 2^T \to 2^T$. For each points $a, b \in X$ we define the set

$$\langle a, b\rangle_{p,f,M} = f^{-1}(p(\{f(a), f(b)\}) \cap M) =$$
$$= \{z \in X \mid f(z) \in p(\{f(a), f(b)\}) \cap M\}.$$

A subset $A \subseteq X$ is called *induced seg-convex with respect to the functions* p, f *and the set* M if $\langle a, b\rangle_{p,f,M} \subseteq A$, for any a, b$\in$A.

E.A. Youness (1999) defined an **E-convex** set in $\mathbf{R}^n$ by modifying the convex combination condition from the definition of the straight-line segment by moving its end points into another place by means of a mapping E: $\mathbf{R}^n \to \mathbf{R}^n$. In this context, the E - straight-line segment determined by the points x and y is defined by $\langle x, y\rangle = \{(1-t) E(x) + t E(y) \mid t \in [0, 1]\}$. A set $A \subseteq \mathbf{R}^n$ is E-*convex* if for each $x \in A$ and $y \in A$, the condition $\langle x, y\rangle \subseteq A$ holds. More comments on this notion are in chapters 7 and 8 of this book. Chapter 7 discusses more convexities defined by extending E-convexity to more dimensions: **(φ, ψ) – convexity, bi – (φ, ψ) – convexity, bi - (φ -1) – convexity, bi – (1-ψ) – convexity** (D. Duca, E. Duca, L. Lupşa and L. R. Blaga (2000)).

1.3.2 Convexity obtained by restricting the straight-line segment to a part of it

The origin of the approach of the convexity for sets by restricting the straight-line segment is in the paper of J.L.W.V. Jensen (1905) on convexity for functions. He defined the convex function, on a subset of a linear space, by means of a condition on the middle point of the straight-line segment, which is chosen to represent the whole segment. It implies that the middle point of each straight-line segment having its endpoints in the domain of the function must be an element of that domain. In fact, this is the origin of the loss of the connectivity happening in the context of convexity for sets. Therefore, the **midpoint convexity** (originated in J.L.W.V. Jensen (1905, 1906)) is defined as it follows.

Let us suppose that $X = \mathbf{R}^n$, $n \in \mathbf{N}$. A set $A \subseteq \mathbf{R}^n$ is *midpoint convex* if for every $x, y \in A$ one has $\frac{x+y}{2} \in A$.

J. W. Green and W. Gustin (1950) extend the midpoint convexity by the replacement of the convex combinations with scalars from the set $\Delta = \{0, ½, 1\}$, representing the straight-line segment, by convex combinations with scalars from a general set Δ with some properties. They obtain the notion called **quasi-convexity**, defined as it follows. Let us suppose that $n \in \mathbf{N}$ and $\Delta \subset [0, 1]$, $\{0, 1\} \subset \Delta$. A set $A \subseteq \mathbf{R}^n$ is said to be *quasi-convex* if for each x,

$y \in A$ one has $\{t x + (1-t) y \mid t \in \Delta\} \subset A$. Sometimes the set Δ is supposed to satisfy more conditions, as:

1. $1 - \Delta = \Delta$;
2. $(\Delta + \Delta) \cap]0, 1[\subset \Delta$;
3. $\Delta / \Delta \cap]0, 1[\subset \Delta$;
4. $\Delta \Delta \subset \Delta$.

For $\Delta = \{0, p, 1\}$, $p \in]0, 1[$, the quasi-convexity becomes **p – convexity** (A. Aleman (1985)). The p - convexity is another non-connected convexity, obtained in the same manner as the midpoint convexity. But, in this case, the requirement that the middle point of a segment belongs to the set under consideration is replaced by a condition on another point of the segment.

A. Aleman (1985) published another notion of non-connected convexity, calling it **weak convexity**. $(X, +, \cdot)$ is supposed to be a real or complex linear space. A set $A \subseteq X$ is said to be *weakly convex* if for each two points $x \in A$ and $y \in A$, there is a number $p \in]0, 1[$ such that the element $(1-p) x + p y$ belongs to A. Therefore, it is enough that the interior of the straight-line segment $<x, y>$ meets the set A at least once.

1.3.3 Convexity obtained by special straight-line segments

The domain of the convexity obtained by choosing special straight-line segments appeared in connection with two classes of problems. First of all it was generated in connection with the optimisation on sets in which the order is defined by means of a cone (conic convexity). The other root lies in the efforts for the elaboration of algorithms to build the convex hull of a set (operation called convexification) and algorithms to detect the convexity property (ortho-convexity).

P.L. Yu (1974[293]) defined the **directional convexity** as it follows: let us consider the fixed element $u \in \mathbf{R}^n \setminus \{0\}$. A set $A \subseteq \mathbf{R}^n$ is called *directionally convex in the direction u*, if $x \in A$, $y \in A$ and $t \in [0,1]$ implies that there is the number $\mu \geq 0$, such that $t x + (1-t) y + \mu u \in A$.

In particular, P.L. Yu (1974 [293]) considers only the directions of a cone and obtains the **cone convexity**. Let us denote by K a convex cone in $\mathbf{R}^n$. A set $A \subseteq \mathbf{R}^n$ is said to be K – *cone convex*, if $A + K$ is a convex set. Chapter 8 contains more types of convexity properties obtained starting from the cone convexity.

G.J.E. Rawlins and D. Wood (1988) consider only the horizontal and vertical directions and obtain the **ortho-convexity**. In $\mathbf{R}^2$ a set A is called *ortho-convex* if for each points x, y $\in$ A, the straight-line segment (in the sense of the Euclidean geometry) <x, y> lies in A if it is either horizontal or vertical.

1.3.4 Convexity obtained by special conditions on straight-line segments

R. Lowen (1980) had to define the convexity of fuzzy sets by modifying the condition of inclusion of the straight-line segment, adjusting it to sets in which the notion of element involves a degree of uncertainty. He defined the **fuzzy convexity.** Let E be the Euclidean plane. A fuzzy set A in E is represented by a characteristic function χ_A: E $\rightarrow$ [0, 1]. A fuzzy set A in E is *fuzzily convex* if for all x $\in$ E, y $\in$ E, and all z $\in$ <x, y> the inequality

$$\chi_A(z) \geq \min\{\chi_A(x), \chi_A(y)\}$$

occurs. Therefore, a medium value condition is supposed to hold. This condition may be formulated in terms of level sets.

In a complementary manner, B. B. Chaudhuri (1992) defined the **fuzzy concavity** by asking that for all x $\in$ E, y $\in$ E, and z $\in$ <x, y> the inequality

$$\chi_A(z) \leq \max\{\chi_A(x), \chi_A(y)\}$$

holds. In chapter 8 we shall present the directional versions of these properties, used in pattern recognition.

More convexity properties have to take into account that it is necessary to look into a set through another one (a network, a grid, etc.). These situations first occur in integer and in rational programming and later in image analysis.

L. Lupşa (1981) defined more convexity notions with respect to a given set in connection with the integer programming. They are: **strong n-convexity with respect to M, slack n-convexity with respect to M, strong convexity with respect to M, slack convexity with respect to M**. Let (V,+,$\cdot$) be a real linear space and M a nonempty subset of V. Let n be a natural number, n > 0. A subset A $\subseteq$ V is called *strongly* n-*convex with respect to* M if A = $\varnothing$ or if for each system of n points $x^1, \ldots, x^n \in$ A and all real non-negative numbers $t_1, \ldots, t_n$ satisfying the condition $t_1 + \ldots + t_n = 1$ and $t_1x^1 + \ldots + t_n x^n \in$ M we have $t_1x^1 + \ldots + t_n x^n \in$ A. A subset A $\subseteq$ V is called *slackly* n-*convex with respect to* M if A = $\varnothing$ or if for each system of n points $x^1, \ldots, x^n \in M \cap A$ and all real non-negative numbers $t_1, \ldots, t_n$ satisfying the condition $t_1 + \ldots + t_n = 1$ and $t_1x^1 + \ldots + t_n x^n \in$ M we have $t_1x^1 + \ldots + t_n x^n \in$ A. A subset A $\subseteq$ V is called *strongly convex with respect to* M if A = $\varnothing$ or if M $\cap$ conv (A) $\subseteq$ A. A subset A $\subseteq$ V is said to be *slackly convex* *with respect to* M if X = ∅ or if M ∩ conv (M∩A) ⊆ A. Chapter 2 is devoted to this type of convexities.

J.M. Chassery (1978, 1983) defined the **discrete convexity** and used it in shape analysis involved in cytology (see chapter 9). In the plane $\mathbf{R}^2$ we consider the chessboard distance $c(x, y) = \max(|x_1 - y_1|, |x_2 - y_2|)$ for $x = (x_1, x_2) \in \mathbf{R}^2$, $y = (y_1, y_2) \in \mathbf{R}^2$. For the real number h > 0 we apply a grid of step h over the plane and denote the set of lattice points by $\mathbf{Z}^2(h) = \{(ih, jh) \mid i \in \mathbf{Z}, j \in \mathbf{Z}\}$. The set $\mathbf{Z}^2(h)$ is a non-connected subset of $\mathbf{R}^2$. Therefore, a convexity in this set of lattice points is a non-connected convexity in $\mathbf{R}^2$. A set $A \subset \mathbf{Z}^2(h)$ is *ε - convex* if for every x, y ∈ A and t ∈]0, 1[, there is a number ε ∈ [h/2, h[and a point $a \in \mathbf{Z}^2(h)$ such that $c(tx + (1 - t)y, a) \le \varepsilon$. A set $A \subset \mathbf{Z}^2(h)$ is *discrete convex* if there is a real number ε ∈ [h/2, h[for which A is ε - convex.

More generalisations of the discrete convexity will be thoroughly discussed in chapters 3, 4 and 5: **α-convexity, convexity with respect to a behaviour** (G. Cristescu (1997 [64])

1.3.5 Convexity obtained by putting the straight-line segments in relation with special external points

J. Perkal (1956) obtains a non-connected convexity by discussing the position of the points of the given set towards its exterior, calling it **ε – convexity.** Let (X, d) be a metric space. The ε – convex hull of A ⊆ X is the set of all points p ∈ X situated at a distance at least ε/2 from each point situated at a distance greater than ε/2 from A. The simbol $C_\varepsilon(A)$ denotes the ε - convex hull of A. The set A ⊆ X is said to be ε – *convex* if $A = C_\varepsilon(A)$.

G. Cristescu (1996 [56], [57]) discussed the behaviour of the points of each straight-line segment determined by elements of an investigated set towards the points of a reference set through which it is looking into the space. For example, such a reference set is considered to contain, in a particular case, the lattice points of a network over a totally bounded metric space. The notions of convexity defined as consequence of this investigation are: **the strong (α, δ) - convexity with respect to a network, the strong convexity with respect to a set and two behaviours, the slack (α, δ) - convexity with respect to a network the slack convexity with respect to a set and two behaviours.** They are thoroughly studied in chapters 4 and 5.

Other convexities defined using this principle are mentioned in chapter 9 in connection with their applications in image analysis.

1.3.6 Weak segmential approach

This approach has the origin in the remark that each convex set in the Euclidean plane lies into a half-space. A problem concerning the generalisation of this property was formulated during the meeting held in Obervolfach in 1974. The result is the general property of **F-convexity** (R. Blind, G. Valette, T. Zamfirescu (1980)).

Let F be a class of sets in $\mathbf{R}^n$. A set $A \subseteq \mathbf{R}^n$ is said to be F *-convex* if, for each $x \in A$, $y \in A$, with $x \neq y$, there is a set $F \in$ F such that $x \in F$, $y \in F$ and $F \subseteq A$.

In the particular case when F is the class of all 2D – rectangles in $\mathbf{R}^n$, the corresponding convexity is called **rectangular convexity** (R. Blind, G. Valette, T. Zamfirescu (1980)).

1.4 Unions of convex sets

A method of avoiding the connectivity consists in taking into account unions of convex sets as another type of convex sets.

The first notion of convexity implying the shape of a union of convex sets was published by F.A. Valentine (1957), when he defined the notion of **three point convexity.**

E_n is supposed to be the n-dimensional Euclidean space and $A \subseteq E_n$. The set A is said to have the three point property P_3 if for each triple of points x, y, z in A at least one of the closed segments $\langle x, y\rangle$, $\langle x, z\rangle$, $\langle y, z\rangle$ is in A. A set A is *three point convex* if A possesses the property P_3.

F.A. Valentine (1964) generalised the above definition to **m-point convexity** as it follows: the set A is said to be m *-point convex* if for each m-member subset of A at least one of the $\binom{m}{2}$ line segments determined by these points lies in A.

The same author took into account the case in which convex arcs are used instead of straight-line segments. He obtains the **arcwise convexity** (F.A. Valentine (1951)) and the **three point arcwise convexity** (F.A. Valentine (1955)). These properties will be described in chapter 8.

More applications of pattern recognition imply directional tests and this is the reason of the necessity of a collinearity condition in the above-mentioned definition. **Collinear 3-point convexity (CP_3 – convexity)** (L. Latecki, A. Rosenfeld, R. Silverman (1995)) is obtained by introducing the condition of collinearity in the definition of the property P_3. Let $X = \mathbf{R}^2$ and $A \subset \mathbf{R}^2$. We say that the set A has the property CP_3 (collinear P_3) if P_3 holds for all collinear triples of points of A. A set A is CP_3 – *convex* if it has the property CP_3.

The publication of these definitions had as consequence a great deal of research on their shape. First of all F.A. Valentine (1957) published the following result.

Theorem 1.4.1. *Suppose that* A *is a closed connected 3-point convex set in* $\mathbf{R}^2$. *Then* A *is expressible as the union of three or fewer closed convex sets having a nonempty intersection. The number three is the best.*

J.F. Lawrence, W.R. Hare Jr. and J.W. Kenelly (1972) characterised the sets that may be represented as finite unions of convex sets. But M. Breen (1974, 1976, 1977) worked in the less restrictive framework, eliminating both the condition of closure and the condition of connectivity that appear in Valentine's theorem 1.4.1. Her main results concerning the shape of 3-point convex sets and m-point convex sets are quoted in the following theorems. Some notions are needed. Let $A \subseteq \mathbf{R}^p$, $p \in \mathbf{N}^*$. A point $x \in A$ is a point of local convexity of A if there is some neighbourhood U of x such that if y, z $\in A \cap U$ then $<y, z> \subseteq A$. If A fails to be locally convex at some point q in A, then q is called a point of local non-convexity (lnc point) of A. Let us denote by Lnc (A) the set of all lnc points of $\overline{A}$.

Theorem 1.4.2. *If* A *is a 3-point convex set, if* $(\overline{A} \setminus A) \cap Fr(\overline{A}) = \emptyset$ *and* card (Lnc (A)) = n *for* n *an odd integer*, n > 1, *then* A *is expressible as a union of four or fewer convex sets.*

Theorem 1.4.3. *If* A *is a 3-point convex set, if* $(\overline{A} \setminus A) \cap Fr(\overline{A}) = \emptyset$ *and* card (Lnc (A)) = n ≥0, *where* n (possibly infinite) *is not an odd integer greater then one, then* A *is expressible as a union of four or fewer convex sets.*

Theorem 1.4.4. *If* A *is an open 3-point convex set, then* A *is expressible as a union of four or fewer convex sets. The result is best possible.*

Theorem 1.4.5. *If* A *is a 3-point convex set and* $\overline{A}$ *is convex, if* $(\overline{A} \setminus A) \subseteq Fr(\overline{A})$, *then* A *is a union of three or fewer convex sets. The bound of three is best possible.*

Theorem 1.4.6. *If* A *is a 3-point convex set, if* $(\overline{A} \setminus A) \subseteq Fr(\overline{A})$, *then* A *is a union of four or fewer convex sets. The number of four is best possible.*

Theorem 1.4.7. *If* A *is a 3-point convex set, then* A *is as a union of six or fewer convex sets. The result is best possible.*

W.I. Stamley and J.M. Marr (1963) and F.A. Valentine (1965) published weaker results. M. Breen (1976) obtained the bound of 2(m-1) convex parts that are disjoint components of a m-point convex set.

1.5 Intersectional approach

It is well known that each convex set in a n-dimensional linear space lies in a half-space. More, each convex set is the intersection of all half-spaces containing it. This is the starting point of the intersectional method of defining the property of convexity.

W. Fenchel (1952) first defined an intersectional convexity, called **even convexity in a locally convex space**. A subset M of a locally convex space X is said to be an open half-space if there is a linear functional $\varphi \in X^* \setminus \{0\}$ and $d \in \mathbf{R}$ such that

$$M = \{x \in X \mid \varphi(x) < d\}.$$

A subset $A \subseteq X$ is said to be *evenly convex* if it is the intersection of a family of open half-spaces.

L. Danzer, B. Grunbaum, V. Klee (1963) defined the general notion of **convexity with respect to a family of sets**. Let X be a nonempty set and a family of subsets $\mathsf{M} \subset 2^X$. A set $A \subseteq X$ is said to be *convex with respect to* M (or, simply, M – *convex*) if there is a subfamily M' of M such that $A = \cap \{M \mid M \in \mathsf{M}'\}$.

More convexity properties are reducible to M – convexity. These convexities are discussed in chapter 8. K. Fan (1963) introduces the notion of **convexity with respect to a set of functions**, which is a particular case of M – convexity. Let X be a nonempty set and W a set of functions, $w: X \to \mathbf{R}$. A set $A \subseteq X$ is said to be *convex with respect to* W (or, simply, W – *convex*) if for each $x \notin A$ there is a function $w \in W$ such that $\sup w(A) < w(x)$. This is a separational approach. I. Singer (1984) extends the notion of W – convexity, allowing functions of type $w: X \to \overline{R} = [-\infty, +\infty]$. He proves that the W – convexity concept obtained in this manner is equivalent to the M – convexity. I. Singer (1997) denotes by $S_d(w) = \{x \in X \mid w(x) \le d\}$. Then a set $A \subseteq X$ is convex with respect to W if and only if

$$A = \cap \{S_d(w) \mid (w, d) \in W \times \mathbf{R}, \sup w(A) \le d\}.$$

It means that, for $\mathsf{M} = \{S_d(w) \mid (w, d) \in W \times \mathbf{R}\}$, a set $A \subseteq X$ is W – convex if and only if $A = \cap \{M \mid M \in \mathsf{M}'\}$ for

$$\mathbf{M}' = \{S_d(w) \mid (w, d) \in W \times \mathbf{R}, \sup w(A) \le d\} \subseteq \mathbf{M},$$

meaning that A is convex with respect to $\mathbf{M}$.

J. Scrader (1975) defined the **convexity with respect to a pair (W, φ)**, which is another particular case of $\mathbf{M}$ – convexity. Let X and W be two arbitrary nonempty sets and a function $\varphi: X \times W \to \overline{R} = [-\infty, +\infty]$. J. J. Moreau (1966-67) called the function φ a coupling function. A set $A \subseteq X$ is said to be *convex with respect to the pair* (W, φ) (or, briefly, (W, φ) – *convex*) if for each $x \notin A$ there is an element $w \in W$ such that

$$\sup \{\varphi(g, w) \mid g \in A\} < \varphi(x, w).$$

This definition is a separational one, but I. Singer (1997) proved that a set A $\subseteq$ X is convex with respect to the pair (W, φ) if and only if

$$A = \cap \{S_d(\varphi(\bullet, w)) \mid (w, d) \in W \times \mathbf{R}, \sup \{\varphi(g, w) \mid g \in G\} \le d\},$$

with $S_d(\varphi(\bullet, w)) = \{x \in X \mid \varphi(x, w) \le d\}$.

As in the previous example, a set $A \subseteq X$ is (W, φ) – convex if and only if

$$A = \cap \{M \mid M \in \mathbf{M}'\}$$

for $\mathbf{M} = \{S_d(\varphi(\bullet, w)) \mid (w, d) \in W \times \mathbf{R}\}$, and

$$\mathbf{M}' = \{S_d(\varphi(\bullet, w)) \mid (w, d) \in W \times \mathbf{R}, \sup \{\varphi(g, w) \mid g \in G\} \le d\} \subseteq \mathbf{M}.$$

It means that A is convex with respect to $\mathbf{M}$.

The convex hull operator g defined by V.P. Soltan (1984) is another intersectional approach of the convexity, but it leads to a notion of straight-line segment.

1.6 Separational approach

The separation properties of convex sets in a linear space are discussed in almost all books dealing with the convexity for sets: see W.W. Breckner (1974), H.D. Eggleston (1959), P.M. Gruber, J.M. Willis (1993), F.A. Valentine (1964), R.J. Webster (1994). The separation properties leading to separational methods of defining a convexity are:

1. A convex set in a linear space lies in a hyperplane. Therefore, the sign of the linear functions that define all hyperplanes containing a convex set characterises the set.
2. The idea of the nearest point, which is thoroughly discussed by R.J. Webster (1994).

3. The possibility of separating each point that does not belong to the interior of a convex set.
4. The existence of the support hyperplane in each point of the boundary of a convex set in a linear space.

The previous paragraph presented some separational convexities that are also intersectional: **convexity with respect to a set of functions** (K. Fan (1963)), **convexity with respect to a pair (W, φ)** (J. Scrader (1975))

L. Hörmander (1989) uses the convex hull operator that does not lead visibly to a straight-line segment, to define the **pseudoconvexity**. In fact, it is a separational definition. Let $A \subseteq \mathbf{C}^n$ be an open set. We recall that a function $u: A \to [-\infty, +\infty]$ is called plurisubharmonic if u is semicontinuous from above and for arbitrary z and w from $\mathbf{C}^n$, the function $\lambda \to u(z+\lambda w)$ is subharmonic in the part of **C** where it is defined. The set of all plurisubharmonic functions on A is denoted by P (A). If K is a compact subset of the open set $A \subseteq \mathbf{C}^n$ then the P(A) – hull of K is defined by

$$\hat{K}_A^P = \{z \mid z \in A, u(z) \leq \sup_K u \text{ for all } u \in P(A)\}.$$

The open set $A \subseteq \mathbf{C}^n$ is *pseudoconvex* if there is a continuous plurisubharmonic function u in A such that the set $A_c = \{z \mid z \in A, u(z) < c\}$ is relatively compact in A (i.e. is contained in a compact subset of A) for every $c \in \mathbf{R}$. This property holds if and only if K relatively compact in A implies $\hat{K}_A^P$ relatively compact in A.

J.E. Martínez- Legaz, I. Singer (1990) used the separational approach in an ordered set, obtaining the notion of **separational multi-order convexity.** A multi-ordered set is an ordered pair (X, O), where X is a set and O is a nonempty family of partial orders on X. Let (X, O) be a multi-ordered set. A set $A \subseteq X$ is said to be O – *separationally convex* (or, briefly, O_{sep} – *convex*) if for each $x_0 \in X \setminus A$, there is an order $\rho = \rho(x_0) \in O$ such that the relation $x_0 \, \bar{\rho} \, g$ occurs for each $g \in X$, where $\bar{\rho}$ is the relation opposite to ρ.

A.M. Rubinov, B. Glover, V. Jeyakumar (1995) defined another separational property, called **H-convexity**. Let X be an arbitrary set and the subset $Z \subseteq X$, $Z \neq \varnothing$. Let H be a set of functions defined on X and mapping into **R**. A set $A \subseteq H$ is called H-*convex on* Z if and only if for any $h' \in H$ such that $h' \notin A$, there is an element $x \in Z$ for which

$$h'(x) > \sup \{h(x) \mid h \in A\}.$$

Chapter 8 contains more convexity properties defined in all the above-described manners, classified according to other criteria.

Chapter 2

Convexity with respect to a set

The necessity of characterising the set of efficient and non-efficient points of a vectorial programming problem in integers (see chapter 12) leads to the approach of the convexity with respect to a set. An analysis of the set SI of those solutions of a linear system of inequations having all the co-ordinates integer numbers shows that this set is not convex. But it has the property that, for any choice of a pair of its points, each convex combination of its elements having all the co-ordinates integer numbers belongs to SI. Starting from this remark, L. Lupşa (1980[150]) first defined the notion of convex set with respect to a given set and studied its properties. This notion corresponds to today's notion of a strong 2-convex set with respect to a given set. The notion of strong convexity with respect to a given set was introduced in L. Lupşa (1981[152]), showing by means of an example that the notions of convexity with respect to a set and strong convexity with respect to a set are distinct. After a deeper investigation of these sets, L. Lupsa (1982) introduced the notion of n-convexity with respect to a set, which is equivalent to the strong n-convexity from this chapter. The paper of L. Lupşa (1986[163]) is a synthesis of four types of convexity properties: strong n-convexity with respect to a set, slack n-convexity with respect to a set, strict convexity with respect to a set and slack convexity with respect to a set. The examples presented in that paper show that the notions are distinct from each other and also distinct both from the classical convexity and from the cone convexity. J.E. Martinez-Legaz and I. Singer (1990) proved, when they defined the segmential convexity, that V = $\mathbf{R}^n$ and M = $\mathbf{Z}^n$, the segmential convexity with respect to $(\mathbf{Z}^n,\tau')$ is equivalent to the strong 2-convexity. In general these notions are different.

This new type of convexity is important because it allows the study of the properties of some subsets of $\mathbf{Z}^n$ with applications, as it will be presented in this book, in the optimisation theory and in pattern recognition.

In this chapter, when we shall have more vectors, we shall upper-script the indices showing the number of the vector, when more vectors will be taken into account. For indicating the number of a scalar from more scalars, we shall subscript the indices. In order to indicate the coordinates of a vector $x^k \in \mathbf{R}^m$, we shall use the notation $x^k = (x_1^k,...,x_m^k)$.

2.1 Types of convexity with respect to a given set

Let $(V,+,\cdot)$ be a real linear space. In the following by M we denote a nonempty subset of V. Let n be a natural number, n>0.

Definition 2.1.1. *A subset* $X \subseteq V$ *is called*:

i) *strongly* n*-convex with respect to* M *if* $X = \varnothing$ *or if for each system* x^1, ..., $x^n \in X$ *and all real numbers* t_1, ..., t_n *satisfying the condition*

(2.1.1) $t_1 \geq 0, ..., t_n \geq 0, t_1 + ... + t_n = 1,$

and

(2.1.2) $t_1x^1 + ... + t_nx^n \in M$

we have

(2.1.3) $t_1x^1 + ... + t_nx^n \in X.$

ii) *slackly* n*-convex with respect to* M *if* $X = \varnothing$ *or if for all system* x^1, ..., $x^n \in M \cap X$ *and all real numbers* t_1, ... , t_n *satisfying conditions* (2.1.1) *and* (2.1.2) *we have*

(2.1.4) $t_1x^1 + ... + t_nx^n \in X.$

iii) *strongly convex with respect to* M *if* $X = \varnothing$ *or if*

(2.1.5) $M \cap \mathrm{conv}\,(X) \subseteq X;$

iv) *slackly convex with respect to* M *if* $X = \varnothing$ *or if*

(2.1.6) $M \cap \mathrm{conv}\,(M \cap X) \subseteq X.$

Example 2.1.1. The set X = {(0, 0), (1, 1), (2, 2)} is strongly 2-convex with respect to $\mathbf{Z}^2$. It is easy to see that X is not classically convex.

Example 2.1.2. The set $X = \{(1, 1), (3, -1)\}$ is not strongly 2-convex with respect to $\mathbf{Z}^2$ because $(1, 0) \in \mathbf{Z}^2$ but $(1, 0) = 0.5\,(1, 1) + 0.5\,(3, -1) \notin X$.

Example 2.1.3. The set $X = \{(0, 2), (-1, 1), (1, 0)\}$ is strongly 2-convex and slackly 2-convex with respect to $\mathbf{Z}^2$. It is neither strongly 3-convex nor slackly 3-convex with respect to $\mathbf{Z}^2$, because $(0, 1) = \frac{1}{3}\,(0, 2) + \frac{1}{3}\,(-1, 1) + \frac{1}{3}\,(1, 0) \in \mathbf{Z}^2$, but $(0, 1) \notin X$.

Example 2.1.4. The set $X = \{(x, y) \in \mathbf{R}^2 \mid xy \leq 2,\ 1 \leq x \leq 4\}$ is not classically convex, but it is slackly n-convex with respect to $\mathbf{Z}^2$ for any number $n \in \mathbf{N}$, $n \geq 2$.

Example 2.1.5. The set $X = \{(x, y) \in \mathbf{R}^2 \mid -\frac{1}{2} \leq x \leq \frac{1}{2},\ y = 1 - 4x^2\}$ is slackly 2-convex with respect to $\mathbf{Z}^2$, since $X \cap \mathbf{Z}^2 = \{(0, 1)\}$. The set X is not strongly 2-convex with respect to $\mathbf{Z}^2$, because $z = \frac{1}{2}\,(-\frac{1}{2}, 0) + \frac{1}{2}\,(\frac{1}{2}, 0) = (0, 0) \in \mathbf{Z}^2$, but $(0, 0) \notin X$.

Example 2.1.6. Let $X = \{(x, y) \in \mathbf{R}^2 \mid 0 \leq x \leq 6y^2,\ 2x + 3y \leq 15,\ y \geq 0\}$.

We have $\text{conv}\,(X) = \{(x, y) \in \mathbf{R}^2 \mid 0 \leq x \leq 6y,\ 2x + 3y \leq 15,\ y \geq 0\}$. Since $\mathbf{Z}^2 \cap \text{conv}\,(X) = \{(0, 0), (0, 1), (0, 2), (0, 3), (0, 4), (0, 5), (1, 1), (1, 2), (1, 3), (1, 4), (2, 1), (2, 2), (2, 3), (3, 1), (3, 2), (3, 3), (4, 1), (4, 2), (5, 1), (6, 1)\} \subseteq X$, the set X is strongly convex with respect to $\mathbf{Z}^2$. But the set X is not classically convex.

Example 2.1.7. The set $X = \{(x, y) \in \mathbf{R}^2 \mid [x] = y\}$ is slackly convex with respect to $\mathbf{Z}^2$ because we have

$$X \cap \mathbf{Z}^2 = \{(x, y) \in \mathbf{Z}^2 \mid x = y\},$$

$$\text{conv}\,(\mathbf{Z}^2 \cap X) = \{(x_1, x_2) \in \mathbf{R}^2 \mid x_1 = x_2\}$$

and

$$\mathbf{Z}^2 \cap \text{conv}\,(\mathbf{Z}^2 \cap X) \subseteq X.$$

But the set X is not convex.

Example 2.1.8. Let be $X = \{(0, 2), (\frac{1}{2}, 0), (1, 3), (2, 3)\}$. Since

$$\mathbf{Z}^2 \cap \text{conv}\,(X) = \{(0, 2), (1, 1), (1, 2), (1, 3), (2, 3)\} \not\subset X,$$

the set X is not strongly convex with respect to $\mathbf{Z}^2$. But the set X is slackly convex with respect to $\mathbf{Z}^2$ because we have

$$\mathbf{Z}^2 \cap \text{conv}\,(\mathbf{Z}^2 \cap X) = \{(0, 2), (1, 3), (2, 3)\} \subseteq X.$$

Example 2.1.9. Let be $M = [0, 1] \times [1, 3]$ and

$$X = \{(x, y) \in \mathbf{R}^2 \mid 0 \le x \le 1, 2 \ge y \ge \sqrt{x}\}.$$

Because $M \cap \text{conv}\,(X) = [0, 1] \times [1, 2] \subseteq X$, the set X is strongly convex with respect to M. We notice that the set X is not classically convex.

Remark 2.1.1. i) Every subset X of V is strongly n-convex and slackly n-convex with respect to any of its subset, for any natural number n.

ii) If X is a convex subset of V, then X is both strongly n-convex and slackly n-convex with respect to any nonempty subset M of V, for any natural number n.

iii) If the subset X of V is a strongly (slackly) n-convex with respect to a nonempty set M of V, then X is strongly (slackly) n-convex with respect to any nonempty subset P of M.

Remark 2.1.2. Let n and k be two natural numbers, $2 \le k < n$. Let X and M be subsets of V. If X is strongly n-convex (slackly n-convex) with respect to M, then X is also strongly k-convex (slackly k-convex) with respect to M. The converse is not always true (see example 2.1.5).

Remark 2.1.3. Let X and M be two nonempty subset of V.

i) If X is strongly n-convex with respect to M, then X is also slackly n-convex with respect to M.

ii) If X is slackly n-convex with respect to M, then X is not always strongly n-convex with respect to M (see example 2.1.5).

Remark 2.1.4. Let be M and X two nonempty subset of V.

i) If $M \subseteq X$, then X is strongly and slackly convex with respect to M.

ii) If X is convex, then X is strongly and slackly convex with respect to M, but, in general, if X is strongly (slackly) convex with respect to M, it is not convex (see examples 2.1.6 - 2.1.8).

iii) If a subset X is strongly convex with respect to M, then X is also slackly convex with respect to M; the converse is not always true (see example 2.1.8).

iv) If $X \subseteq M$, then X is strongly convex with respect to M if and only if it is slackly convex with respect to M.

Remark 2.1.5. It is easy to see that:

i) A set $X \subseteq V$ is slackly n-convex with respect to M if and only if the set $X \cap M$ is strongly n-convex with respect to M.

ii) A set $X \subseteq V$ is slackly convex with respect to M if and only if the set $X \cap M$ is strongly convex with respect to M.

Theorem 2.1.1. *If* $X \subseteq V$ *is strongly convex with respect to* M *then* X *is strongly* n-*convex with respect to* M, *for any natural number* n, $n \geq 2$.

Proof. Let n be a natural number, $n \geq 2$, $x^1, \dots, x^n \in X$ and $t_1, \dots, t_n$, real numbers satisfying the conditions (2.1.1) and (2.1.2). Thus, by Carathéodory's theorem, we get

(2.1.7) $\quad t_1 x^1 + \dots + t_n x^n \in \text{conv}(X).$

From (2.1.7) and (2.1.2) it follows that

(2.1.8) $\quad t_1 x^1 + \dots + t_n x^n \in M \cap \text{conv}(X).$

Because X is strongly convex with respect to M, we get that $\sum_{j=1}^{n} t_j x^j \in X$.

This implies that X is strongly n-convex with respect to M.

Theorem 2.1.2. *If* V *is a finite dimensional linear space, a set* $X \subseteq V$ *is strongly convex with respect to* M *if and only if it is strongly* n-*convex with respect to* M, *where*

(2.1.9) $\quad n = 1 + \dim X.$

Proof. The necessity is implied by theorem 2.1.1.

Sufficiency. Now let X be a strong n-convex set with respect to M and let $z \in M \cap \text{conv}(X)$. Because $\dim X = n-1$, by Carathéodory's theorem there are n elements $x^1, \dots, x^n$ of X and n nonnegative real numbers $t_1, \dots, t_n$, with $t_1 + \dots + t_n = 1$, such that $z = t_1x^1 + \dots + t_nx^n$. From the strong n-convexity of set X and as $z \in M$, it follows $z \in X$. But z was arbitrarily chosen. Therefore we have $M \cap \text{conv}(X) \subseteq X$, which implies that X is strongly convex with respect to M.

From theorem 2.1.1 and remark 2.1.5, we obtain

Corollary 2.1.1. *A set* $X \subseteq V$ *is slackly convex with respect to* M *if and only if* X *is slackly* n-*convex with respect to* M, *where* $n = 1 + \dim X$.

The notion of convexity with respect to a given set is a new type of general convexity. From examples 2.1.1 - 2.1.8 it results that this convexity is different from the classical convexity.

In the particular case $V = \mathbf{R}^n$ and $M = \mathbf{Z}^n$ the strong 2-convexity is equivalent to the segmential convexity with respect to $(\mathbf{Z}^n, \tau')$ defined by J. E. Martinez-Legaz and I. Singer (1990). From theorem 2.1.1 and example

2.1.3 it follows that, in general, the notions of integer convex set and segmential convex set are different.

2.2 Properties of strong n-convex sets and of slack n-convex sets with respect to a given set

Proposition 2.2.1. *If* X_k, $k \in K$, *is a family of strong (slack)* n-*convex sets with respect to* M, *then the set* $X = \bigcap_{k \in K} X_k$ *is also strongly (slackly)* n-*convex with respect to* M.

Proof. Let $x^1, ..., x^n \in X$, and let $t_1, ..., t_n$ be real numbers satisfying the conditions (2.1.1) and (2.1.2). Because each set X_k, $k \in K$, is strongly n-convex with respect to M, we have $t_1 x^1 + ... + t_n x^n \in X_k$, for all $k \in K$; thus $t_1x^1 + ... + t_nx^n \in \bigcap_{k \in K} X_k$. Hence X is strongly n-convex with respect to M. The proof is similar when the sets X_k, $k \in K$, are slackly convex.

From remark 2.1.2 and proposition 2.2.1 we obtain:

Corollary 2.2.1. *If* X_k, $k \in K$, *is strongly (slackly)* n_k-*convex with respect to* M, *then the set* $X = \bigcap_{k \in I} X_k$ *is strongly (slackly)* n-*convex with respect to* M, *where* $n = \min \{n_k \mid k \in K\}$.

Definition 2.2.1. *Let be* $r \in \mathbf{R}^*$. *A nonempty subset* U *of* V *is called* r-*set if* $\frac{u}{r} \in U$ *for each* $u \in U$.

Example 2.2.1. If $C \subseteq V$ is a cone, then C is a r-set, for each $r \in \mathbf{R}^*$.

Example 2.2.2. The set **Z** of integer numbers is a (-1)-set.

Example 2.2.3. The set **Q** of rational numbers is a r-set for all $r \in \mathbf{Q}^*$.

Proposition 2.2.2. *If* $M \subseteq V$ *is a nonempty* r-*set and* $X \subseteq V$ *is strongly (slackly)* n-*convex with respect to* M, *then* rX *is strongly (slackly)* n-*convex with respect to* M.

Proof. Let $x^1, ..., x^n \in rX$ and let $t_1, ..., t_n$ be real numbers satisfying the conditions (2.1.1) and (2.1.2). For each x^k, $k \in \{1, ..., n\}$, there is $y^k \in X$ such that $x^k = r\, y^k$. Taking into account that $r\,(t_1 y^1 + ... + t_n y^n) = t_1 x^1 + ... + t_n x^n$, from (2.1.2) we get that $r\,(t_1 y^1 + ... + t_n y^n) \in M$. Because M is a r-set it follows that $t_1 y^1 + ... + t_n y^n \in M$. But X is strongly n-convex with respect to

M, and $y^k \in X$, for each $k \in \{1, \dots, n\}$. It follows that $t_1 y^1 + \dots + t_n y^n \in X$. Then

$$t_1 x^1 + \dots + t_n x^n = r\,(t_1 y^1 + \dots + t_n y^n) \in r\,X.$$

Hence the set rX is strongly n-convex. Analogously one can prove a similar property for the case when X is slackly n-convex with respect to M.

Remark 2.2.1. Let $n \in \mathbf{N}^*$. From example 2.1.8 and proposition 2.1.2 it results that if $X \subseteq V$ is strongly (slackly) n-convex with respect to **Z**, then the set (-X) is also strongly (slackly) n-convex with respect to **Z**.

Remark 2.2.2. If X and Y are nonempty subsets of V, which are strongly (slackly) n-convex with respect to M, where M is a nonempty subset of V. Then X +Y is not always strongly (slackly) n-convex with respect to M.

Example 2.2.4. The sets $X = \{(2, 1), (0, 2)\}$ and $Y = \{(1, 3), (0, 2)\}$ are strongly (slackly) 2-convex with respect to $\mathbf{Z}^2$, but the set

$$X + Y = \{(3, 4), (2, 3), (1, 5), (0, 4)\}$$

is not strongly (slackly) 2-convex with respect to $\mathbf{Z}^2$, because

$$\frac{1}{3}(3, 4) + \frac{2}{3}(0, 4) = (1, 4) \in \mathbf{Z}^2, \text{ but } (1, 4) \notin X + Y.$$

Proposition 2.2.3. *Let* M *and* X *be two nonempty subsets of* V, $n \in \mathbf{N}^*$, *and let* $Y \subseteq V$ *be a classical convex set such that* $M - Y \subseteq M$. *If* X *is strongly (slackly)* n*-convex with respect to* M, *then* X + Y *is also strongly (slackly)* n*-convex with respect to* M.

Proof. Let $z^1, \dots, z^n$, be n elements of X + Y and $t_1, \dots, t_n$ be real numbers satisfying (2.1.1) and

$$(2.2.1) \qquad t_1 x^1 + \dots + t_n x^n \in M.$$

For each $k \in \{1, \dots, n\}$, there are $x^k \in X$, and $y^k \in Y$, such that $z^k = x^k + y^k$. Then from (2.1.6) we get

$$(2.2.2) \qquad (t_1 x^1 + \dots + t_n x^n) + (t_1 y^1 + \dots + t_n y^n) \in M.$$

But Y is classically convex. Then

$$(2.2.3) \qquad t_1 y^1 + \dots + t_n y^n \in Y.$$

Because $M - Y \subseteq M$, (2.2.3) and (2.2.2) imply that $t_1 x^1 + \dots + t_n x^n \in M$. Since X is strongly n-convex with respect to M, it results that

$$(2.2.4) \qquad t_1 x^1 + \dots + t_n x^n \in X.$$

From (2.2.3) and (2.2.4) we get that

$$t_1 z^1 + ... + t_n z^n = (t_1 x^1 + ... + t_n x^n) + (t_1 y^1 + ... + t_n y^n) \in X + Y.$$

Hence X + Y is strongly n-convex with respect to M.

Remark 2.2.3. Let $n \in \mathbf{N}^*$ and $z \in \mathbf{Z}$. If the set X is strongly (slackly) n-convex with respect to $\mathbf{Z}^n$, then, from proposition 2.2.3 and example 2.2.2 it follows that the sets X + {z} and X – {z} are strongly (slackly) n-convex with respect to M.

2.3 Properties of strong convex sets and of slack convex sets with respect to a given set

Theorem 2.3.1. *The intersection of a family* X_k, $k \in K$, *of strong convex sets with respect to* M, *is strongly convex set with respect to* M.

Proof. We have

(2.3.1) $\quad M \cap \text{conv}(X_k) \subseteq X_k$, for all $k \in K$.

Because $\bigcap_{k\in K} X_k \subseteq X_k$, for each $k \in K$, we get $\text{conv}(\bigcap_{k\in K} X_k) \subseteq \text{conv}(X_k)$, for all $k \in K$. Hence

(2.3.2) $\quad M \cap \text{conv}(\bigcap_{k\in K} X_k) \subseteq M \cap \text{conv}(X_k)$, for all $k \in K$.

From (2.3.1) and (2.3.2) it results that $M \cap \text{conv}(\bigcap_{k\in K} X_k) \subseteq X_k$, for all $k \in K$. Hence $M \cap \text{conv}(\bigcap_{k\in K} X_k) \subseteq \bigcap_{k\in K} X_k$. It follows that $\bigcap_{k\in K} X_k$ is strongly convex with respect to M.

From remark 2.1.5 and theorem 2.1.1 we get

Corollary 2.3.1. *The intersection of a family* X_k, $k \in K$, *of slackly convex sets with respect to* M, *is a slackly convex set with respect to* M.

Theorem 2.3.2. *The set* $X \subseteq V$ *is strongly convex with respect to* M *if and only if* $M \cap \text{conv}(X) = M \cap X$.

Proof. Necessity. Assume that X is strongly convex with respect to M. From the definition it results that $M \cap \text{conv}(X) \subseteq X$. But $M \cap \text{conv}(X) \subseteq M$. Hence

(2.3.3) $\quad M \cap \text{conv}(X) \subseteq M \cap X$.

Because $X \subseteq \text{conv}(X)$, we have

(2.3.4) $M \cap \operatorname{conv}(X) \supseteq M \cap X$.

From (2.3.3) and (2.3.4) we obtain $M \cap X = M \cap \operatorname{conv}(X)$.

Sufficiency. It is evident.

From remark 2.1.5 and theorem 2.3.2 we obtain

Corollary 2.3.2. *A set* $X \subseteq V$ *is slackly convex with respect to* M *if and only if* $M \cap \operatorname{conv}(M \cap X) = M \cap X$.

Theorem 2.3.3. *If* M *is a cone and* X *is strongly convex with respect to* M, *then we have*

(2.3.5) $M \cap (a X + b X) = M \cap (a + b) X$,

for every $a, b \in [0, +\infty[$.

Proof. If $a = 0$ or $b = 0$, then (2.3.5) is true. Suppose that $a \neq 0$ and $b \neq 0$.

Let $z \in (a X + b X) \cap M$. Then there are two points $u \in X$ and $v \in X$, such that $z = a u + b v$. Because M is a cone, we have $\frac{1}{a+b} z \in M$. Certainly,

$$\frac{1}{a+b} z \in \frac{a}{a+b} X + \left(1 - \frac{a}{a+b}\right) X.$$

Hence

$$\frac{1}{a+b} z \in M \bigcap \left(\frac{a}{a+b} X + \left(1 - \frac{a}{a+b}\right) X \right) \subseteq M \cap \operatorname{conv}(X).$$

Then, theorem 2.3.2 gives $\frac{1}{a+b} z \in M \cap X$. This implies $z \in (a + b) X$, and $z \in (a + b) M = M$. Hence, $z \in M \cap (a + b) X$. Because z was arbitrarily chosen, we have

(2.3.6) $(a X + b X) \cap M \subseteq (a + b) X \cap M$.

Now let $z \in (a + b) X \cap M$. Then there is $x \in X$, such that

$z = (a + b) x$. But $(a + b) x = a x + b x$. Hence $z \in a X + b X$, and since

$z \in M$, we have $z \in M \cap (a X + b X)$. As z was arbitrarily chosen, it results that

(2.3.7) $M \cap (a + b) X \subseteq M \cap (a X + b X)$.

From (2.3.6) and (2.3.7) we obtain (2.3.5) which was proved.

In what follows we denote by

i) $\Lambda(X)$ the set of all the subsets S of V which are strongly convex with respect to M and which contain X, and

ii) $\Gamma(X)$ the set of all the subsets U of V which are slackly convex with respect to M and which contain $X \cap M$.

Definition 2.3.1. i) *The intersection of all the elements of* $\Lambda(X)$ *is called the strong convex hull of* X *generated by* M *and denoted by* $\mathrm{tconv}_M X$

ii) *The intersection of all elements of* $\Gamma(X)$ *is called the slackly convex hull of* X *generated by* M *and denoted by* $\mathrm{sconv}_M X$.

Remark 2.3.1. i) We have both $X \subseteq \mathrm{tconv}_M X$ and $X \cap M \subseteq \mathrm{sconv}_M X$.

ii) From theorem 2.3.1 and corollary 2.3.1 we get that the strong convex hull of a set $X \subseteq V$ generated by M is strongly convex with respect to M and the slackly convex hull of a set $X \subseteq V$ generated by M is slackly convex with respect to M.

Theorem 2.3.4. i) *A set* $X \subseteq V$ *is strongly convex with respect to* M *if and only if*

(2.3.8) $\quad X = \mathrm{tconv}_M X.$

ii) *A set* $X \subseteq V$ *is slackly convex with respect to* M *if and only if*

(2.3.9) $\quad X \cap M = \mathrm{sconv}_M X.$

Proof. i) **Necessity**. If X is strongly convex with respect to M, then $X \in \Lambda$; this implies that $\mathrm{tconv}_M X \subseteq X$. But, from remark 2.3.1, we have $X \subseteq \mathrm{tconv}_M X$. Hence $X = \mathrm{tconv}_M X$.

Sufficiency. Because the set $\mathrm{tconv}_M X$ is strongly convex with respect to M, from $X = \mathrm{tconv}_M X$ it results that X is strongly convex with respect to M.

ii) **Necessity**. Applying theorem 2.3.1, we get that the set $X \cap M$ is slackly convex with respect to M. Then $X \cap M \in \Gamma$. It follows that $\mathrm{sconv}_M X \subseteq X \cap M$. But, from remark 2.3.1 we have $X \cap M \subseteq \mathrm{sconv}_M X$. Hence $X \cap M = \mathrm{sconv}_M X$.

Sufficiency. Let $X \cap M = \mathrm{sconv}_M X$. From remark 2.3.1 it results that $X \cap M$ is slackly convex with respect to M. Then, from the definition of slackly convex set, we get that $M \cap \mathrm{conv}\,(M \cap (M \cap X)) \subseteq X \cap M$. But $M \cap (M \cap X) = M \cap X$. Hence we have $M \cap \mathrm{conv}\,(M \cap X) \subseteq M \cap X \subseteq X$. It follows that X is slackly convex with respect to M.

Theorem 2.3.5. *For each* $X \subseteq V$ *we have*

(2.3.10) $\quad \mathrm{tconv}_M X = (M \cap \mathrm{conv}\,(X)) \cup X.$

and

(2.3.11) $\text{sconv}_M X = M \cap \text{conv}(X \cap M)$.

Proof. i) First we prove that the set $(M \cap \text{conv}(X)) \cup X$ is strongly convex with respect to M. We remark that

(2.3.12) $(M \cap \text{conv}(X)) \cup X = \text{conv}(X) \cap (M \cup X)$.

From remark 2.3.1, it follows that $M \cup X$ and conv (X) are strongly convex with respect to M. By theorem 2.3.1, we conclude that $\text{conv}(X) \cap (M \cup X)$ is strongly convex with respect to M. From (2.3.12) it follows that $(M \cap \text{conv}(X)) \cup X$ is strongly convex with respect to M.

Now we prove that (2.3.10) is true. Because $X \subseteq X \cup (M \cap \text{conv}(X))$, we get that $(M \cap \text{conv}(X)) \cup X \in \Lambda(X)$. Then

(2.3.13) $\text{tconv}_M X \subseteq (M \cap \text{conv}(X)) \cup X$.

Let $z \in (M \cap \text{conv}(X)) \cup X$. Two cases may obtain:

a) If $z \in X$, then, from remark 2.3.1, we have $z \in \text{tconv}_M X$.

b) Let be $z \in (M \cap \text{conv}(X))$. Since $X \subseteq S$ for every $S \in \Lambda(X)$, we have $\text{conv}(X) \subseteq \text{conv}(S)$, for all $S \in \Lambda(X)$. Because every $S \in \Lambda(X)$ is strongly convex with respect to M, we have $M \cap \text{conv}(X) \subseteq M \cap \text{conv}(S) \subseteq S$, for every $S \in \Lambda(X)$. It follows that $z \in \bigcap_{S \in \Lambda} S = \text{tconv}_M X$.

Therefore, in both cases $z \in \text{tconv}_M X$. As z was arbitrarily chosen, we have

(2.3.14) $(M \cap \text{conv}(X)) \cup X \subseteq \text{tconv}_M X$.

From (2.3.13) and (2.3.14) it results (2.3.10).

ii) Applying remark 2.1.1 and theorem 2.3.1, one gets that $M \cap \text{conv}(X \cap M)$ is slackly convex with respect to M. Because $X \cap M \subseteq M \cap \text{conv}(X \cap M)$, we obtain that $M \cap \text{conv}(X \cap M) \in \Gamma(X)$. Then

(2.3.15) $\text{sconv}_M X \subseteq M \cap \text{conv}(X \cap M)$.

Since $X \cap M \subseteq Y$ then for every $Y \in \Gamma(X)$, we have $X \cap M \subseteq Y \cap M$, for all $y \in \Gamma(X)$. So, from slackly convexity of $Y \in \Gamma(X)$, we get that $M \cap \text{conv}(X \cap M) \subseteq M \subseteq \text{conv}(Y \cap M) \subseteq Y$, for all $Y \in \Gamma(X)$. Therefore

(2.3.16) $M \cap \text{conv}(X \cap M) \subseteq \bigcap_{Y \in \Gamma} Y = \text{sconv}_M X$.

From (2.3.15) and (2.3.16) it results (2.3.11).

Remark 2.3.2. In view of theorems 2.1.1 and 2.3.4 and applying Carathéodory's theorem, we get that:

i) if X is a subset of V and $n = 1+\dim X$, then for each point $z \in \text{tconv}_M X$ there are n points $x^k \in X$, $k \in \{1, ..., n\}$, such that

$$z \in \text{conv}(\{x^1, ..., x^n\}).$$

ii) if X is a subset of V and $p = 1 + \dim(X \cap M)$, then for each point $z \in \text{sconv}_M X$ there are p points $x^k \in X \cap M$, $k \in \{1, ..., p\}$, such that

$$z \in \text{conv}(\{x^1, ..., x^n\}).$$

Theorem 2.3.6. *Let* m *be a natural number,* $m \geq n$. *If* $M \subseteq \mathbf{R}^n$ *is a nonempty convex set,* X_k, $k \in \{1, ..., m\}$ *are strongly convex subsets of* $\mathbf{R}^n$ *with respect to* M *satisfying the condition that*

$$(2.3.17) \qquad \left(\bigcap_{j=1}^{n+1} X_{k_j}\right) \cap M \neq \varnothing, \textit{ for any } k_1, ..., k_{n+1} \in \{1, ..., m\},$$

then $M \cap \left(\bigcap_{k=1}^{m} X_k\right) \neq \varnothing$.

Proof. The proof is by induction on m. For $m = n+1$, the assertion of the theorem is trivial. Suppose that $m > n+1$ and that the conclusion of the theorem is true for $m - 1$ sets. Then there are m points $x^1, ..., x^m$ such that

$$(2.3.18) \qquad x^k \in \left(\bigcap_{\substack{j=1\\ j\neq k}}^{m} X_j\right) \cap M, \text{ for each } k \in \{1, ..., m\}.$$

Because

$$(2.3.19) \qquad \begin{cases} \sum_{k=1}^{m} t_k x^k = 0 \\ \sum_{k=1}^{m} t_k = 0 \end{cases}$$

is a homogeneous system of n+1 linear equations and $m \geq n+1$ unknowns, t_j, $j \in \{1, ..., m\}$, it has a nontrivial solution $s_1,...,s_m$. Then there are $k \in \{1, ..., m\}$ and $j \in \{1, ..., m\}$, $k \neq j$, such that $s_k > 0$ and $s_j < 0$. Without restriction of generality, we assume that $s_1 > 0$, $s_2 \geq 0$, ..., $s_r \geq 0$ and $s_{r+1} < 0$, $s_{r+2} \leq 0$, ..., $s_m \leq 0$. Let $s = (s_1 + ... + s_r)^{-1}$ and

(2.3.20) $$y = \sum_{i=1}^{r} s_i s x^i$$

Because for any i ∈ {1, ..., r} we have $x^i \in \bigcap_{j=r+1}^{m} X_j$, and $x^i \in M$, in view of the convexity of the set M, from (2.3.20) we get

(2.3.21) $$y \in \mathrm{conv}\,(\bigcap_{j=r+1}^{m} X_j), \text{ and } y \in M.$$

Because the sets X_j, j ∈ {r+1, ..., m} are strongly convex with respect to M, the set $\bigcap_{j=r+1}^{m} X_j$ is also strongly convex with respect to M. Then, from (2.3.21) we get that

(2.3.22) $$y \in (\bigcap_{j=r+1}^{m} X_j), \text{ and } y \in M.$$

Taking into account that $s_1 + ... + s_r = (-s_{r+1}) + ... + (-s_m)$ and putting $\alpha_j = s_j/\alpha$, for all j ∈ {r + 1, ..., m}, where $\alpha = s_{r+1} + ... + s_m$, (2.3.19) and (2.3.20) imply

$$y = s^{-1}\cdot(-s_{r+1}x^{r+1} - ... - s_m x^m) =$$
$$= -\alpha^{-1}\cdot(-s_{r+1}x^{r+1} - ... - s_m x^m) = \alpha_1 x^{r+1} + ... + \alpha_m x^m.$$

Then, similarly one can show that

(2.3.23) $$y \in (\bigcap_{j=1}^{r} X_j), \text{ and } y \in M.$$

From (2.3.22) and (2.3.23) it follows that $M \cap (\bigcap_{k=1}^{m} X_k) \neq \emptyset$.□

2.4 Topology with respect to a given set

Let (V, || ||) be a normed linear space and let M be a nonempty subset of V. By μ (M) we denote the real number given by

(2.4.1) $$\mu(M) = \inf\{\|x - y\| \mid x \in M, y \in M, x \neq y\}.$$

Let be $X \subseteq V$.

Definition 2.4.1. *We say that a point* $x \in M$ *is an interior point of* X *with respect to* M *if there is a real number* r, $r > \mu(M)$ *such that* $B(x, r) \cap M \subseteq X$, *where* $B(x, r) = \{y \in V \mid \|x - y\| < r\}$. *The set of all the interior points of* X *with respect to* M *is called the interior of* X *with respect to* M *and we denote it by* $I_M(X)$.

Definition 2.4.2. *We say that a point* $x \in M$ *is a boundary point of* X *with respect to* M *if for each real number* r, $r > \mu(M)$, *we have*

$$B(x, r) \cap M \cap X \neq \varnothing \text{ and } B(x, r) \cap M \cap (V \setminus X) \neq \varnothing.$$

The set of all the boundary points of X *with respect to* M *is called the boundary set of* X *with respect to* M *and is denoted by* $\partial_M(X)$.

Definition 2.4.3. *A point* $x \in M$ *is called an outer point of* X *with respect to* M *if there is a real number* r, $r > \mu(M)$, *such that* $B(x, r) \cap X \cap M = \varnothing$. *The set of all outer points of* X *with respect to* M *is called the exterior of* X *with respect to* M *and we denote it by* $E_M(X)$.

Definition 2.4.4. *A point* $x \in M$ *is said to be an isolated point of X with respect to* M *if there is a real number* r, $r > \mu(M)$, *such that*

$$B(x, r) \cap X \cap M = \{x\}.$$

Definition 2.4.5. *We say that a point* $x \in M$ *is an accumulation point of* X *with respect to* M *if for each real number* r, $r > \mu(M)$, *we have*

$$(B(x, r) \setminus \{x\}) \cap X \cap M \neq \varnothing.$$

The set of all accumulation points of X *with respect to* M *is called the derived of* X *with respect to* M *and we denote it by* $(X')_M$.

Definition 2.4.6. *We say that a point* $x \in M$ *is an adherence point of* X *with respect to* M *if for each real number* r, $r > \mu(M)$, *we have*

$$(B(x, r) \setminus \{x\}) \cap X \cap M \neq \varnothing.$$

The set of all adherence points of X *with respect to* M *is called the adherence of* X *with respect to* M *and we denote it by* $cl_M(X)$.

Example 2.4.1. Let $V = \mathbf{R}$, $M = \mathbf{Q}$, and $X = \{x \in \mathbf{Q} \mid -\sqrt{2} \leq x \leq \sqrt{2}\}$. We have $\mu(M) = 0$, $I_M(X) = X$, $\partial_M(X) = \varnothing$, $E_M(X) = \mathbf{Q} \setminus [-\sqrt{2}, \sqrt{2}]$, and $(X')_M = X$, $cl_M(X) = X$..

Example 2.4.2. Let $V = \mathbf{R}$, $M = \mathbf{Q}$, and $X = \{0, 1, 3\}$. We have $\mu(M)=0$, $I_M(X) = \varnothing$, $\partial_M(X) = \{0, 1, 3\}$, $E_M(X) = Q \setminus \{0, 1, 3\}$, $(X')_M = \varnothing$, $cl_M(X) = X$. The points $x = 0$, $x = 1$ and $x = 3$ are isolated points of X.

Example 2.4.3. Let $V = \mathbf{R}$, $M = \mathbf{Z}$, and $X = \{0, 1, 2, 3, 5\}$. We have $\mu(M) = 1$, $I_M(X) = \{1, 2\}$, $\partial_M(X) = \{-1, 0, 3, 4, 5, 6\}$, $(X')_M = \{-1, 0, 1, 2, 3, 4, 6\}$, $cl_M(X) = \{-1, 0, 1, 2, 3, 4, 5, 6\}$. The point 5 is an isolated point of X with respect to M.

Example 2.4.4. Let $V = \mathbf{R}^2$, $M = \mathbf{Z}^2$, and $X = \{(0,2), (-1,1), (1,0), (0,1)\}$. We have $I_M(X) = \varnothing$, and $\partial_M(X) = X \cup \{(-2, 1), (-1, 0), (-1, 2), (0, 0), (0, 3), (1, -1), (1, 1), (1, 2), (2, 0)\}$ and $(X')_M = \{(0,2), (-1,1), (0,1), (-2, 1), (-1, 0), (-1, 2), (0, 0), (0, 3), (1, -1), (1, 1), (1, 2), (2, 0)\}$. Obviously, $cl_M(x) = \{(0,2), (-1,1), (1,0), (0,1), (-2, 1), (-1, 0), (-1, 2), (0, 0), (0, 3), (1, -1), (1, 1), (1, 2), (2, 0)\}$. The point (1,0) is an isolated point of X with respect to M.

Example 2.4.5. Let $V = \mathbf{R}$, $M = \{-3/2, -1, 0, 1\}$ and $X = \{0, 1, 2\}$. We have $\mu(M) = 1/2$, $I_M(X) = \{0, 1\}$, $\partial_M(X) = \varnothing$, $E_M(X) = \{-3/2, -1\}$, $(X')_M = \varnothing$, and $cl_M(X) = \{0, 1\}$. We remark that the point $x = 2$ is neither interior point nor exterior point nor bounded point of X with respect to M. The points $x=0$ and $x = 1$ are simultaneously interior and isolated point of X with respect to M.

Example 2.4.6. Let $V = \mathbf{R}$, $M = \{0\} \cup \{1/n \mid n \in \mathbf{N}^*\}$ and $X = Q \cap]0, \sqrt{2}\,]$. We have $\mu(M) = 0$, $I_M(X) = \{1/n \mid n \in \mathbf{N}^*\}$, $\partial_M(X) = \{0\}$, $E_M(X) = \varnothing$, $(X')_M = M$ and $cl_M(X) = M$.

Remark 2.4.1. If M and X are nonempty sets of V, then the following assertions are true:

i) if $x \in X$ is an isolated point of X, then $x \in \partial_M(X)$;

ii) $\partial_M(X) \subseteq cl_M(X)$;

iii) $I_M(X) \subseteq cl_M(X)$;

iv) $cl_M(X) = \partial_M(X) \cup I_M(X)$;

v) $I_M(X) \cap \partial_M(X) = \varnothing$, $I_M(X) \cap E_M(X) = \varnothing$, $\partial_M(X) \cap E_M(X) = \varnothing$.

vi) $I_M(X) \cup \partial_M(X) \cup E_M(X) = M$.

Remark 2.4.2. Let M be a nonempty subset of V. It is easy to see that:

(a) $I_M(X) \subseteq X \cap M$, for any $X \subseteq V$;

(b) if $M = V$, then $I_M(X) = \text{int } X$, $\partial_M(X) = Fr(X)$, $E_M(X) = Ext(X)$ and $\partial_M(X) = \overline{X}$, for any $X \subseteq V$;

(c) $I_M(M) = M$, $\partial_M(M) = \varnothing$, $E_M(M) = \varnothing$, $cl_M(M) = M$;

(d) $I_M(\varnothing) = \varnothing$, $\partial_M(\varnothing) = \varnothing$, $E_M(\varnothing) = M$ and $cl_M(\varnothing) = \varnothing$;

(e) $I_M(V) = M$, $\partial_M(V) = \varnothing$, $E_M(V) = \varnothing$ and $cl_M(V) = M$.

Definition 2.4.7. *We say that* $X \subseteq V$ *is*

i) *an open set with respect to* M *if* $I_M(X) = X \cap M$;

ii) *a closed set with respect to* M *if* $cl_M(X) = X \cap M$.

From remark 2.4.2 it follows that, if M is a non-empty set of V, then the sets $\varnothing$, M and V are simultaneously open and closed with respect to M.

The sets X from examples 2.4.1 and 2.4.5 are simultaneously open and closed with respect to M. The set X from example 2.4.2 is closed with respect to M. The set X from examples 2.4.3 and 2.4.4 are neither open nor closed with respect to M. The set X from example 2.4.6 is open with respect to M and it is not closed with respect to M.

Theorem 2.4.1. *If* G_k, $k \in K$, *is a family of subsets of* V, *open with respect to* M, *then the set* $G = \bigcup_{k \in K} G_k$ *is open with respect to* M.

Proof. Let $x \in G$. There is $j \in K$ such that $x \in G_j$. Because G_j is open with respect to M, there is $r > \mu(M)$ such that $B(x, r) \cap M \subseteq G_j$. But $G_j \subseteq G$. It follows that $B(x, r) \cap M \subseteq G$. Hence $x \in I_M(G)$. Therefore $I_M(G) = G$. The set G is open with respect to M.

Theorem 2.4.2. *If* G_k, $k \in \{1, ..., p\}$, *are subsets of* V, *open with respect to* M, *then the set* $H = \bigcap_{k=1}^{p} G_k$ *is open with respect to* M.

Proof. Let $x \in H$. Then $x \in G_k$, for each $k \in \{1, ..., p\}$. As G_k, $k \in \{1, ..., p\}$, is open with respect to M, there are $r_k > \mu(M)$, $k \in \{1, ..., p\}$ such that $B(x, r_k) \cap M \subseteq G_k$, for each $k \in \{1, ..., p\}$. If we take $r = \min\{r_k \mid k \in \{1, ..., p\}\}$, then we have $r > \mu(M)$, and $B(x, r) \cap M \subseteq B(x, r_k) \cap M \subseteq G_k$, for each $k \in \{1, ..., p\}$. It follows that $B(x, r) \cap M \subseteq \bigcap_{k=1}^{p} G_k = H$. Hence $x \in I_M(H)$. Then $H \subseteq I_M(H)$. From remark 2.4.1 we have $I_M(H) \subseteq H$. It follows that $I_M(H) = H$. Therefore the set H is open with respect to M.

From theorems 2.4.1, 2.4.2 and remark 2.4.2 we get

Corollary 2.4.1. *If* M *is a nonempty set of* V, *then the set*

$$T = \{G \subseteq V \mid G \text{ has the I-property with respect to } M\}$$

is a topology on V *which we call topology with respect to* M.

Definition 2.4.8. *Let* p *be a natural number,* $p \geq 2$. *We say that*

i) *a point* $x \in X \cap M$ *is a* p-*extremal point of* X *with respect to* M *if there are no* p *points* $x^1, ..., x^p$ *of* $((X \setminus \{x\}) \cap M)$ *and* p *nonnegative real numbers* $t_1, ..., t_p$ *such that the following three conditions are simultaneously true*:

$$t_1+...+t_p=1, \quad t_1x^1+...t_px^p \in M \quad \textit{and} \quad t_1x^1+...+t_px^p = x;$$

ii) a point $x \in X \cap M$ *is an extremal point* of X *with respect to* M *if*

$$x \notin M \cap \text{conv}\,((X \setminus \{x\}) \cap M).$$

The set of all p-extremal points of X with respect to M we denote by $Exr_{M,p}(X)$. The set of all extremal points of X with respect to M we denote by $Exr_M(X)$.

Remark 2.4.3. If $X \subseteq V$ is a nonempty set, then it is easy to see that $Exr_M(X) \subseteq Exr_{M,p}(X) \subseteq Exr_{M,s}(X)$, for any natural numbers p and s, $2 \le s < p$.

Example 2.4.7. If $V = \mathbf{R}$, $M = \mathbf{Q}$, and $X = \{x \in \mathbf{Q} \mid -\sqrt{2} \le x \le \sqrt{2}\}$, then $Exr_M(X) = Exr_{M,n}(X) = \varnothing$, for all natural numbers n, $n \ge 2$.

Example 2.4.8. If $V = \mathbf{R}$, $M = \mathbf{Q}$, and $X = \{0, 1, 3\}$, then $Exr_M(X) = Exr_{M,n}(X) = \{0, 3\}$, for all natural number n, $n \ge 2$.

Example 2.4.9. If $V = \mathbf{R}^2$, $M = \mathbf{Z}^2$, and $X = \{(0, 2), (-1, 1), (1, 0), (0, 1)\}$, then $Exr_{M,2}(X) = X$, and $Exr_M(X) = Exr_{M,n}(X) = Exr_{M,3}(X) = \{(0, 2), (-1, 1), (1, 0)\}$, for all natural numbers n, $n \ge 3$.

Theorem 2.4.3. *If* $X \subseteq V$ *is slackly convex with respect to* M, $x \in X$, *and* p *is a natural number*, $p \ge 2$, *then the following assertions are equivalent*:

i) $x \in Exr_p(X)$;

ii) *if there are* p *elements* $x^1, ..., x^p \in X \cap M$ *with the property that* $x \in M \cap \text{conv}\,\{x^1, ..., x^p\}$, *then* $x^1 = ... = x^p = x$;

iii) *the set* $X \setminus \{x\}$ *is slackly* p-*convex.*

Proof. Suppose that i) is true and ii) is false. Then there are $x^1, ..., x^p$, elements of $X \cap M$, and $t_1, ..., t_p$, nonnegative real numbers with $t_1 + ... + t_p = 1$, such that $x = t_1x^1 + ... + t_px^p$. Let

$$P = \{j \in \{1, ..., p\} \mid t_j \neq 0\}, \text{ and } J = \{j \in P \mid x^j = x\}.$$

If $J = P$, then ii) is true. If we suppose that $J \neq P$, then $(1 - \sum_{j \in J} t_j)\, x = \sum_{j \in P \setminus J} t_j\, x^j$ and dividing by $(1 - \sum_{j \in J} t_j)$ which is a strictly positive number, we get $x = \sum_{j\in P\setminus J} s_j x^j$, where $s_j = \frac{t_j}{1-\sum_{j\in J} t_j}$, for all $j \in P \setminus J$. Because $\sum_{j\in P\setminus J} s_j = 1$, and $s_j > 0$, for all $j \in P \setminus J$, it follows that

$$x \in M \cap \text{conv}\{x^j \mid j \in P \setminus J\}.$$

Then $x \notin \text{Exr}_q(X)$, where $q = \text{card}(P \setminus J)$. But, in view of remark 2.4.2, $\text{Exr}_p(X) \subseteq \text{Exr}_q(X)$. Hence $x \notin \text{Exr}_p(X)$. That contradicts the fact that $x \in \text{Exr}_p(X)$. Hence $J = P$. Therefore ii) is true.

Now suppose that ii) is true and iii) is false. Then there are p elements, $x^1, ..., x^p$ of $(X \setminus \{x\}) \cap M$ and p nonnegative real numbers $t_1, ..., t_p$ with $t_1 + ... + t_p = 1$ such that $u = t_1x^1 + ... + t_px^p \in M$ and $u \notin X \setminus \{x\}$. From the slack p-convexity of X we get $u \in X$. It follows that $u = x$. Applying ii) it results that $x^1 = ... = x^p = x$. This contradicts the fact that $x^k \in X \setminus \{x\}$, for all $k \in \{1, ..., p\}$. Therefore iii) is true.

The implication iii) => i) is immediate.

Definition 2.4.9. *We say that a point* $x \in X$ *has the* ci-*property if there is a real number* r, $r > 0$, *such that* $B(x, r) \subseteq \text{conv}(X)$. *By* ci(X) *we denote the set of all points of* X *which have the* ci-*property.*

Definition 2.4.10. *We say that a subset* X *of* V *has the* ci-*property with respect to* M *if* $M \cap \text{ci}_M(X) \subseteq X$.

Example 2.4.10. Let $V = \mathbf{R}^2$, $M = \mathbf{Z}^2$, $A = \{(-1, 0), (0, 1), (0, 2), (1, 0)\}$ and $B = \{(-1, -1), (0, 1), (0, 2), (1, -1)\}$. It is easy to see that $\text{ci}(A) = \{(0, 1)\}$ and $\text{ci}(B) = \{(0, 1)\}$. We have $A \cap B = \{(0, 1)\}$ and $\text{ci}(A \cap B) = \emptyset$. The sets A and B have the ci- property.

Theorem 2.4.4. *If* G_k, $k \in K$, *is a family of subsets of* V *with the* ci-*property, then the set* $G = \bigcup_{k\in K} G_k$ *has also the* ci-*property.*

Proof. Let $x \in G$. There is $j \in K$ such that $x \in G_j$. Because G_j has the ci-property, there is $r > 0$ such that $B(x, r) \subseteq \text{conv}(G_j)$. But $\text{conv}(G_j) \subseteq \text{conv}(G)$. It follows that $B(x, r) \subseteq \text{conv}(G)$. Hence $x \in \text{ci}(G)$. Therefore $\text{ci}(G) = G$; the set G has the ci-property.

From example 2.4.10 it results that the intersection of a finite family of subsets of V, which has the ci-property, is not a set with ci-property.

2.5 The problem of the best approximation

In this section the problem of the best approximation of an element of the space $\mathbf{R}^n$ by elements of a strongly or slackly convex subset of $\mathbf{R}^n$ is discussed.

Let be $X \subseteq \mathbf{R}^n$ and let $y \in \mathbf{R}^n$. We remember that a point $\alpha \in X$ is called element of best approximation of y by elements of X, if

$$\| y - \alpha \| \leq \|y - x\|, \text{ for all } x \in X.$$

Theorem 2.5.1. *If* M *is a convex subset of* $\mathbf{R}^n$, X *is a strongly convex set with respect to* M, $y \in M$, *and there is* $\alpha \in X \cap \operatorname{int} M$ *such that* α *is an element of best approximation of* y *by elements of* X, *then* α *is the single element of best approximation of* y *by elements of* X.

Proof. If $y \in X$, then the theorem is true. Let now $y \in \mathbf{R}^n \setminus X$. Suppose that there is a point $u \in X$, $u \neq \alpha$, which is an element of best approximation of y by elements of X. Since $\alpha \in X \cap \operatorname{int} M$, there is a real number $r > 0$ such that $\{x \in \mathbf{R}^n \mid \|x - \alpha\| < r\} \subseteq M$. Then there is a real number t, $0 < t < 1$ such that $z = t\alpha + (1-t)u \in M$. But X is strongly convex with respect to M and $z \in M \cap \operatorname{conv}(X)$. It follows that $z \in X$.

We have $\|z - y\| = \|t\alpha + (1-t)u - y\| = \|t\alpha + (1-t)u - ty + (1-t)y\|$. Hence

$$(2.5.1) \qquad \|z - y\| \leq t\|\alpha - y\| + (1-t)\|u - y\|.$$

We prove that in (2.5.1) we cannot have equality. Suppose that $\|t\alpha + (1-t)u - ty\| = t\|\alpha - y\| + (1-t)\|u-y\|$. Then, from the Cauchy-Schwartz inequality, it follows that there are two real numbers a and b, with $|a| + |b| \neq 0$, such that

$$(2.5.2) \qquad at(\alpha-y) + b(1-t)(u-y) = 0.$$

As $\alpha \neq y$ and $u \neq y$, it follows that $a \neq 0$ and $b \neq 0$. Two cases may appear: $a \cdot b > 0$ and $a \cdot b < 0$.

Case I. If $a \cdot b > 0$, then from (2.5.2), we get

$$y = \frac{at}{at + (1-t)b} \cdot \alpha + \frac{b(1-t)}{at + (1-t)b} u.$$

As at

$$(at + (1-t)b)^{-1} > 0, \; b(1-t)(at + (1-t)b)^{-1} > 0$$

and at

$$(at + (1-t)b)^{-1} + b(1-t)(at + (1-t)b)^{-1} = 1,$$

we have $y \in \text{conv}(\{\alpha,u\})$ and $y \in M$. It follows that $y \in M \cap \text{conv}(X)$. Because X is strongly convex with respect to M, we have $y \in X$, which is a contradiction.

Case II. If $a \cdot b < 0$, then we get from (2.5.2)

$$(2.5.3) \qquad \|\alpha - y\| = \left|\frac{b}{a}\right| \cdot \frac{1-t}{t} \cdot \|u - y\|.$$

Because α and u are elements of best approximation of y by elements of X, we have $\|\alpha - y\| = \|u - y\|$. Then, from (2.5.3), it follows that $t = |b|\,(|a|+|b|)^{-1}$ Replacing t in (2.5.2), we get

$$(2.5.4) \qquad a|b|\alpha + b|a|u - (a|b| + b|a|)y = 0$$

As $a \cdot b < 0$, we obtain $a|b| + |b|a = 0$. Then, from (2.5.4) we get $\alpha = u$, thus contradicting the assumption that $u \neq \alpha$.

Because in both cases we obtain a contradiction, we cannot have equality in (2.5.2). Hence

$$(2.5.5) \qquad \|z - y\| < t \cdot \|\alpha - y\| + (1-t) \cdot \|u - y\|.$$

As α and u are elements of best approximation of y by elements of X, (2.5.5) implies

$$(2.5.6) \qquad \|z - y\| < \|\alpha - y\|.$$

The strict inequality (2.5.6) contradicts the hypothesis that α is an element of best approximation of y by elements of X. It follows that there is no $u \in X$, $u \neq \alpha$, which is an element of best approximation of y by elements of X.

Let $y \in \mathbf{R}^n$ and $X \subseteq \mathbf{R}^n$. By A (X, y) we denote the set of all elements of best approximation of y by elements of X.

Theorem 2.5.2. *Let* p *be a natural number*, $p \geq 2$.

i) *If* $X \subseteq \mathbf{R}^n$ *is* p-*strongly convex with respect to* M, *then the set* A (X, y) *is strongly* p-*convex with respect to* M.

ii) *If* $X \subseteq \mathbf{R}^n$ *is* p-*slackly convex with respect to* M, *then the set* A (X, y) *is slackly* p-*convex with respect to* M.

Proof. i) If A (X, y) = $\varnothing$ or card A (X, y) = 1, the conclusion of the theorem is true. Suppose that card A (X, y) ≥ 2. Let $x^* \in$ A (X, y). We denote by $a = \|y - x^*\|$. Obviously we have $a = \|y - u\|$ for each $u \in$ A (X, y).

Let $x_1, ..., x_p$ be elements of A (X, y) and $t_1, ..., t_p$ be nonnegative real numbers with $t_1 + ... + t_p = 1$ and $t_1x_1 + ... + t_px_p = z \in M$. Because X is strongly p-convex with respect to M, we get $z \in X$. Then we have

$$(2.5.7) \quad \|z - y\| = \|t_1(x_1 - y) + ... + t_p(x_p - y)\| \le t_1a + ... + t_pa = a.$$

But $\|x - y\| \ge a$, for all $x \in X$. Hence

$$(2.5.8) \quad \|z - y\| \ge a.$$

From (2.5.7) and (2.5.8) it results that $\|z - y\| = a$. It follows that $z \in$ A (X, y). Therefore A (X, y) is strongly p-convex with respect to M.

ii) The proof is by analogy with i).

Corollary 2.5.1. i) *If* $X \subseteq \mathbf{R}^n$ *is strongly convex with respect to* M, *then the set* A (X, y) *is strongly convex with respect to* M.

ii) *If the set* X *is slackly convex with respect to* M, *then the set* A (X, y) *is slackly convex with respect to* M.

Proof. i) Let dim X = p. Because X is strongly convex with respect to M, X is also strongly (p+1)-convex with respect to M. Applying theorem 2.5.1, it results that the set A (X, y) is (p+1) strongly convex with respect to M. But dim A (X, y) $\le$ p. Then, in view of theorem 2.1.1, we get that A (X, y) is strongly convex with respect to M.

Similarly we can prove ii).

Theorem 2.5.3. *If* X *is strongly convex with respect to* $\mathbf{Z}^n$, *and* $y \in \mathbf{Z}^n$, *then* I_{Z^n} (A (X, y)) $= \varnothing$.

Proof. Suppose that I_{Z^n} (A (X, y)) $\neq \varnothing$, and let $x \in I_{Z^n}$ (A (X, y)). Then the points $u = x + e^1$ and $v = x - e^1$, where $e^1 = (1, 0, \dots, 0) \in \mathbf{Z}^n$, are elements of A (X, y), which implies

$$(2.5.9) \quad \|y - u\| = \|y - v\| = \|y - x\|.$$

On the other hand, we have $x = 2^{-1} u + 2^{-1} v$. Hence

$$(2.5.10) \quad \|x-y\|^2 = 4^{-1}\|(u-y) + (v-y)\|^2 <$$
$$< 4^{-1}(\|(u-y) + (v-y)\|^2 + \|(u-y) - (v-y)\|^2).$$

From the properties of a norm, we get

$$(2.5.11) \quad \|(u-y) + (v-y)\|^2 + \|(u-y) - (v-y)\|^2 = 2\|u-y\|^2 + 2\|v-y\|^2.$$

(2.5.9), (2.5.10) and (2.5.11) imply

$$\|x - y\|^2 < 4^{-1} \cdot 4 \cdot \|x - y\|^2,$$

which is a contradiction. The assumption that I_{Z^n} (A (X, y)) $\neq \varnothing$ is false. Hence I_{Z^n} (A (X, y)) $= \varnothing$, which was to be proved.

2.6 Separation of strong and slack convex sets

In what follows we suppose that a scalar product denoted by (|) is defined in the linear space V.

Let $a \in \mathbf{R}^n \setminus \{0\}$, $b \in \mathbf{R}$, $A \subseteq V$ and $B \subseteq V$. We remember what we said previously:

i) The hyperplane $H = \{x \in V \mid (a \mid x) = b\}$ separates the nonempty sets A and B if

$$\sup \{(a \mid x) \mid x \in A\} \leq b \leq \inf \{(a \mid x) \mid x \in B\};$$

in this case the sets A and B are called separable.

ii) The hyperplane $H = \{x \in V \mid (a \mid x) = b\}$ strictly separates the nonempty set A and B if

$$\sup \{(a \mid x) \mid x \in A\} < b < \inf \{(a \mid x) \mid x \in B\};$$

in this case the sets A and B are called strictly separated.

Theorem 2.6.1. *If* M *is a nonempty subset of* $\mathbf{R}^n$, X *is a compact subset of* $\mathbf{R}^n$ *which is strongly convex with respect to* M *and* $z \in M \setminus X$, *then there exists a hyperplane* H *which strictly separates* {z} *and* X.

Proof. Because X is strongly convex with respect to M, we have $z \notin$ conv (X). As X is compact, conv (X) is a closed convex set. It results that there is a vector $a \in \mathbf{R}^n \setminus \{0_n\}$ and a number $b \in \mathbf{R}$, such that $(a \mid x) > b$, for all $x \in$ conv (X) and $(a \mid z) < b$. Taking $c = (a \mid z) + \frac{b-(a|z)}{2}$, we have $(a \mid z) < c < (a \mid x)$, for all $x \in$ conv (X). Since $X \subseteq$ conv (X), it follows that the hyperplane

$$H = \{x \in \mathbf{R}^n \mid (a \mid x) = c\}$$

strictly separates the sets {z} and X.

Corollary 2.6.1. *If* X *is a compact subset of* $\mathbf{R}^n$ *which is slackly convex with respect to* M, *and* $z \in M \setminus X$, *then there is a hyperplane* H *which strictly separates* {z} *and* $X \cap M$.

Proof. Because X is slackly convex with respect to M, the set $X \cap M$ is strongly convex with respect to M. Applying proposition 2.6.1, we get that there is a hyperplane H which strictly separates $X \cap M$ and $\{z\}$.

Corollary 2.6.2. *If* X *and* Y *are nonempty subsets of* $\mathbf{R}^n$ *such that the set* $Y - X$ *is compact and strongly convex with respect to* M *and if* $0_n \in M \setminus (Y - X)$, *then the sets* X *and* Y *can be strictly separated by a hyperplane.*

Proof. Applying proposition 2.6.1 we get that there are $a \in \mathbf{R}^n$, $a \neq 0_n$, and $b \in \mathbf{R}$ such that the hyperplane $H = \{x \in \mathbf{R}^n \mid (a \mid x) = b\}$ strictly separates the sets $\{0_n\}$ and Y - X. Then

$$(2.6.1) \qquad \inf \{(a \mid z) \mid z \in Y - X\} > b > 0.$$

From (2.6.1) it follows $(a \mid x) + b < (a \mid y)$, for all $(x, y) \in X \times Y$. Hence

$$(2.6.2) \qquad \sup \{(a \mid x) \mid x \in X\} + b \leq \inf \{(a \mid y) \mid y \in Y\}.$$

Since $b > 0$, we have

$$(2.6.3) \qquad \sup \{(a \mid x) \mid x \in X\} < \sup \{(a \mid x) \mid x \in X\} + b/2 <$$

$$< \sup \{(a \mid x) \mid x \in X\} + b.$$

Taking $c = \sup \{(a \mid x) \mid x \in X\} + b/2$, from (2.6.2) and (2.6.3) it results that there is a hyperplane H which strictly separates the sets X and Y.

Example 2.6.1. Let us consider the following sets: $V = \mathbf{R}^2$, $M = \mathbf{Z}^2$, $X = \{(1, 1), (2, 1)\}$, and $Y = \{(0, 0), (0, 1), (1, 0)\}$. The set

$$Y - X = \{(-1, -1), (-1, 0), (0, -1), (-2, -1), (-2, 0)\}$$

is compact and strongly convex with respect to $\mathbf{Z}^2$ and $0_2 \in \mathbf{Z}^2 \setminus (Y - X)$. By corollary 2.6.2 it results that there is a hyperplane H which strictly separates the sets X and Y.

Theorem 2.6.2. *If* X *is strongly convex with respect to* M *and* $z \in M \setminus X$, *then there exists a hyperplane* H *which separates* $\{z\}$ *and* X.

Proof. Because X is strongly convex with respect to the set M, we have $z \notin \text{conv}(X)$. The set $\{z\}$ and conv (X) are convex. Then, it results that there is a vector $a \in \mathbf{R}^n \setminus \{0_n\}$ and a real number b, such that the hyperplane $H = \{x \in \mathbf{R}^n \mid (a \mid x) = b\}$, separates the sets $\{z\}$ and conv (X). Since we have obviously $X \subseteq \text{conv}(X)$, it follows that the hyperplane H also separates the sets $\{z\}$ and X.

Corollary 2.6.3. *If the set* $Y - X$ *is strongly convex with respect to* M *and* $0_n \notin M \setminus (Y - X)$, *then the sets* X *and* Y *can be separated by a hyperplane.*

Proof. Applying theorem 2.6.2 it results that there are $a \in \mathbf{R}^n \setminus \{0_n\}$ and $b \in \mathbf{R}$, such that

$$(2.6.4) \quad 0 \le b \le \inf \{(a \mid z) \mid z \in Y - X\}.$$

Taking $c = \sup \{(a \mid y) \mid y \in X\} + b$, we have $\sup \{(a \mid x) \mid x \in X\} \le c \le \inf \{(a \mid y) \mid y \in Y\}$. The hyperplane $H^* = \{x \in \mathbf{R}^n \mid (a \mid x) = c\}$ separates the sets X and Y.

Corollary 2.6.4. *If* X *is slackly convex with respect to* M *and* $z \in M \setminus X$, *then there is a hyperplane* H *which separates* $\{z\}$ *and* $X \cap M$.

Proof. Because X is slackly convex with respect to M, $X \cap M$ is strongly convex with respect to M. Applying theorem 2.6.2, we get that there is a hyperplane H which separates $\{z\}$ and $X \cap M$.

Definition 2.6.1. *We say that the hyperplane* H *separates the sets* A *and* B *with respect to* M, *if* H *separates the sets* $A \cap M$ *and* $B \cap M$.

Example 2.6.2. Let $V = \mathbf{R}^2$, $M = \mathbf{Z}^2$,

$$A = \{(x, y) \in \mathbf{R}^2 \mid x \in [-2, \frac{1}{2}], 0 \le y \le \frac{1}{2}(x - \frac{1}{2})^2\} \text{ and}$$

$$B = \{(x, y) \in \mathbf{R}^2 \mid x \in [-\frac{1}{2}, 2], 0 \le y \le \frac{1}{2}(x + \frac{1}{2})^2\}$$

We have $A \cap \mathbf{Z}^2 = \{(-2, 0), (-2, 1), (-2, 2), (-2, 3), (-1, 0), (-1, 1), (0, 0)\}$ and $B \cap \mathbf{Z}^2 = \{(2, 0), (2, 1), (2, 2), (2, 3), (1, 0), (1, 1), (0, 0)\}$. It is easy to see that the hyperplane $H = \{(x, y) \in \mathbf{R}^2 \mid x = 0\}$ separates $A \cap \mathbf{Z}^2$ and $B \cap \mathbf{Z}^2$. It follows that the hyperplane H separates the sets A and B with respect to $\mathbf{Z}^2$.

Theorem 2.6.3. *If* M *is convex,* X *and* Y *are two nonempty strongly convex sets with respect to* M *and* $X \cap Y \cap M = \emptyset$, *then there is a hyperplane* H *which separates the sets* X *and* Y *with respect to* M.

Proof. Since M is a convex set, the sets $M \cap \text{conv}(X)$ and $M \cap \text{conv}(Y)$ are also convex. As X and Y are strongly convex with respect to M, we have $M \cap \text{conv}(X) = M \cap X$, and $M \cap \text{conv}(Y) = M \cap Y$. Then

$$(M \cap \text{conv}(X)) \cap (M \cap \text{conv}(Y)) = X \cap Y \cap M = \emptyset.$$

It results that there is a hyperplane H, which separates the sets $M \cap \text{conv}(X)$ and $M \cap \text{conv}(Y)$. Then H separates the sets $M \cap X$ and $M \cap Y$. Therefore H separates the sets X and Y with respect to M.

Corollary 2.6.5. *If* M *is convex,* X *and* Y *are two nonempty slackly convex sets with respect to* M *and* $X \cap Y \cap M = \emptyset$, *then there is a hyperplane* H *which separates the sets* X *and* Y *with respect to* M.

Proof. Because X and Y are slackly convex with respect to M, the sets $M \cap \text{conv}(X)$ and $M \cap \text{conv}(Y)$ are strongly convex with respect to M. Applying theorem 2.6.2, we get that there is a hyperplane H which separates the sets $(X \cap M)$ and $(Y \cap M)$ with respect to M. Therefore H separates the sets X and Y with respect to M. .

2.7 Integer convex sets and discrete polyhedral sets

An important particular case of convexity with respect to a given set is when $V = \mathbf{R}^n$ and $M = \mathbf{Z}^n$.

Definition 2.7.1. *A subset* $X \subseteq \mathbf{Z}^n$ *is called an integer convex set if* X *is strongly convex with respect to* $\mathbf{Z}^n$.

From the properties of the strongly convex sets we obtain:

Proposition 2.7.1. *The intersection of a family* $(X_k)_{k \in K}$ *of integer convex sets is integer convex.*

Proposition 2.7.2. *A set* $X \subseteq \mathbf{Z}^n$ *is integer convex if and only if*

$$X = \mathbf{Z}^n \cap \text{conv}(X).$$

Proposition 2.7.3. *A set* $X \subseteq \mathbf{Z}^n$ *is an integer convex set if and only if for each* $x_1, \ldots, x_{n+1}$ *elements of* X, *and each nonnegative numbers* $t_1, \ldots, t_{n+1}$, *satisfying the conditions* $t_1 + \ldots + t_{n+1} = 1$ *and* $t_1x_1 + \ldots + t_{n+1}x_{n+1} \in \mathbf{Z}^n$, *we have* $t_1x_1 + \ldots + t_{n+1}x_{n+1} \in X$.

In $\mathbf{R}^n$ we consider the Euclidean norm i.e. $\|u - v\| = \sqrt{\sum_{j=1}^{n}(u_j - v_j)^2}$, for all $u = (u_1, \ldots, u_n) \in \mathbf{R}^n$, and all $v = (v_1, \ldots, v_n) \in \mathbf{R}^n$.

We specify that $e_k = (0, \ldots, 0, 1, 0, \ldots, 0) \in \mathbf{R}^n$, where 1 is on k-position, for each $k \in \{1, \ldots, n\}$.

Let $X \subseteq \mathbf{Z}^n$. In what follows we denote:

- by tconv(X), the convex hull of X with respect to $\mathbf{Z}^n$, i.e. $\text{tconv}(X) = \text{tconv}_{\mathbf{Z}^n}(X)$;

- by I(X), the interior set of X with respect to $\mathbf{Z}^n$; if $\| \ \|$ is the Euclidean norm, i.e. $I(X) = I_{\mathbf{Z}^n}(X)$;

- by B(X), the boundary set of X with respect to $\mathbf{Z}^n$, if $\| \ \|$ is the Euclidean norm, i.e. $B(X) = B_{Z}{}^{n}(X)$;

- by E(X), the outside set of X with respect to $\mathbf{Z}^n$, i.e. $E(X) = E_{Z}{}^{n}(X)$;

- by $Exr_p(X)$, the set of p- extremal points of X with respect to $\mathbf{Z}^n$, where p is a natural number, $p \geq 2$.

- by Exr(X), the set of extremal points of X with respect to $\mathbf{Z}^n$.

Proposition 2.7.4. *If* $X \subseteq \mathbf{Z}^n$, *then*

$$\text{tconv}(X) = X \cup \{x \in \mathbf{Z}^n \mid x \in \text{conv}(X)\}.$$

The integer convex sets have also essential features.

Proposition 2.7.5. *If a set* $X \subseteq \mathbf{Z}^n$ *is integer convex, then for any two elements* $x = (x_1, \ldots, x_n)$ *and* $y = (y_1, \ldots, y_n)$ *of* X *and for any natural number* r, r *common division of the numbers* $|x_j - y_j|$, $j \in \{1, \ldots, n\}$, *we have*

$$(2.7.1) \qquad \frac{s}{r}x + (1 - \frac{s}{r})y \in X, \textit{ for each } s \in \{0, 1, \ldots, r\}.$$

Proof. Let r be a common divisor of the numbers $|x_j - y_j|$, $j \in \{1, \ldots, n\}$, and $s \in \{0, 1, \ldots, r\}$. As $x \in \mathbf{Z}^n$, $y \in \mathbf{Z}^n$, it follows that

$$\frac{s}{r}x + (1 - \frac{s}{r})y = y + \frac{s}{r}(x - y) \in \mathbf{Z}^n,$$

and because X is an integer convex set, we get that (2.7.1) is true.

Proposition 2.7.6. *If* $X \subseteq \mathbf{Z}^n$ *and* $x \in X$, *then* x *is an interior point of* X *if and only if* $x + e^k \in X$ *and* $x - e^k \in X$ *for each* $k \in \{1, \ldots, n\}$.

Proof. Necessity. Because $\mu(\mathbf{Z}^n) = 1$, if $x \in I(X)$, then there is a real number r, $r > 1$, such that $B(x, r) \cap \mathbf{Z}^n \subseteq X$. But, for each $k \in \{1, \ldots, n\}$, we have $x + e^k \in B(x, r) \cap \mathbf{Z}^n$, and $x - e^k \in B(x, r) \cap \mathbf{Z}^n$. Hence $x + e^k \in X$ and $x - e^k \in X$ for each $k \in \{1, \ldots, n\}$.

Sufficiency. If we take $r \in \mathbf{R}$, $r > 1$, then

$$B(x, r) \cap \mathbf{Z}^n =$$

$$= \{x\} \cup \{x + e_k \mid k \in \{1, \ldots, n\}\} \cup \{x - e_k \mid k \in \{1, \ldots, n\}\} \subseteq X.$$

Hence $x \in I(X)$.

Remark 2.7.1. If the Euclidean norm is replaced by Chebicheff's norm, namely $\|x - y\| = \max\{|x_j - y_j| \mid j \in \{1,\ldots,n\}\}$, for all $x, y \in \mathbf{R}^n$, then we have $B(x, r) =]x_1 - r, x_1 + r[\times \ldots \times]x_n - r, x_n + r[$. Therefore, in the particular case $n = 2$, if $X \subseteq \mathbf{Z}^2$, then a point $x \in I(X)$ if and only if is an interior point according to the definition of A. Rosenfeld (1970, 1974) and J.M. Chassery (1979). In this case, the topology with respect to $\mathbf{Z}^2$ reduces to the known digital topology used in pattern recognition.

From proposition 2.7.6 and remark 2.4.1 it results that

Corollary 2.7.1. *If* $X \subseteq \mathbf{Z}^n$ *then a point* $x \in \mathbf{Z}^n$ *is an outer point of* X *if and only if* $x + e_k \in \mathbf{Z}^n \setminus X$ *and* $x - e_k \in \mathbf{Z}^n \setminus X$, *for each* $k \in \{1, ..., n\}$.

Corollary 2.7.2. *If* $X \subseteq \mathbf{Z}^n$ *then a point* $x \in \mathbf{Z}^n$ *is a boundary point of* X *if and only if there are* $j, k \in \{1, ..., n\}$ *such that*

$$x + e^j \in X \text{ or } x - e^j \in \mathbf{Z}^n \setminus X, \text{ and } x + e^k \in \mathbf{Z}^n \setminus X \text{ or } x - e^k \in \mathbf{Z}^n \setminus X.$$

Proposition 2.7.7. *If* $X \subseteq \mathbf{Z}^n$ *is an integer convex set, then* I(X) *is also integer convex set.*

Proof. Let $x \in \mathbf{Z}^n \cap \text{conv}\,(I(X))$. From theorem 2.1.2 it follows that there are n + 1 elements $x_1, ..., x_{n+1}$ of I(X) and n + 1 nonnegative real numbers $t_1, ..., t_{n+1}$ with $t_1 + ... + t_{n+1} = 1$, such that $x = t_1x_1 + ... + t_{n+1}x_{n+1}$. In view of proposition 2.7.6, we have $x_j + e^k \in X$, and $x_j - e^k \in X$, for each $k \in \{1, ..., n+1\}$ and each $k \in \{1, ..., n\}$. Then

$$x + e^k = t_1x_1 + ... + t_{k-1}x_{k-1} + t_k(x_k + 1) + t_{k+1}x_{k+1} + ... + t_{n+1}x_{n+1} \in$$
$$\in \mathbf{Z}^n \cap \text{conv}\,(X),$$

and

$$x - e^k = t_1x_1 + ... + t_{k-1}x_{k-1} + t_k(x_k - 1) + t_{k+1}x_{k+1} + ... + t_{n+1}x_{n+1} \in$$
$$\in \mathbf{Z}^n \cap \text{conv}\,(X),$$

for all $k \in \{1,...,n\}$. Then, due to proposition 2.7.6 we get that $x \in I(X)$. Because x was arbitrarily chosen in $\mathbf{Z}^n \cap \text{conv}\,I(X)$, it follows that

$$\mathbf{Z}^n \cap \text{conv}\,(I(X)) \subseteq I(X);$$

therefore I(X) is an integer convex set.

Proposition 2.7.8. *If* $X \subseteq \mathbf{Z}^n$ *is an integer convex set,* $x \in I(X)$, $y \in \partial(M)$ *and* $z \in [x, y] \cap M$, *then* $z \in I(X)$ *or* $z \in \partial(X)$.

Proof. As X is an integer convex set it results $z \in X$. Applying remark 2.4.2 it follows that $z \in I(X)$ or $z \in \partial(X)$.

We remark that in the classical case if X is a convex set, $x \in \text{Int}\,(X)$ and $z \in \text{Fr}(X)$, then $\langle x, y \rangle \subseteq \text{Int}\,(X)$. In the case of an integer convex set, this property is not true.

Example 2.7.1. Let X = {(0, 0), (0, 1), (0, 2), (1, 0), (1, 1), (1, 2), (2, 0), (2, 1), (2, 2), (3, 1), (4, 1)}. It is easy to see that X is an integer convex set.

$(1, 1) \in I(X)$, and $(4, 1) \in \partial(X)$. We have $(2, 1) \in \langle(1, 1), (4, 1)\rangle$ and $(2, 1) \in I(X)$. We also have $(3, 1) \in \langle(1, 1), (4, 1)\rangle$ and $(3, 1) \in \partial(X)$.

Proposition 2.7.9. *If* $X \subseteq \mathbf{Z}^n$ *is an integer convex set and* $x \in Exr(X)$, *then* $x \notin I(X)$.

Proof. Suppose that $x \in I(X)$. Let be $k \in \{1, ..., n\}$. Then, $x + e^k \in X$, and $x - e^k \in X$. Because $x = \frac{1}{2}(x + e^k) + \frac{1}{2}(x - e^k)$, it follows that $x \notin Exr_2(X)$, and then, in view of remark 2.7.1 we get $x \notin Exr(X)$. This contradicts the hypothesis. Hence $x \notin I(X)$.

Corollary 2.7.3. *If* $X \subseteq \mathbf{Z}^n$ *is an integer convex set, then*

$$Exr_{n+1}(X) \subseteq \partial(X), \text{ and } Exr(X) \subseteq \partial(X).$$

Proposition 2.7.10. *If* $X \subseteq \mathbf{Z}^n$ *is an integer convex set and* $x \in X$, *then the following assertions are equivalent*:

i) $x \in Exr(X)$;

ii) *if, for a natural number* p, $p \geq 2$, *there are* p *elements* $x_1, ..., x_p$ *of* X *such that* $x \in M \cap conv\{x_1,...,x_p\}$, *then* $x_1 = ... = x_p = x$;

iii) *the set* $X \setminus \{x\}$ *is integer convex.*

Proof. Let i) be true, i.e. $x \in Exr(X)$. Suppose that there are: a natural number p, $p \geq 2$, and p elements $x_1, ..., x_p$ of X such that $x \in M \cap conv(\{x_1, ..., x_p\})$. Let $J = \{j \in \{1, ..., p\} \mid x_j = x\}$. It follows that

$$x \in M \cap conv(\{x_k \mid k \in \{1, ..., p\} \setminus J\}) \subseteq M \cap conv(X \setminus \{x\})$$

which contradicts i). Hence ii) is true.

Now let ii) be true. If we suppose that $X \setminus \{x\}$ is not integer, then $M \cap conv(X \setminus \{x\}) \not\subset X \setminus \{x\}$. It follows that there is a natural number p, $p \geq 2$, p elements $x_1, ..., x_p$ of $X \setminus \{x\}$ and p nonnegative real numbers $t_1, ..., t_p$ with $t_1 + ... + t_p = 1$, such that $u = t_1x_1 + ... + t_px_p \in M$ and $u \notin X \setminus \{x\}$. But X is an integer set. Then $u \in X$. This implies that $x = u$. Applying ii) we get that $x_1 = ... = x_p = x$ which contradicts the assumption that $x_k \in X \setminus \{x\}$ for all $k \in \{1, ..., n\}$. Therefore $X \setminus \{x\}$ is integer set.

In view of the definition of the extremal point with respect to $\mathbf{Z}^n$, it is easy to see that if iii) is true, then i) is also true.

Proposition 2.7.11. *If* $X \subseteq \mathbf{Z}^n$ *is a finite integer convex set and* $x^0 \in \mathbf{Z}^n \setminus X$, *then there is a hyperplane that strictly separates* $\{x^0\}$ *and* X.

Proof. As $X \subseteq \mathbf{Z}^n$, it follows that X is a compact set. Then conv (X) is also compact set. Because $x^0 \in \mathbf{Z}^n \setminus X$, and X is integer convex set, we get that $x^0 \notin \text{conv}(X)$. It results that there is a hyperplane H, which strictly separates the set $\{x^0\}$ and conv (X). But $X \subseteq \text{conv}(X)$. Therefore H strictly separates the sets $\{x^0\}$ and X.

Proposition 2.7.12. *If* $X \subseteq \mathbf{Z}^n$ *is an integer convex set and* $x^0 \in \mathbf{Z}^n \setminus X$, *then there is a hyperplane which separates* $\{x^0\}$ *and* X.

Proof. In view of the integer convexity of X, from $x^0 \in \mathbf{Z}^n \setminus X$ we get that $x^0 \notin \text{conv}(X)$. Then the sets $\{x^0\}$ and X are two convex and mutually disjoint sets. It results that there is a hyperplane H, which separates the set $\{x^0\}$ and conv (X). But $X \subseteq \text{conv}(X)$. Therefore H separates the sets $\{x^0\}$ and X.

We remember that a subset P of $\mathbf{R}^n$ is said to be a polyhedral set if it is the intersection of a finite number of closed half-spaces or if $P = \mathbf{R}^n$. Every polyhedral set P has a representation

$$(2.7.2) \qquad P = \bigcap_{i=1}^{m} K(a^i, b_i),$$

where each $K(a^i, b_i)$ is a closed half-space, i.e.

$$K(a^i, b_i) = \{x \in \mathbf{R}^n \mid (a^i \mid x) \geq b_i\},$$

$a^i \in \mathbf{R}^n$, $b_i \in \mathbf{R}$, $i \in \{1, ..., m\}$, being given vectors and numbers .

Definition 2.7.2. *A polyhedral set is called an integer polyhedral set if in representation* (2.7.2) *we have* $a^i \in \mathbf{Z}^n$, $b_i \in \mathbf{Z}$, *for all* $i \in \{1, ..., n\}$.

Definition 2.7.3. *We shall call a representation* (2.4.1) *of* P *irreducible if*

$$m = 1 \textit{ or if } m > 1 \textit{ and } Q \not\subset \bigcap_{\substack{i=1 \\ i \neq j}}^{m} K(a^i, b_j), \textit{for all } j \in \{1, ..., m\}.$$

We remember that for a classical closed convex set C, a convex subset F of C is called a face of C if for any two distinct points $y, z \in C$, such that $[y, z] \cap F \neq \varnothing$, we have $[y, z] \subseteq F$. A k - face is a face F of C having dim F = k. Both $\varnothing$ and C are faces of C, called improper faces; all the other faces are called proper faces. A facet of C is a face F of C with $0 \leq \dim F = \dim C - 1$. A point $x \in C$ is said to be an extreme point of C if $\{x\}$ is a face of C. In what follows, the set of extreme points of a polyhedral set P is denoted by exr(P). If P is a polyhedral set then exr(P) = Exr(P). A. Brondsted (1983) defines a polytope as a set which is the convex hull of a nonempty finite set $\{x_1,...,x_p\}$ and proves that:

Proposition 2.7.13. *Let* $P \neq \mathbf{R}^n$ *be a polyhedral set with* dim P = n. *If* (2.7.2) *is a representation of* P, *then the following holds*:

(a) $\operatorname{bd} P = \bigcap_{i=1}^{m} H(a^i, b_i) \cap P$, *where* $H(a^i, b_i) = \{x \in \mathbf{R}^n \mid \langle a^i, x\rangle = b_i\}$;

(b) *each facet of* P *is of the form* $H(a^i, b_i) \cap P$;

(c) *a set* $H(a^i, b_i) \cap P$ *is a facet of* P *if and only if the representation* (2.7.2) *is irreducible.*

Proposition 2.7.14. *Let* P *be a polytope, and* $\{x^1, ..., x^p\}$ *a finite subset of* P. *Then the following two conditions are equivalent:*

(a) $P = \operatorname{conv}(\{x^1, ..., x^p\})$;

(b) $\operatorname{ext} P \subseteq \{x^1, ..., x^p\}$.

Corollary 2.7.4. *If* P *is a polytope, then we have* $P = \operatorname{conv}(\operatorname{exr}(P))$.

Proposition 2.7.15. *A subset* P *of* $\mathbf{R}^n$ *is a polytope if and only if it is a bounded polyhedral set.*

Proposition 2.7.16. *If* P *is a polytope in* $\mathbf{R}^n$ *and* F *is a proper face of* P, *then* F *is also a polytope and* $\operatorname{exr}(F) = F \cap \operatorname{exr}(P)$.

Now we prove a very interesting property, for our purpose, of integer sets

Theorem 2.7.1. *If* $X \subseteq \mathbf{Z}^n$ *is a finite convex set, then there are* $m \in \mathbf{N}^*$, $a^i \in \mathbf{Z}^n$, $i \in \{1, ..., m\}$ *and* $b_i \in \mathbf{Z}$, $i \in \{1, ..., m\}$ *such that*

(2.7.3) $X = \{x \in \mathbf{Z}^n \mid (a^i \mid x) \geq b_i$, *for each* $i \in \{1, ..., m\}\}$.

Proof. Let $n = \dim X$. Because X is a finite set, the set $P = \operatorname{conv}(X)$ is a bounded polytope. Then, there are $m \in \mathbf{N}^*$, $a^i \in \mathbf{R}^n$, $i \in \{1, ..., m\}$ and $b_i \in \{1, ..., m\}$ such that

(2.7.4) $P = \{x \in \mathbf{R}^n \mid (a^i \mid x) \geq b_i$, for each $i \in \{1, ..., m\}\}$.

It remains to prove that $a^i \in \mathbf{Z}^n$ and $b_i \in \mathbf{Z}$, for all $i \in \{1, ..., m\}$.

Let $i \in \{1, ..., m\}$. In view of proposition 2.7.9, the set F_i,

(2.7.5) $F_i = \{x \in P \mid (a^i \mid x) = b_i\}$,

is a facet of P. Applying propositions 2.7.16 and 2.7.14, we get that

(2.7.6) $\operatorname{exr}(F_i) = F_i \cap \operatorname{exr}(P) \subseteq X$.

Since F_i is a facet of P, we have $\dim F_i = n - 1$. As $F_i = \operatorname{aff}(\operatorname{exr}(F_i))$, it follows that $\dim F_i = \dim(\operatorname{aff}(\operatorname{exr}(F_i)))$. From (2.7.6) it results that there are $x^1, ..., x^n \in F_i \cap \mathbf{Z}^n$, with $x^1,...,x^n$ affinely independent. Then the points

$$x^1 - x^n, ..., x^{n-1} - x^n,$$

are linearly independent. Hence

$$\text{rank}\begin{pmatrix} x_1^1 - x_1^n & \dots & x_1^{n-1} - x_1^n \\ \dots & \dots & \dots \\ x_1^n - x_n^n & \dots & x_n^{n-1} - x_n^n \end{pmatrix} = n - 1.$$

Consider the system

(2.7.7) $\quad (a^i \mid x) - b_i = 0, \; i \in \{1,\dots,n\}.$

Because $x^k \in F_i$, for each $k \in \{1, \dots, n\}$, this system is compatible. On the other hand, because the points $x^1, \dots, x^n$ exist, we can consider a^i, b_i, in the system (2.7.7) as unknowns. We have grade $(x^1, \dots, x^n, -e^0) = n$, where by e^0 we denote the vector $e^0 = (1, \dots, 1)$. Two cases are possible:

I) $D = \begin{vmatrix} x_1^1 & \dots & x_n^1 \\ \dots & \dots & \dots \\ x_1^n & \dots & x_n^n \end{vmatrix} \neq 0$. Then a solution of the system (2.7.7) is given by

$a_j^i = D_j$ $(j=1,\dots,n)$, $b_i = D$, where $D_j = \begin{vmatrix} x_1^1 & \dots x_{j-1}^1 & 1 & x_{j+1}^1 \dots & x_n^1 \\ \dots & \dots & \dots & \dots & \dots \\ x_1^n & \dots x_{j-1}^n & 1 & x_{j+1}^n \dots & x_n^n \end{vmatrix}$, for all $j \in \{1,\dots,n\}$.

II) $D = 0$, but there is a $k \in \{1, \dots, n\}$ such that

$$\Delta_k = \begin{vmatrix} x_1^1 & \dots & x_{k-1}^1 & x_{k+1}^1 & \dots & x_n^1 & -1 \\ \dots & \dots & \dots & \dots & \dots & \dots & \dots \\ x_1^n & \dots & x_{k-1}^n & x_{k+1}^n & \dots & x_n^n & -1 \end{vmatrix} \neq 0.$$

Then $a_k^i = D_k$, $a_j^i = D_{kj}$, for $j \in \{1, \dots, n\} \setminus \{k\}$, $b_i = D_{kk}$ is a solution of the system (2.7.7), where

$$\Delta_{kj} = \begin{vmatrix} x_1^1 & \dots & x_{j-1}^1 & x_k^1 & x_{j+1}^1 & \dots & x_{k-1}^1 & x_{k+1}^1 & \dots & x_n^1 & -1 \\ \dots & \dots & \dots & \dots & \dots & \dots & \dots & \dots & \dots & \dots & \dots \\ x_1^n & \dots & x_{j-1}^n & x_k^n & x_{j+1}^n & \dots & x_{k-1}^n & x_{k+1}^n & \dots & x_n^n & -1 \end{vmatrix} = (-1)^{k-1-j}\Delta_j,$$

and

$$\Delta_{kk} = \begin{vmatrix} x_1^1 & \dots & x_{k-1}^1 & x_{k+1}^1 & \dots & x_n^1 & x_k^1 \\ \dots & \dots & \dots & \dots & \dots & \dots & \dots \\ x_1^n & \dots & x_{k-1}^n & x_{k+1}^n & \dots & x_n^n & x_k^n \end{vmatrix} = (-1)\cdot(-1)^{n-k}D_k = (-1)^{n+1-k}D_k.$$

Obviously, as $x^h \in \mathbf{Z}^n$, in both cases we have $a^i_j \in \mathbf{Z}$ ($j \in \{1, ..., n\}$), $b_i \in \mathbf{Z}$. Hence

(2.7.8) $\quad \text{conv}(X) = \{x \in \mathbf{R}^n \mid (a^i \mid x) \geq b_i, \text{ for each } i \in \{1, ..., m\}\}$,

where $a^i \in \mathbf{Z}^n$, and $b_i \in \mathbf{Z}$, for each $i \in \{1, ..., m\}$. But, because X is an integer convex set, we have $X = \mathbf{Z} \cap \text{conv}(X)$. From (2.7.8) we get that

$$X = \{x \in \mathbf{Z}^n \mid (a^i \mid x) \geq b_i, \text{ for each } i \in \{1, ..., m\}\}$$

Definition 2.7.4. *A subset* Y *of* $\mathbf{Z}^n$ *is called a discrete polyhedral set if there is an integer polyhedral set* P *such that* $Y = P \cap \mathbf{Z}^n$. *A bounded discrete polyhedral set is called a discrete polytope.*

Corollary 2.7.5. *Every finite integer set is a discrete polytope.*

2.8 Convexity space with respect to a given set

In this paragraph we present the extension of the notion of convexity with respect to a given set for subsets of a convexity space, given by L. Blaga and L. Lupşa (1998).

Let X be an arbitrary set. We recall that:

Definition 2.8.1. *A pair* (X, G), *with* $G \subseteq 2^X$, *is called a convexity space on* X *if*

1. $X \in G$ *and* $\emptyset \in G$,
2. $\bigcap_{i \in I} A_i \in G$, *for any arbitrary family* $(A_i)_{i \in I}$, *with* $A_i \in G$.

Definition 2.8.2. *A subset* $A \subseteq X$ *is called convex in the convexity space* (X, G) *if* $A \in G$.

Definition 2.8.3. *A mapping* $g: 2^X \to 2^X$ *is called the convex hull operator on* X *if:*

1. $A \subseteq g(A)$,
2. $g(g(A)) = g(A)$,
3. $g(B) \subseteq g(A)$, *for any* $A, B \in 2^X$ *with* $B \subseteq A$.

Remark 2.8.1. If g is a convex hull operator defined on X, and $G = \{A \subseteq X \mid g(A) = A\}$, then the pair (X, G) is a convexity space on X, called the convexity space generated by g.

Remark 2.8.2. If (X, G) is a convexity space on X, then a convex hull operator on X is defined by the mapping g: $2^X \to 2^X$ given by

(2.8.1) $g(A) = \cap \{C \in G \mid A \subseteq C\}$, for all $A \in 2^X$.

This mapping is called the convex hull operator generated by G.

Remark 2.8.3. If g is a convex hull operator defined on X then, for every two points $x \in X$ and $y \in X$, the set $<x, y> = g(\{x, y\})$ is said to be the straight-line segment determined by x and y. The set $<xy = \{z \in X \mid y \in <x, z>\}$ is called the half-line containing y and opposite to the point x.

Let X and M be two arbitrary sets and let g be a convex hull operator on X. Let k be a natural number, $k \geq 2$.

Definition 2.8.4. *A subset* A *of* X *is called*:

i) k-*strongly convex with respect to* g *and* M *if for any* $a_1, \dots, a_k$ *elements of* A *we have*

(2.8.2) $g(\{a_1, \dots, a_k\}) \cap M \subseteq A$;

ii) *strongly convex with respect to* g *and* M *if*

(2.8.3) $g(A) \cap M \subseteq A$;

iii) *slackly* k-*convex with respect to* g *and* M *if for any* $a_1, \dots, a_k$ *elements of* A *we have*

(2.8.4) $g(\{a_1, \dots, a_k\} \cap M) \cap M \subseteq A$;

iv) *slackly convex with respect to* g *and* M *if*

(2.8.5) $g(A \cap M) \cap M \subseteq A$.

Theorem 2.8.1. *If* g *is a convex hull operator on* X *and* M *is an arbitrary subset of* X, *then the following statements are true*:

a) *The set* $G_{st,k,g,M}$ *of the subsets of* X *which are strongly* k-*convex with respect to* g *and* M *is a convexity space.*

b) *The set* $G_{st,g,M}$ *of the subsets of* X *are strongly convex with respect to* g *and* M *is a convexity space.*

c) *The set* $G_{sl,k,g,M}$ *of the subsets of* X *which are slackly* k-*convex with respect to* g *and* M *is a convexity space.*

d) *The set* $G_{sl,g,M}$ *of the subsets of* X *which are slackly convex with respect to* g *and* M *is a convexity space.*

The proof is immediate.

Remark 2.8.4. Let g be a convex hull operator on X, M be an arbitrary subset of X. If $A \in 2^X$ and $M \subseteq A$, then $A \in G_{st,g,M}$, $A \in G_{sl,g,M}$ and for any natural k, $k \geq 2$ we have $A \in G_{st,k,g,M}$ and $A \in G_{sl,k,g,M}$.

Theorem 2.8.2. *Let* g *be a convex hull operator defined on* X *and let* G *be the convexity space generated by* g.

a) *If* k *and* n *are natural numbers,* $2 \leq k < n$, *then the following inclusions are true:*

$$G \subseteq G_{st,g,M} \subseteq G_{st,n,g,M} \subseteq G_{st,k,g,M};$$

$$G \subseteq G_{sl,g,M} \subseteq G_{sl,n,g,M} \subseteq G_{sl,k,g,M}.$$

b) *If* A *and* B *are two sets such that* $A \subseteq B$ *and* $A \neq B$, *and if* n *is a natural number,* $n \geq 2$, *then the following inclusions are true*:

$$G_{st,g,B} \subseteq G_{st,g,A}; \quad G_{st,n,g,B} \subseteq G_{st,n,g,A};$$

$$G_{sl,g,B} \subseteq G_{sl,g,A}; \quad G_{sl,n,g,B} \subseteq G_{sl,n,g,A}.$$

c) *If* n *is a natural number,* $n \geq 2$, *then*

$$G_{st,g,M} \subseteq G_{sl,g,M}; \quad G_{st,n,g,M} \subseteq G_{sl,n,g,M}.$$

Proof. a) Let A be a subset of G. Then $g(A) \subseteq A$, which implies

$$g(A) \cap M \subseteq A \cap M \subseteq A.$$

So, $A \in G_{st,g,M}$.

Let now $A \in G_{st,g,M}$. From the third property of a convex hull operator we have

(2.8.6) $g(\{a_1,...,a_n\}) \subseteq g(A)$.

But $A \in G_{st,g,M}$, so

(2.8.7) $g(A) \cap M \subseteq A$.

From the above inclusions, we get $g(\{a_1,...,a_n\}) \cap M \subseteq A$. Hence

$$A \in G_{st,n,g,M}.$$

Let $a_1,...,a_k$ be elements of A and $A \in G_{st,n,g,M}$. Putting $a_{k+1}=...=a_n=a_1$ and taking into account the definition of $G_{st,n,g,M}$, we have

$$g(\{a_1,...,a_k\}) \cap M = g(\{a_1,...,a_n\}) \cap M \subseteq A.$$

This implies that $A \in G_{st,k,g,M}$.

The other inclusions can be proved similarly.

Examples 2.1.1, 2.1.3 show that there are situations when the above inclusions are strict. In these examples we consider the convexity space with $X = \mathbf{R}^2$, $M = \mathbf{Z}^2$, g (A) = conv (A) and the convexity notion is strong convexity with respect to M.

Theorem 2.8.2, example 2.1.1 and example 2.1.3 show that if G is the set of all the convex subsets of $\mathbf{R}^2$, then $G_{st,3,g,Z^2}$ and $G_{st,2,g,Z^2}$ are convexity spaces which do not coincide with G; moreover

$$G \subset G_{st,3,g,Z^2} \subset G_{st,2,g,Z^2}.$$

This points out that the notions introduced by definition 2.8.4 are consistent.

Theorem 2.8.3. *If* $A \subseteq X$ *then* A *is strongly convex with respect to* g *and* M *if and only if*

$$M \cap g(A) = M \cap A.$$

Proof. Necessity is immediate from definition of the strong convexity with respect to g and M.

In order to prove the sufficiency let us suppose that the equality is true. It is obvious that $M \cap g(A) \subseteq M$, so the equality implies

$$M \cap g(A) \subseteq A.$$

Hence A is strongly convex with respect to M.

Corollary 2.8.1. *If* $A \subseteq X$ *and* $B \subseteq X$ *are two strongly convex sets with respect to* g *and* M, *then* $A \cap B$ *is also strongly convex with respect to* M.

Theorem 2.8.4. *If* $A \subseteq X$, *then the set* $(M \cap g(A)) \cup A$ *is strongly convex with respect to* g *and* M.

Proof. We have

$$(M \cap g(A)) \cup A = (A \cup g(A)) \cap (M \cup A).$$

From the first property of the convex hull operator it follows that $A \cup g(A) = g(A)$. Hence

$$(M \cap g(A)) \cup A = g(A) \cap (M \cup A). \tag{2.8.8}$$

Because g (g (A)) = g (A), it results that $g(A) \in G$. In view of theorem 2.8.2 we get $g(A) \in G_{st,g,M}$. Remark 2.8.4 implies that $M \cap A \in G_{st,g,M}$. Applying theorem 2.8.3, we have $g(A) \cap (M \cup A) \in G_{st,g,M}$. Then from (2.8.8) it results that $(M \cap g(A)) \cup A$ is strongly convex with respect to g and M.

In view of theorem 2.8.1, $G_{st,g,M}$ is a convexity space. We denote its convex hull operator by $stconv_{g,M}$.

Theorem 2.8.5. *For all* $A \in 2^X$ *we have*

(2.8.9) $\quad stconv_{g,m}(A) = (M \cap g(A)) \cup A.$

Proof. From remark 2.8.2 we have

$$stconv_{g,M}(A) = \cap \{Y \in G_{st,g,M} \mid A \subseteq Y\}.$$

Because $A \subseteq (M \cap g(A)) \cup A$ and $(M \cap g(A)) \cup A$ is strongly convex with respect to g and M, we get that

(2.8.10) $\quad stconv_{g,M}(A) \subseteq (M \cap g(A)) \cup A.$

Now let $z \in (M \cap g(A)) \cup A$. Two cases are possible:

a) $z \in A$; then obviously $z \in stconv_{g,M}(A)$.

b) $z \in M \cap g(A)$. Let

$$S = \{B \in G_{st,g,M} \mid A \subseteq B\}$$

and let $Y \in S$. Because g is a convex hull operator, we have $g(A) \subseteq g(Y)$. Since Y is strongly convex with respect to g and M, we get

$$M \cap g(A) \subseteq M \cap g(Y) \subseteq Y.$$

Hence $z \in Y$. But Y was arbitrarily chosen in S. Then $z \in \cap\{Y \mid Y \in S\} = stconv_{g,M}(A)$. Because z was also arbitrarily chosen and in both cases we get $z \in stconv_{g,M}(A)$, it follows that

(2.8.11) $\quad stconv_{g,M}(A) \supseteq (M \cap g(A)) \cup A.$

Now, (2.8.9) results from (2.8.10) and (2.8.11).

It is obvious that all theorems 2.8.1 – 2.8.5 remain valid for any convex hull operator g. The particular choice of the given set M and of the convex hull operator g allows us to obtain other results concerning the properties of the convex sets with respect to a given set in the form of separation theorems and to give some topological properties. The next example is in connection with the problem of the determination of some properties that can be related to problems of multiple criteria decision making. Our convexity language leads to some properties of the non-dominated and dominated points, which cannot be obtained otherwise.

Example 2.8.1. Let $(X, +, \cdot, K)$ be a finite dimensional real or complex linear space and let $\succ$ be a binary relation on X with the properties

if $x \succ y$ then $x + u \succ y + u$ for all $u \in X$;

if $x \succ y$ then $tx \succ ty$ for all $x \in X$ and for all $t \in K$ with $t \geq \theta_K$.

Let M be a non-void subset of X, let $A \in 2^X$ and let k be a natural number $k \geq 2$. Let

$$ED(A, \succ) = \{x \in A \mid \exists\, a \in A, a \succ x\}$$

be the set of dominated points of A. The following statements are true:

a) If $A \in G_{st,k,g,M}$ then $ED(A, \succ) \in G_{st,k,g,M}$.

b) If $A \in G_{st,g,M}$ then $ED(A, \succ) \in G_{st,g,M}$.

c) If $A \in G_{sl,k,g,M}$ then $ED(A, \succ) \in G_{sl,k,lg,M}$.

d) If $A \in G_{sl,g,M}$ then $ED(A, \succ) \in G_{sl,g,M}$.

Chapter 3

Behaviours. Convexity with respect to a behaviour

The concept of behaviour of an object is largely used in order to make difference between things. As a mathematical concept, Elena Popoviciu used the term behaviour for the first time in 1965, on the occasion of a conference held at Cluj-Napoca "Babeş-Bolyai" University, without publishing it. The discussions on this theme have been reconsidered during the meetings of the Interdisciplinary Researches Laboratory of Cluj-Napoca "Babeş-Bolyai" University between 1980-1983. In this period, she wrote especially about the particular case of behaviours that are shapes of functions, in order to generalise the property of convexity of a function. The main results in this direction are written in the papers of E. Moldovan (1955, 1958, 1959) and E. Popoviciu (1969, 1972, 1983). Her disciples, M. Ivan (1975, 1980, 1982) and R. Precup (1984, 1995) gave applications of this concept in interpolation and approximation.

The starting point of this chapter can be found in our public and private discussions with Elena Popoviciu and in the above mentioned papers of M. Ivan and R. Precup on the notion of allure and its generalisations. The results concerning the notion of behaviour are quoted from our previous papers G. Cristescu (1996[55], 1997[61]). The aim of the first part of this chapter is to point the sets of behaviours that may be endowed with a mathematical structure and to provide more examples from various domains of mathematics.

After introducing the notion of behaviour, we study the manner in which it leads to a new notion of convex set that does not imply the connectivity. The behaviours provide a new tool of obtaining new convexity type properties, to build new theories of convexity and to retrieve the known ones. The examples presented in this chapter will show the wide range of domains of mathematics in which this notion penetrates as an approximation of the classical convexity. Convexity properties of this type appear starting with the constructive theory of functions, in the integer programming and in the rational programming. They play an important role in the shape recognition, including, as a particular case, the digital convexity. They extend the domain of convexity to fractals, for which such a property becomes possible. The basic results in this direction have been published by G. Cristescu (1997 [65], 2000[71]).

3.1 The notion of behaviour

Let A and B be two nonempty sets, $\mathcal{T}$ a collection of transformations T:A→B, D a nonempty part of B and a ∈ A.

Definition 3.1.1. i) *If* T (a) ∈ D *then the pair* (D, T) *is said to be a strict behaviour of the element* a.

ii) *If* T (a) ∈ D *for every* T ∈ $\mathcal{T}$ *then the pair* (D, $\mathcal{T}$) *is said to be a strong behaviour of the element* a.

iii) *If there is a transformation* T ∈ $\mathcal{T}$ *such that* T (a) ∈ D, *then the pair (D, $\mathcal{T}$) is said to be a weak behaviour of the element* a.

In order to avoid the confusion the symbol s - (D, $\mathcal{T}$) will be used to denote a strong behaviour and w - (D, $\mathcal{T}$) will denote a weak behaviour.

Definition 3.1.2. i) *If there is an element* a ∈ A *such that* T (a) ∈ D *then the pair* (D, T) *is said to define a behaviour in* A *by means of* B.

ii) *If there is an element* a ∈ A *such that for every* T ∈ $\mathcal{T}$, T (a) ∈ D, *then the pair* (D, $\mathcal{T}$) *is said to define a behaviour in* A *by means of* B.

iii) *If there is an element* a ∈ A *such that there is a* T ∈ $\mathcal{T}$ *having the property that* T (a) ∈ D, *then the pair* (D, $\mathcal{T}$) *is said to define a behaviour in* A *by means of* B.

Therefore, there are three ways of defining a behaviour in the set A by means of a set B: using the strict behaviour, the strong behaviour and the weak behaviour of some elements of A.

Example 3.1.1. Let us suppose that A is the set of all the human beings on the Earth at a certain moment t, and B = **R.** The function T: A → B is defined by

$$T(a) = \text{the age of the person a},$$

where a ∈ A. If D =]0, 14] ⊂ B, then the strict behaviour (D, T) of one person means that that person is a child.

Example 3.1.2. For the same set A as in the example 3.1.1 let us consider the set M of all the medical investigations that might be applied to a person. Take B = {0, 1} and the set of functions

$$\mathrm{T} = \{T(m) \mid T(m) : A \to B, T(m)(a) =$$

$$= \begin{cases} 1 & \text{if a has the investigation m normal} \\ 0 & \text{if a has not a normal m} \end{cases}, a \in A, m \in M\}.$$

Then the strong behaviour ({1}, T) of a person a ∈ A means that that person is healthy. The weak behaviour ({0}, T) of a person b ∈ A means that at least one investigation applied to b has not a normal answer, therefore b is not healthy.

Obviously, every strict behaviour is a strong behaviour, but we prefer to keep both notions for historical reasons and because, sometimes, it is easier to present the proofs in a simpler case, the extension for the general one being obvious. The strict behaviours and the strong behaviours have been defined first by Elena Popoviciu in 1965 and called simply, behaviours as in E Popoviciu (1997). Then,

$$\text{(3.1.1)} \quad \mathrm{Comp}\,(A, B) = \{ (D, T) \mid D \subseteq B, T : A \to B \} \cup$$

$$\cup \{ s\text{-}(D, \mathrm{T}) \mid D \subseteq B, \mathrm{T} \subset \{ f \mid f : A \to B \} \} \cup$$

$$\cup \{ w\text{-}(D, \mathrm{T}) \mid D \subseteq B, \mathrm{T} \subset \{ f \mid f : A \to B \} \}$$

will be called the set of behaviours of A by means of B. The elements of the set Comp(A, B) will be denoted either by C = (D, T) or C = s - (D, T) or C = w - (D, T).

For a behaviour C = (D, T) ∈ Comp (A, B), the set $C(A) = \{x \in A \mid T(x) \in D\}$ is called the set of those elements of A having the behaviour C. If C = s-(D, T) ∈ Comp(A, B) then the set of the elements of A having the behaviour C is $C(A) = \{ x \in A \mid T(x) \in D, \forall T \in \mathrm{T} \}$. If C = w - (D, T) ∈ Comp (A, B) then the set of the elements of A having the behaviour C is also denoted by $C(A) = \{x \in A \mid \exists T \in \mathrm{T}, T(x) \in D\}$.

Now, two behaviours, C_1 and $C_2 \in$ Comp (A, B) are considered.

Definition 3.1.3. i) *The behaviour* C_1 *is said to be identical to the behaviour* C_2 *if and only if* $C_1(A) = C_2(A)$.

ii) *The behaviour* C_1 *weakly precedes the behaviour* C_2 *if and only if* $C_2(A) \subseteq C_1(A)$.

iii) *The behaviour* C_1 *strictly precedes the behaviour* C_2 *if and only if* $C_2(A) \subset C_1(A)$.

In what follows, the situation when C_1 weakly precedes C_2 will be denoted by $C_1 \leq C_2$, and $C_1 < C_2$ will signify that C_1 strictly precedes C_2. If C_1 is identical with C_2 then the notation is $C_1 = C_2$.

Proposition 3.1.1. i) *The identity of behaviours is a relation of equivalence in* Comp (A, B).

ii) *The weak precedence is an order relation in* Comp (A, B).

iii) *The strict precedence is a preference relation in* Comp (A, B).

Proof. It is obvious, but, for example, iii) will be proved.

Asymmetry: If $C_1, C_2 \in$ Comp (A, B) and $C_1 < C_2$ then $C_2(A) \subset C_1(A)$ without the possibility of equality and so it follows that the relation $C_2 < C_1$ does not take place.

Transitivity: If $C_1, C_2, C_3 \in$ Comp (A, B), $C_1 < C_2$ and $C_2 < C_3$ then $C_2(A) \subset C_1(A)$ and $C_4(A) \subset C_2(A)$. This means that $C_4(A) \subset C_1(A)$ and therefore $C_1 < C_3$.

Theorem 3.1.1. (Comp (A, B), $\leq$) *is a preinductive semilattice with respect to the union.*

Proof. Let $C_1, C_2 \in$ Comp (A, B) and first $C_1 = (D, T)$, $C_2 = (E, S)$, $D, E \subseteq B$, $T: A \rightarrow B$, $S: A \rightarrow B$ is supposed. Then the behaviour w-$(D \cup E, T)$, denoted by $C_1 \vee C_2$, is the supremum of C_1 and C_2, where T is the collection of functions $f: A \rightarrow B$ having the following properties:

a) $f(a) = m$, if $T(a) \notin D$ and $S(a) \notin E$, where $m \in B - (D \cup E)$;

b) $f(a) = T(a)$ or $f(a) = S(a)$ if $T(a) \in D$ and $S(a) \in E$;

c) $f(a) = T(a)$ if $T(a) \in D$ and $S(a) \notin E$;

d) $f(a) = S(a)$ if $T(A) \notin D$ and $S(a) \in E$.

In order to verify that the behaviour $C_1 \vee C_2$ is the supremum of C_1 and C_2 it is enough to prove that $(C_1 \vee C_2)(A) = C_1(A) \cup C_2(A)$.

First, let $x \in (C_1 \vee C_2)(A)$. Then there is a transformation $f \in \mathcal{T}$ such that $f(x) \in D \cup E$. Two cases are possible:

1) $f(x) = T(x)$ and then $x \in C_1(A)$, therefore $x \in C_1(A) \cup C_2(A)$;

2) $f(x) = S(x)$ and then $x \in C_2(A)$, therefore $x \in C_1(A) \cup C_2(A)$.

It follows that $(C_1 \vee C_2)(A) \subseteq C_1(A) \cup C_2(A)$.

Now, $x \in C_1(A) \cup C_2(A)$ is supposed. Then $x \in C_1(A)$ or $x \in C_2(A)$, following that $T(x) \in D$ or $S(x) \in E$. This means that $f(x) = T(x) \in D \cup E$ or that $f(x) = S(x) \in D \cup E$ and therefore $x \in (C_1 \vee C_2)(A)$, so that $C_1(A) \cup C_2(A) \subseteq (C_1 \vee C_2)(A)$.

This proves that in this case $(D \cup E, T) = C_1 \vee C_2 = \sup\{C_1, C_2\}$.

The proof is also valid in the following cases:

$C_1 = (D, T)$, $C_2 = s - (E, F)$;

$C_1 = (D, T)$, $C_2 = w - (E, F)$;

$C_1 = s - (D, F_1)$, $C_2 = s - (E, F_2)$;

$C_1 = s - (D, F_1), C_2 = w - (E, F_2);$
$C_1 = w - (D, F_1), C_2 = w - (E, F_2),$
where F, F_1, F_2 are sets of transformations $g : A \to B$.

A similar reasoning might be used in order to define the supre num of a finite number of behaviours from Comp (A, B). Indeed, if $C_1 = (D_1, T_1)$, $C_2 = (D_2, T_2)$, ..., $C_n = (D_n, T_n)$, are behaviours from Comp (A, B), where $D_i \subseteq B$ and T_i: $A \to B$ for i = 1, 2, ..., n, then the weak behaviour w-($D_1 \cup D_2 \cup ... \cup D_n$, T) is the supremum of the behaviours $C_1, C_2, ..., C_n$, where T is the set of all the transformations $f : A \to B$ having the properties:

a) $f(a) = m \in B - \bigcup_{i=1}^{n} D_i$ if $T_i(A) \not\subset D_i$, i = 1, 2, ..., n, where m is arbitrary chosen;

b) $f(a) = T_k(a)$ where k is any of the indices from {1, 2, ..., n} for which $T_k(a) \in D_k$ takes place.

Denoting by

$$(3.1.2) \qquad \bigvee_{i=1}^{n} C_i = w - (\bigcup_{i=1}^{n} D_i, T)$$

and showing that

$$(3.1.3) \qquad (\bigvee_{i=1}^{n} C_i)(A) = \bigcup_{I=1}^{n} C_i(A)$$

one can prove by induction that the supremum of the n behaviours was defined.

Example 3.1.3. Let $a \in \mathbf{R}$, $b \in \mathbf{R}$, $A = \{f \mid f : [a, b] \to \mathbf{R}\}$, $B = 2^{\mathbf{R}}$, the set of all the subsets of **R**, $D_1 = \{M \subset \mathbf{R} \mid M$ bounded in $\mathbf{R}\}$ and $D_2 = \{\{m\} \mid m \in \mathbf{R}\}$. The following transformations are taken into account:

$T_1: A \to B$, $T_1(f) = \text{Im } f = \{f(x) \mid x \in [a, b]\}$;

$T_2: A \to B$, $T_2(f) = \bigvee_a^b f$ = the total variation of the function f on [a, b];

$$T_3^{x_0} : A \to B, \quad T_3^{x_0}(f) = \lim_{x \to x_0} f(x);$$

$$T_4^{x_0} : A \to B, \quad T_4^{x_0}(f) = \lim_{x \to x_0} \frac{f(x) - f(x_0)}{x - x_0}.$$

The following sets of transformations are also considered: $\mathsf{T} = \{ T_3^{x_0} \mid x_0 \in [a, b]\}$, $\mathsf{F} = \{T_4^{x_0} \mid x_0 \in [a, b]\}$. The following behaviours are defined using these transformations:

(D_1, T_1) = the behaviour of bounded function on [a, b];

(D_2, T_2) = the behaviour of function with bounded variation on [a, b];

$(D_2, T_3^{x_0})$ = the behaviour of continuous function on x_0;

$(D_2, T_4^{x_0})$ = the behaviour of function having the first derivative finite on the point x_0;

w-(D_2, T) = the behaviour of continuous function on [a, b];

w-(D_2, F) = the behaviour of differentiable function on [a, b].

The following basic results of the mathematical analysis are retrieved:

(3.1.4) $(D_1, T_1) < (D_2, T_2) < w - (D_2, \mathsf{T}) < w - (D_2, \mathsf{F})$;

(3.1.5) $(D_2, T_2) < (D_2, T_3^{x_0}) < (D_2, T_4^{x_0})$;

(3.1.6) $(D_2, T_3^{x_0}) < w - (D_2, \mathsf{T})$;

(3.1.7) $(D_2, T_4^{x_0}) < w - (D_2, \mathsf{F})$.

Example 3.1.4. Let (X, d) be a metric space, $A = X \times 2^X$, $B = \mathbf{R}$, $D = [0, \varepsilon]$, $T: X \times 2^X \to \mathbf{R}$, $T(a, M) = d(a, M) = \inf\{d(a, x) \mid x \in M\}$, where $(a, M) \in X \times 2^X$. If $T(a, M) \in [0, \varepsilon]$ then the point a belongs to an ε-neighbourhood of the set M. It means that the pair (a, M) has the behaviour $([0, \varepsilon], T)$.

Example 3.1.5. Let (X, d) be a metric space, Y a nonempty part of X, $A = X \times Y$, $B = \mathbf{R}$, $D = [0, \varepsilon[$ for $\varepsilon > 0$ and $f: A \to B$, $f(x, y) = d(x, y)$. The behaviour $([0, \varepsilon[, f)$ in X means that the element y is an ε-approximation of x by elements from Y.

Example 3.1.6. Let X be a real linear space, $A = 2^X$, $B = 2^X$, $D = \{\phi\}$, $T: 2^X \to 2^X$, $T(M) = \text{conv}(M) - M$, where $M \in 2^X$ and by conv (M) the union of all closed straight line segments determined by pairs of points of M is denoted. Then the behaviour $(\{\phi\}, T)$ means the convexity property of the subsets of X.

Example 3.1.7. Let (X, d) be a metric space, s(X) the set of the sequences of elements of X, $A = s(X) \times X$, $B = s(\mathbf{R})$ and $D = s(\mathbf{R}_+)$. For every sequence of nonnegative real numbers, $(\varepsilon_n)_{n \in N}$ convergent to zero the transformation $f(\varepsilon_n)$: $s(X) \times X \to s(\mathbf{R})$, $f(\varepsilon_n)((x_n), a) = \varepsilon_n - d(x_n, a)$, for every $((x_n), a) \in s(X) \times X$, is defined.

Let us consider the set $T = \{f(\varepsilon_n) \mid (\varepsilon_n) \in D \text{ and } \varepsilon_n \to 0\}$ and the weak behaviour w - (D, T). A pair $((x_n), a) \in s(X) \times X$ has the behaviour w - (D, T) if and only if there is a sequence $(\varepsilon_n)_{n\in N}$, of nonnegative real numbers convergent to zero, such that $f(\varepsilon_n)((x_n), a) \in D$ is convergent to zero. Therefore, $d(x_n, a) \le \varepsilon_n$, $n \in \mathbf{N}$, implying that the sequence (x_n) is convergent in X, having the limit a.

Example 3.1.8. Let X be a real linear space and K a wedge in X, i.e. a set having the property that for each $x \in K$, $y \in K$ and $\alpha \in \mathbf{R}$ we have $x + y \in K$ and $\alpha x \in K$. Let us take $A = X^2$, $B = X$ and the transformation $f: A \to B$, $f(x, y) = y - x$. The behaviour (K, f) of a pair of elements $(x, y) \in X^2$ means that $y - x \in K$, therefore $x \le_K y$, where $\le_K$ is the pre-order relation induced by the wedge K in X.

Example 3.1.9. Let $A \neq \varnothing$, $B = \{0, 1\}$, $D = \{1\}$ and $T: A \to B$, defined by $T(a) = 1$ for each $a \in A$. The behaviour (D, T) is the universal behaviour in A by means of B.

Example 3.1.10. Let X be a nonempty set and let To (X) be the set of all the topologies on X. The elements of every topology $\tau \in$ To (X) are open sets, and (X, τ) is a topological space. If $\tau \in$ To (X), then a set $\beta \subset \tau$ is a basis for the topology τ if every set from τ is the union of a family of elements of β.

Theorem 3.1.2. *The property of the pair* $(\tau, \beta) \in$ To (X) $\times\, 2^{2^X}$ *to associate a basis* β *to the topology* τ *is a strict behaviour.*

Proof. It is well known that a set $\beta \subset 2^X$ is a basis for a topology $\tau \in$ To(X) if and only if the following conditions are satisfied

b1) $\bigcup_{B\in\beta} B = X$;

b2) every finite intersection of elements of β is a union of the elements of a part of β, including the empty part.

On another hand, to every set $\beta \subset 2^X$ satisfying b1) and b2) it corresponds a unique topology $\tau \in$ To(X) having the basis β. Let us denote now

$$\mathsf{B} = \{\beta \subset 2^X \mid \beta \text{ satisfies b1) and b2)}\},\ S = \text{To}(X) \times \mathsf{B}$$

and let us define the transformation $T\colon \mathrm{To}\,(X) \times 2^{2^X} \to 2^{2^X}$ by:

$$T(\tau, \beta) = \begin{cases} \tau - \left\{ \bigcup\limits_{G\in\beta'} G \,\middle|\, \beta' \subseteq \beta \right\}, & \text{for } (\tau, \beta) \in S \\ X, & \text{for } (\tau, \beta) \notin S \end{cases}$$

Consider the behaviour $(\varnothing, T)$. It also follows that this behaviour describes the unique topology τ having the basis β.

If (X, τ) is a topological space and $x \in X$ then the set $\mathcal{V}(x)$ of all the neighbourhoods of the point x has the following basic properties:

v1) if $V \in \mathcal{V}(x)$ then $x \in V$;

v2) if $V \in \mathcal{V}(x)$ and $V \subset U$ then $U \in \mathcal{V}(x)$;

v3) if $V_i \in \mathcal{V}(x)$, $i = 1, 2, ..., n$, then $\bigcap\limits_{i=1}^{n} V_i \in \mathcal{V}(x)$, for every $n \in \mathbf{N}$;

v4) if $V \in \mathcal{V}(x)$ then there is $W \in \mathcal{V}(x)$ such that from $y \in W$ it follows $V \in \mathcal{V}(y)$.

For every topology τ on X it is possible to generate the set

$$\mathcal{V}(\tau) = \bigcup_{x\in X} \{\mathcal{V}(x)\},$$

called the set of the neighbourhoods corresponding to the topology τ.

Theorem 3.1.3. *The property of a pair* $(\tau, \mathcal{V}) \in \mathrm{To}\,(X) \times 2^{2^X}$ *to associate to the topology* τ *the set* $\mathcal{V}$ *of the corresponding neighbourhoods is a strict behaviour.*

Proof. Let us denote by $\mathcal{M}$ the set of those subsets $\mathcal{V} \subseteq 2^X$ that have the property that for every point $x \in X$ there is $\mathcal{V}(x) = \{V(x)\} \subset \mathcal{V}$ that satisfies the properties v1) - v4) and $\mathcal{V} = \bigcup\limits_{x\in X} \{V(x)\}$. For every $\mathcal{V} \in \mathcal{M}$ we define the set $\tau(\mathcal{V}) \subseteq 2^X$ by $\tau(\mathcal{V}) = \{ A \subset X \mid \forall\, x \in A \Rightarrow A \in \mathcal{V}(x) \}$. It is well known that $\tau(\mathcal{V})$ is the unique topology such that for every $x \in X$, $\mathcal{V}(x)$ is the set of the neighbourhoods of the point x.

Now, let us define the transformation $T\colon \mathrm{To}\,(X) \times 2^{2^X} \to 2^{2^X}$ by

$$T(\tau, \mathcal{V}) = \begin{cases} \tau - \tau(\mathcal{V}), & \text{if } \mathcal{V} \in \mathcal{M} \\ X & \text{if } \mathcal{V} \notin \mathcal{M}, \end{cases}$$

for every $(\tau, V) \in \mathrm{To}(X) \times 2^{2^X}$. It follows that V is the set of the neighbourhoods corresponding to the topology τ if and only if the pair (τ, V) has the behaviour $(\varnothing, T)$.

Let W(x) be a set of neighbourhoods of the point x. The set W(x) is a fundamental system of neighbourhoods for the point x if for every neighbourhood $V \in V(x)$ there is $W \in W(x)$ with the property that $W \subset V$.

Theorem 3.1.4. *The property of a set* $W(x) \subset V(x)$ *of being a fundamental system of neighbourhoods of the point* x *is a weak behaviour in* 2^X.

Proof. Consider the set of functions $F = \{ f_U : V(x) \to U \mid U \in 2^{V(x)}\}$ and the set of transformations

$$T = \{T: 2^{V(x)} \to 2^X \mid T(W) = \bigcup_{A \in V(x)} [f_W(A) - A],\ f_W \in F\}.$$

Let us consider the behaviour w-$(\varnothing, T)$. A subset $W(x) \subset V(x)$ has the behaviour w-$(\varnothing, T)$ if and only if there is a transformation $T \in T$ such that $T(W(x)) = \varnothing$, which means that there is $f_W \in F$ such that $f_W(A) \subseteq A$, for every $A \in V(x)$, so W(x) is a fundamental system of neighbourhoods for the point x.

Therefore, the notion of behaviour studied in this paragraph makes possible to treat by a unitary language the topology theory, various chapters of the theory of functions, convexity, convergence, and approximation. This proves that all these domains are dealing with the study of various behaviours of some mathematical objects.

3.2 Properties of classes of behaviours

In a real linear space $(A, +, \cdot)$, a nonempty set B is considered together with a behaviour $C \in \mathrm{Comp}(A, B)$. The structure of B will be described according to the necessity.

In this paragraph it is studied the way in which the properties of the transformation T (or of the transformations of the set T), together with the algebraic structure of the sets B and D, lead to an algebraic structure of the set C(A).

Definition 3.2.1. *The behaviour* C *is said to be a translation invariant if for every element* $x \in C(A)$ *we have* $x + a \in C(A)$.

All the behaviours described in the examples 3.1.3, 3.1.6 and 3.1.8 have the property of translation invariance. If the sets X considered in the examples 3.1.4, 3.1.5, 3.1.7 are subsets of a real linear space then the behaviours described in these examples are translation invariants too.

Definition 3.2.2. *The behaviour* C *is said to be stable with respect to the addition if for each pair of elements* x *and* y *having this behaviour, the element* x + y *has the same behaviour.*

The behaviours described in the examples 3.1.3, 3.1.6 and 3.1.8 have this stability. Also, the behaviour described in the example 3.1.7 has this property if the set X is a part of a real linear space.

Proposition 3.2.1. *Let* (B, ∗) *a group,* T: A → B *a function with the property that, for every* a, b ∈ A, T (a + b) = T (a) ∗ T (b), *and a behaviour* C = (D, T) *with* D ⊂ B. *Suppose that* D = T (C (A)). *Then* D *is a stable part of the group* (B, ∗) *if and only if the behaviour* C *is stable with respect to the addition.*

Proof. Suppose that D is stable in the group (B, ∗) and let x ∈ C (A) and y ∈ C (A). Then T (x) ∈ D and T (y) ∈ D. Therefore T (x + y) = T (x) ∗ T(y) ∈ D and then x + y ∈ C (A).

Conversely, let us suppose that the behaviour C is stable with respect to the addition. Then, for a ∈ D and b ∈ D there are x ∈ A and y ∈ A such that T (x) = a and T (y) = b, so a ∗ b = T (x) ∗ T (y) = T (x + y) ∈ D and then D is stable in the group (B, ∗).

Proposition 3.2.2. *Let* (B, ∗) *be a group,* D ⊂ B, T *a set of transformations* T: A → B *having the property that for every pair of points* a, b ∈ A, T (a + b) = T (a) ∗ T (b), *and* C = s - (D, T) *a strong behaviour. Suppose that, for every transformation* T ∈ T, D = T (C (A)). *Then, the necessary and sufficient condition for* D *to be a stable part of the group* (B, ∗) *is the stability of the behaviour* C *with respect to the addition.*

The proof follows the same method as the previous one.

Definition 3.2.3. a) *The behaviour* C *is every positive scale change invariant if for every* x ∈ C (A) *and number* $\alpha \geq 0$ *one has* α x ∈ C (A).

b) *The behaviour* C *is every scale change invariant if for every* x ∈ C(A) *and number* $\alpha \in \mathbf{R}$ *one has* αx ∈ C (A).

The behaviours described in examples 3.1.3 and 3.1.6 are every scale change invariants, and the behaviour from example 3.1.8 is every positive scale invariant change. Also, in case when the set X is a subset of a real linear space, then the behaviour described in example 3.1.7 is an every scale change invariant.

Proposition 3.2.3. *Let* (B, +, ·) *be a real linear space,* D ⊂ B, T: A → B *a positively homogeneous and additive transformation, and* C = (D, T) *a strict behaviour. Let us suppose that* D = T (C (A)). *Then the necessary and sufficient condition for* D *to be a wedge in* B *is the stability of the behaviour* C *with respect to the addition and its invariance to every positive scale change.*

Proof. Suppose that D is a wedge in B. Then D is a stable part (B, +) and, according to proposition 3.2.1, the behaviour C is stable with respect to the addition. Let now x ∈ C (A) and $\alpha \in \mathbf{R}_+$. Then T (x) ∈ D holds. But T (αx) = α T (x) ∈ D, because D is a wedge and, therefore, αx ∈ C (A). So, C is every positive scale change invariant.

Conversely, if C is stable with respect to the addition and with respect to every positive scale change then, according to proposition 3.2.1, D is a stable part of the group (B, ·). Let a ∈ D and $\alpha \in \mathbf{R}_+$. Then there is x ∈ C(A) such that T (x) = a and so αa = α T (x) = T (α x) ∈ D. Therefore D is a wedge in B.

Proposition 3.2.4. *Let* (B, +, ·) *be a real linear space,* D ⊂ B, T *a set of transformations* T: A → B *positively homogeneous and additive and* C = s - (D, T) ∈ Comp (A, B). *Suppose that* D = T (C (A)) *for every* T ∈ T. *Then the set* D *is a wedge in* B *if and only if* C *is stable with respect to addition and every positive scale change invariant.*

The proof is similar to that of propositions 3.2.2 and 3.2.3.

Proposition 3.2.5. *Let* (B, +, ·) *be a real linear space,* D ⊂ B, T: A → B *a linear transformation, and* C = (D, T) ∈ Comp (A, B) *a strict behaviour. Suppose that* D = T (C (A)). *Then the set* D *is a linear subspace of* B *if and only if* C *is stable with respect to addition and an every scale change invariant.*

Proof. Let us suppose that D is a linear subspace of B. Due to the linearity of T, for every x, y ∈ C (A) and $\alpha \in \mathbf{R}$ the following relations take place:

T (x + y) = T (x) + T (y) ∈ D and T (α x) = α T (x) ∈ D.

As consequence, the elements x + y and α x have the behaviour C and therefore C is stable with respect to addition and every scale change invariant.

Conversely, supposing that C is stable with respect to addition and an every scale change invariant, proposition 3.2.1 gives that D is a stable part of (B, ·). If a ∈ D and $\alpha \in \mathbf{R}$ then there is x ∈ C (A) such that T (x) = a and so αa = α T (x) = T (α x) ∈ D, therefore D is a linear subspace of (B, +, ·).

Proposition 3.2.6. *Let* (B, +, ·) *be a real linear space,* D ⊂ B, T *a set of linear transformations* T: A → B, *and* C = s - (D, T) ∈ Comp (A, B). *Let us suppose that* D = T (C (A)), *for every* T ∈ T. *Then the set* D *is a linear subspace of* B *if and only if* C *is stable with respect to the addition and every scale change invariant.*

The proof is similar to that of propositions 3.2.2 and 3.2.5.

From definitions 3.1.3, 3.2.2 and 3.2.3 the following properties regarding the structure of some sets of elements having special behaviours are easy to derive.

Corollary 3.2.1. *The behaviour* C *is stable with respect to the addition if and only if* C (A) *is a stable part of the group* (A, +).

Corollary 3.2.2. *The behaviour* C *is stable with respect to the addition and every positive scale change invariant if and only if* C (A) *is a wedge in* A.

Corollary 3.2.3. *The behaviour* C *is stable with respect to the addition and every scale change invariant if and only if* C (A) *is a linear subspace of* A.

Corollary 3.2.4. *Let us consider* C_1, C_2 ∈ Comp (A,B) *such that* $C_1 \leq C_2$. *If* C_1 *and* C_2 *are stable with respect to the addition then* C_2 (A) *is a stable part of the group* (C_1 (A), +). *If* 0 ∈ C_2 (A) ⊆ C_1 (A) *then* (C_2 (A), +) *is a subgroup of the al group* (C_1 (A), +).

Corollary 3.2.5. *Let it be considered* C_1, C_2 ∈ Comp (A, B) *such that* $C_1 \leq C_2$. *If* C_1 *and* C_2 *are stable with respect to the addition and every scale change invariants then* C_2 (A) *is a linear subspace of* C_1 (A).

These general structure properties are retrieved in the case of example 3.1.3 where every functions set involved in precedence relations (3.1.4), (3.1.5), (3.1.6) and (3.1.7) is a linear subspace of the space it is embedded in.

The next properties refer to the behaviours defined by means of subsets D_i of B, i = 1, 2 which are involved in various relations or operations.

Proposition 3.2.7. *Let* $D_1 \subseteq D_2 \subset$ B, C_1 ∈ Comp (A, B), C_1 = (D_1, T) *(strict, strong, weak) and* C_2 ∈ Comp (A, B), C_2 = (D_2, T) *(respectively strict, strong, weak), both defined by means of the same transformations. Then we have* C_1 (A) ⊆ C_2 (A).

The proof is immediate, considering all the types of behaviours.

Proposition 3.2.8. *If* $D_1 \subset$ B, $D_2 \subset$ B, $D_1 \cap D_2 = \varnothing$, C_1 ∈ Comp (A, B), C_1 = (D_1, T) *(strict, strong),* C_2 ∈ Comp (A, B), C_2 = (D_2, T) *(strict, strong),* *both of them being defined by means of the same transformations, then* $C_1(A) \cap C_2(A) = \varnothing$.

Proof. Suppose that both behaviours are strict, so $C_1 = (D_1, f)$, $C_2 = (D_2, f)$ with f: A → B. If there is an element $x \in C_1(A) \cap C_2(A)$ then $f(x) \in D_1$ and $f(x) \in D_2$, therefore $f(x) \in D_1 \cap D_2$, which is a contradiction with the hypothesis that $D_1 \cap D_2 = \varnothing$. The proof is similar for the strong behaviours.

Proposition 3.2.9. *If* $D_1 \subset B$, $D_2 \subset B$, $C_1 \in$ Comp (A, B), $C_1 = (D_1, T)$ *(strict, strong, weak),* $C_2 \in$ Comp (A, B), $C_2 = (D_2, T)$ *(strict, strong, weak),* $C_4 = (D_1 \cup D_2, T)$, *all of them being defined by means of the same transformations, then* $C_1(A) \cup C_2(A) \subseteq C_3(A)$.

Proof. Let us suppose that the three behaviours are strict and the function f: A → B defines them. So we have $C_1 = (D_1, f)$, $C_2 = (D_2, f)$. If the element $x \in C_1(A) \cup C_2(A)$ then $x \in C_1(A)$ or $x \in C_2(A)$. It means that either $f(x) \in D_1$ or $f(x) \in D_2$. Therefore $f(x) \in D_1 \cup D_2$ and so $x \in C_3(A)$ fulfils.

Conversely, if $x \in C_3(A)$ then $f(x) \in D_1 \cup D_2$ so $f(x) \in D_1$ or $f(x) \in D_2$ and then $x \in C_1(A)$ or $x \in C_2(A)$. Therefore $x \in C_1(A) \cup C_2(A)$, and the required condition takes place in this case.

If all the behaviours are strong the reasoning described above remains valid for every transformation $f \in T$, obtaining the inclusion of the left side in the right one.

The converse inclusion is not valid, as it will be proved by means of example 3.2.1.

Suppose now that the three behaviours are weak. If $x \in C_1(A) \cup C_2(A)$ then either $x \in C_1(A)$ or $x \in C_2(A)$ or both relations are valid. It means that there is $f \in T$ such that $f(x) \in D_1$ or there is $g \in T$ such that $g(x) \in D_2$. In the first case, from $f(x) \in D_1$ it follows that $f(x) \in D_1 \cup D_2$ and so $x \in C_3(A)$. In the second case, $f(x) \in D_2 \subset D_1 \cup D_2$ fulfils and so $x \in C_3(A)$. Therefore $C_1(A) \cup C_2(A) \subseteq C_3(A)$ is valid.

Conversely, if $x \in C_3(A)$ then is $h \in T$ such that $h(x) \in D_1 \cup D_2$, meaning that $h(x) \in D_1$ or $h(x) \in D_2$, so $x \in C_1(A)$ or $x \in C_2(A)$ and therefore $x \in C_1(A) \cup C_2(A)$, as required.

Example 3.2.1. Let X be a real linear space, Y a linear subspace of X, S a maximal proper linear subspace of Y and $y_o \in Y - S$ a fixed point.

Definition 3.2.4. *A linear operator* U: X → X *is said to be an interpolation operator with respect to the subspace* $Y \subset X$ *if it satisfies the conditions:*

(I1) $U x \in Y$, $x \in X$;

(I2) $U x = x$, $x \in Y$.

In this case the subspace Y *is said to be an interpolation one with respect to the operator* U.

Definition 3.2.5. *A linear functional* F: X → **R** *is called the divided difference attached to the interpolation operator* U *with respect to the subspace Y if it satisfies the following conditions:*

(DD1) FU=F;

(DD2) F (S) = {0};

(DD3) F (y_0) = 1.

The existence and the uniqueness of such a functional, denoted by [U;], is proved by M. Ivan (1980). Such a functional can be obtained, for example, by the association between every function defined on the set of n + 1 points and the coefficient of the term of highest degree of the Lagrange polynomial of n + 1 degree that interpolates the functions on the given points.

If we take B = **R**, $D_1 = (-\infty, 0)$, $D_2 = (0, +\infty)$ and the divided difference [U; ·]: X → **R** is attached to every operator U belonging to a set U of interpolation operators with respect to the subspace Y, the following strong behaviours are defined:

(3.2.1) $C_1 = s - ((-\infty, 0), T_U)$

(3.2.2) $C_2 = s - ((0, +\infty), T_U)$,

where $T_U = \{[U,.] \mid U \in U\}$.

An element x ∈ X belongs to C_1 (X) if [U; x] < 0 for every U ∈ U. This property was called by M. Ivan (1982) the U-*concavity property*. The property x ∈ C_2 (X), meaning that [U; x] > 0 for every U ∈ U was called in M. Ivan (1982) U-*convexity property*. The following partition is obvious:

$$X = C_1(X) \cup C_2(X) \cup [X - (C_1(X) \cup C_2(X))].$$

The U-*convexity* property was described like a behaviour of a special type by R.Precup (1985), without the use of the divided differences attached to the interpolating operators, as it follows: the maximality of S in Y means that Y can be partitioned into three parts:

(3.2.3) $Y = Y^- \cup S \cup Y^+$,

where

$$Y^- = \{y \in Y \mid \exists\, a \in S \text{ and } \exists\, \lambda < 0,\ y = a + \lambda y_0\},$$

$$Y^+ = \{y \in Y \mid \exists\, a \in S \text{ and } \exists\, \lambda > 0,\ y = a + \lambda y_0\}.$$

Definition 3.2.6 *The element* x ∈ X *is said to have the shape* (Y, U) *(or* (Y^+, U)) *if for every* U ∈ U *the condition* Ux ∈ Y^- *(or* Ux ∈ Y^+*) is satisfied.*

The relationship between the shapes defined by R. Precup (1985) and the behaviours (3.2.1) and (3.2.2) is established in the following theorem, equivalent to proposition 1.1.2 from R. Precup (1985).

Theorem 3.2.1. (1) *The element* x ∈ X *has the behaviour* C_1 = s - ((-∞, 0), T_U) *if and only if it has the shape* (Y^-, U);

(2) *An element* x ∈ X *has the behaviour* C_2 = s - ((0, +∞), T_U) *if and only if it has the shape* (Y^+, U).

Proof. (1) Let x ∈ X be an element having the shape (Y^-, U). Then, for every operator U ∈ U , we have U x ∈ Y^-. But U x ∈ Y, S is maximal in Y and y_0 ∈ Y - S, involve that U x = a + λy_0, with a ∈ S and $\lambda < 0$, this representation of U x being unique. Now, using (DD1) - (DD3), one obtains

$$[U; x]= [U; U x]= [U; a + \lambda\, y_0] = [U; a] + \lambda\, [U; y_0] = \lambda < 0,$$

therefore the shape (Y^-, U) is equivalent to the behaviour C_1 = s-((-∞, 0), T_U). In a similar way, (2) can be proved.

Coming back to proposition 3.2.9, the set C_1 (X) contains all the U – concave elements of X and the set C_2 (X) contains all the U – convex elements of X. Therefore, $C_1(X) \cup C_2(X)$ is the set of all the elements of X which are either U – convex or U – concave. On the other hand, $D_1 \cup D_2$ = **R** \ {0} and the behaviour C_3 = ($D_1 \cup D_2$, T_U) (strict, strong, weak) describes the elements of X with the following property: either [U, x]≠ 0 for every U ∈ U if the behaviour is strict or strong, or there is U ∈ U such that [U, x] ≠ 0. It is easy to see that $C_3(X) \not\subset C_1(X) \cup C_2(X)$.

Proposition 3.2.10. *If* $D_1 \subset B$, $D_2 \subset B$, $D_1 \cap D_2 \neq \varnothing$, C_1 ∈ Comp (A, B), C_1 = (D_1, .) *(strict, strong, weak),* C_2 ∈ Comp (A, B), C_2 = (D_2, .) *(strict, strong, weak),* $C_1(A) \cap C_2(A) \neq \varnothing$, C_3 = ($D_1 \cap D_2$,.) *(strict, strong, weak), all of them being defined by means of the same transformations, then* $C_1(A) \cap C_2(A) \subseteq C_3(A)$.

The proof follows the same stages as the previous one.

3.3 Sequences of behaviours

Let X be a nonempty set and the ascending sequence $(X_n)_{n=1,2,...}$ of nonempty subsets of X,

(3.3.1) $$X_1 \subset X_2 \subset \ldots \subset X_n \subset \ldots \subset \bigcup_{n=1}^{\infty} X_n = X.$$

Also, let $B \neq \varnothing$ and $D \subset B$, $D \neq \varnothing$. Let us consider the transformations $T_n : X_n \to B$ and denote by T_k the restriction of T_n to X_k. suppose that

(3.3.2) $$T_n\big|_{X_k} = T_k, \text{ for } n \geq k.$$

These transformations might be, in particular, restrictions of a transformation $f: X \to B$. Let be the behaviours

(3.3.3) $$C_n = (D, T_n),\ n = 1, 2, \ldots.$$

Lemma 3.3.1. $\bigcup_{n=1}^{\infty} C_n (X_n) = C(X)$, *where* C *is the strong behaviour* $C = s\text{-}(D,\mathrm{T})$, *with* $\mathrm{T} = \{ T:X \to B \mid T|_{X_n} = T_n,\ n=1,2,\ldots\}$.

Proof. Since $X_1 \subseteq X_2 \subseteq \ldots \subseteq X$ it follows that

(3.3.4) $$C_1 (X_1) \subseteq C_2 (X_2) \subseteq \ldots \subseteq C_n (X_n) \subseteq \ldots \subseteq C(X).$$

Hence $\bigcup_{n=1}^{\infty} C_n (X_n) \subseteq C(X)$.

To prove the converse inclusion, let $x \in C(X)$. Then $T(x) \in D$ for every $T \in \mathrm{T}$. But for every $T \in \mathrm{T}$, $T|_{X_n} = T_n$, therefore there is a number $p \in \mathbf{N}^*$ such that $T_n(x) \in D$ for $n \geq p$. This means that $x \in C_n(X)$, $n \geq p$, hence $x \in \bigcup_{n=1}^{\infty} C_n (X_n)$. So, we have $C(X) \subseteq \bigcup_{n=1}^{\infty} C_n (X_n)$ and the equality is completely proved.

For a sequence of strict behaviours $C_n = (D, T_n)$, $n = 1, 2, \ldots$, that satisfies (3.3.1), (3.3.2) and (3.3.3) the inclusion (3.3.4) justifies the term of *a descending sequence of behaviours.* In order to thoroughly justify this terminology one can notice that if the set of transformations T_n, $T_n^* : X \to B$, $n = 1, 2, \ldots$, with the properties

(3.3.5) $$T_n^*\big|_{X_n} = T_n \text{ and } T_n^*(x) \in B - D \text{ for } x \in X - X_n,\ n = 1, 2, \ldots,$$

is considered then the strong behaviours $C_n^* = s - (D, T_n)$, $n = 1, 2, \ldots$, have the property that $C_n^*(X_n) = C_n(X_n) = C_n(X)$, $n = 1, 2, \ldots$. Then

(3.3.6) $$C_1^* (X_1) \subseteq C_2^* (X_2) \subseteq \ldots \subseteq C_n^* (X_n) \subseteq \ldots \subseteq C(X).$$

Since the behaviours C_k^* are defined on the same set X and according to the definition 3.1.3. ii), it follows that (3.3.6) is equivalent to

(3.3.7) $\quad \ldots \leq C_n^* \leq \ldots \leq C_2^* \leq C_1^*$,

meaning that the sequence $(C_n^*)_{n=1,2,\ldots}$ is descending with respect to the non-strict precedence relation.

So, it was shown that every descending sequence of behaviours in X_n, $C_n = (D, T_n)$, n = 1, 2, ..., has a corresponding sequence $(C_n^*)_{n=1,2,\ldots}$ of behaviours in X, descending with respect to the non-strict precedence relation.

Definition 3.3.1. *If* $(C_n)_{n=1,2,\ldots}$, $C_n = (D, T_n)$, n = 1, 2, ..., *is a descending sequence of behaviours then the strong behaviour* C = s - (D, T), *with* $T = \{T: X \to B \mid T|_{X_n} = T_n,\ n=1,2,\ldots\}$ *is said to be the limit of the sequence of behaviours* $(C_n)_{n=1,2,\ldots}$ *and is denoted by* $C = \lim_{n\to\infty} C_n$.

The example 3.2.1 contains behaviours that, if both the number of knots in the construction of the Lagrange polynomial and the number of values on the knots are successively increased by the addition of an element then the result contains descending sequences of behaviours: concavities, non-convexities, polynomialities, non-concavities, convexities.

Property 3.3.1. *Let be the sequence of additive groups* $(X_k)_{k=1,2,\ldots}$ *with the property (3.3.1) and the sequence of behaviours* $(C_n)_{n=1,2,\ldots}$ *defined by (3.3.2) and (3.3.3). If each behaviour of the sequence is stable with respect to the addition, then its limit is stable with respect to the addition.*

Proof. Let be $x \in C(X)$ and $y \in C(X)$. Then, according to (3.3.1) and to lemma 3.3.1, there is a natural number p such that $x \in C_n(X_n)$ and $y \in C_n(X_n)$ for $n \geq p$. But C_n is stable with respect to the addition for every n, therefore $x + y \in C_n(X_n)$, $n \geq p$. Then, according to lemma 3.3.1, it follows that $x + y \in C(X)$, therefore C is stable with respect to the addition.

Property 3.3.2. *Let the sequence of real linear spaces* $(X_k)_{k=1,2,\ldots}$ *with the property (3.3.1) and the sequence of behaviours* $(C_n)_{n=1,2,\ldots}$ *defined by (3.3.2) and (3.3.3). If each behaviour of the sequence is every positive scale change (every scale change) invariant then its limit is also every positive scale change (every scale change) invariant.*

Proof. Let $x \in C(X)$ and $\alpha > 0$. Then there is a natural number p such that $x \in C_n(X_n)$ for $n \geq p$. This implies, due to the every positive scale change invariance of the behaviour C_n, that $\alpha x \in C_n(X_n)$ and, according to definition 3.3.1, it follows that $\alpha x \in C(X)$. For the every scale change invariance the proof is similar.

Let now Y be a nonempty set, another nonempty set B and its nonempty subsets

(3.3.8) $$\ldots \subseteq D_n \subseteq D_{n-1} \subseteq \ldots \subseteq D_2 \subseteq D_1 \subset B$$

such that

(3.3.9) $$\bigcap_{n=1}^{\infty} D_n = D \subset B,\ B \neq \varnothing.$$

Let the sequence of behaviours $(C_n)_{n=1,2,\ldots}$, where $C_n = (D_n, T)$, n = 1, 2, ..., with $T: Y \to B$.

Lemma 3.3.2. $\bigcap_{n=1}^{\infty} C_n(Y) = C(Y)$, *where* C *is the behaviour* C=(D,T).

Proof. From (3.3.9) it follows

(3.3.10) $$C(Y) \subseteq \ldots \subseteq C_n(Y) \subseteq \ldots \subseteq C_2(Y) \subseteq C_1(Y),$$

therefore $C(Y) \subseteq \bigcap_{n=1}^{\infty} C_n(Y)$ takes place.

Now, let $x \in \bigcap_{n=1}^{\infty} C_n$. Then $x \in C_n(Y)$ for every nonzero natural number n, meaning that $T(x) \in D_n$, for every n = 1, 2, ... Hence, $T(x) \in \bigcap_{n=1}^{\infty} D_n = D$ and therefore $x \in C(Y)$. It follows that $\bigcap_{n=1}^{\infty} C_n(Y) \subseteq C(Y)$ and, together with the conclusion on the converse inclusion we get the desired result.

The sequence of behaviours $(C_n)_{n=1,2,\ldots}$ satisfying (3.3.10) has, therefore, the monotony property

(3.3.11) $$C_1 \leq C_2 \leq \ldots \leq C_n \leq \ldots,$$

being an ascending sequence of behaviours with respect to a non-strict precedence relation of behaviours.

Definition 3.3.2. *If* $(C_n)_{n=1,2,\ldots}$, $C_n = (D_n, T)$, n = 1, 2,..., *is an ascending sequence of behaviours satisfying (3.3.8) and (3.3.9), then the behaviour* C = (D, T) *is called the limit of the sequence of behaviours* $(C_n)_{n=1,2,\ldots}$ *and is denoted by* $C = \lim_{n\to\infty} C_n$.

Definition 3.3.3. *A sequence of behaviours that is either ascending or descending is said to be a monotone sequence of behaviours.*

Example 3.3.1. Coming back to example 3.1.5, let us consider the metric space (X, d), and a nonempty subset M of X. Let now $A = X \times M$, $B = \mathbf{R}$ and $(\varepsilon_n)_{n=1,2,...}$ be a decreasing sequence of nonnegative real numbers, convergent to zero. Then the sequence of intervals $D_n = [0, \varepsilon_n]$, n = 1, 2,..., is convergent to $D = \{0\}$. Define the function f: $A \to B$ by f (x, y) = d (x, y). Then the sequence of behaviours $C_n = (D_n, f)$ is ascending, and its limit is the behaviour C = (D, f), with the significance discussed in example 3.1.5. The convergence of this sequence means that the element $y \in M$ is an ε_n-approximation of $x \in X$, for every n = 1, 2, ... , if and only if d(x, y) = 0, therefore x = y.

Further results concerning the stability with respect to the addition and the every positive scale change (every scale change) invariance, similar to properties 3.3.1 and 3.3.2, are easy to obtain for ascending sequences of behaviours.

3.4. Convexity with respect to a behaviour

In this paragraph X is assumed to be a nonempty set in which the notion of straight-line segment, denoted by <x, y> for $x, y \in X$, is defined. A behaviour (strict, weak or strong) C = (D, T) is defined in $X \times 2^X$, with $T \in T \subset \{T \mid T: X \times 2^X \to B\}$, for the nonempty set B and its nonempty subset $D \subset B$

Definition 3.4.1. *The set* $Y \subset X$ *is called convex with respect to the behaviour* C *if it is empty or if for every two points* $x, y \in Y$ *and for every* $t \in$ <x, y>, *the element* $(t, Y) \in C\,(X \times 2^X)$.

In what follows G_C will mean the set of all the subsets of X that are convex with respect to the behaviour C.

Theorem 3.4.1. *If* $C_1, C_2 \in$ Comp $(X \times 2^X)$, B) *are two behaviours such that* $C_1 \le C_2$ *then* $G_{C_2} \subseteq G_{C_1}$ *holds.*

Proof. Supposing that $A \in G_{C_2}$ we obtain the validity of the relation $(t, A) \in C_2\,(X \times 2^X)$, for every two points $x, y \in A$ and for every $t \in$ <x, y>. But the condition $C_1 \le C_2$ implies $(t, A) \in C_1\,(X \times 2^X)$, meaning that $A \in G_{C_1}$.

Some interesting particular cases will be discussed the next chapter, establishing a connection with the problem of the convexity recognition in chapter 9.

In order to obtain more properties of this general type of convex sets we shall suppose that (X, +, .) and (V, +, .) are real linear spaces. The subsets $B \subseteq V$ and $D \subseteq B$ are supposed to be nonempty. Also, we consider the behaviour $C = (D, \mathsf{T}) \in \mathrm{Comp}\,(X \times 2^X, B)$, which might be either strong, or weak or strict. The straight-line segment in X is defined classically, by means of the convex combinations of the extremities. Let us consider a point $s \in X$ and a scalar $\alpha \in \mathbf{R}$. Starting from the behaviour C we define the behaviours $s + C$ and αC as it follows: to every $T \in \mathsf{T}$ we associate the transformations $(s + T)\colon X \times 2^X \to B$ and $(\alpha T)\colon X \times 2^X \to \alpha B$ by

(3.4.1) $$(s + T)\,(s + a, s + A) = T\,(a, A)$$

(3.4.2) $$(\alpha T)\,(\alpha a, \alpha A) = \alpha T\,(a, A)$$

for $a \in X$ and $A \subset 2^X$. The above mentioned behaviours are defined by $s + C = (D, s + \mathsf{T})$ and $\alpha C = (\alpha D, \alpha \mathsf{T})$ where $s + \mathsf{T} = \{s + T \mid T \in \mathsf{T}\}$ and $\alpha\mathsf{T} = \{(\alpha T) \mid T \in \mathsf{T}\}$.

Property 3.4.1. *If the set* $Y \subset X$ *is convex with respect to the behaviour* C, *if* $s \in X$ *and* $\alpha \in \mathbf{R}$ *then:*

i) *the set* $s + Y$ *is convex with respect to the behaviour* $s + C$;

ii) *the set* αY *is convex with respect to the behaviour* αC.

Proof. Let us suppose that C is a strong behaviour. Assuming that $x \in Y$ and $y \in Y$ and $t \in \langle x, y\rangle$, the condition $(t, Y) \in C\,(X \times 2^X)$ implies $T\,(t, Y) \in D$ for every $T \in \mathsf{T}$. Therefore, according to (3.4.1), the relation $(s + T)\,(s + t, s + Y) = T\,(t, Y) \in D$, for every $T \in \mathsf{T}$, means that $s + Y$ is convex with respect to the behaviour $s + C$. According to (3.4.2) it follows that $(\alpha T)\,(\alpha t, \alpha Y) = \alpha T\,(t, Y) \in \alpha D$, for every $T \in \mathsf{T}$, signifying that αY is convex with respect to the behaviour αC.

Property 3.4.2. *Let us suppose that* (B, +, .) *is a real linear space and* D *is a subspace of* B. *The strict or strong behaviour* $C \in \mathrm{Comp}\,(X \times 2^X, B)$ *is supposed to be stable with respect to the addition and invariant to every change of scale. If the sets* $Y_1 \subset X$ *and* $Y_2 \subset X$ *are convex with respect to the behaviour* C *then for every scalars* $\alpha \in \mathbf{R}$ *and* $\beta \in \mathbf{R}$ *the set* $\alpha Y_1 + \beta Y_2$ *is convex with respect to the behaviour* C.

Proof. Let us consider the points x and y from $\alpha Y_1 + \beta Y_2$ and $t \in \langle x, y\rangle$. Then, there are the points $x_1, y_1 \in Y_1$, $x_2, y_2 \in Y_2$, $t_1 \in \langle x_1, y_1\rangle$, $t_2 \in \langle x_2, y_2\rangle$ such that $x = \alpha x_1 + \beta x_2$, $y = \alpha y_1 + \beta y_2$, $t = \alpha t_1 + \beta t_2$. From the convexity of Y_1 with respect to the behaviour C it results that $(t_1, Y_1) \in C\,(X \times 2^X)$. The convexity of Y_2 with respect to the behaviour C yields $(t_2, Y_2) \in C\,(X \times 2^X)$. From the invariance of the behaviour C to every change of scale we obtain that $(\alpha t_1, \alpha Y_1) \in C(X \times 2^X)$ and $(\beta t_2, \beta Y_2) \in C(X \times 2^X)$. The stability with respect to the addition yields $(\alpha t_1, \alpha Y_1) + (\beta t_2, \beta Y_2) = (\alpha t_1 + \beta t_2, \alpha Y_1 + \beta Y_2) = (t, \alpha Y_1 + \beta Y_2) \in C(X \times 2^X)$, meaning the convexity of $\alpha Y_1 + \beta Y_2$ with respect to the behaviour C.

Corollary 3.4.1. (B, +, .) *is supposed to be a real linear space and* D *is a subspace of* B. *The strict or strong behaviour* $C \in Comp(X \times 2^X, B)$ *is supposed to be stable with respect to the addition and invariant to every change of scale. The sets* $Y_1, Y_2, \ldots, Y_n \subset X$ *are supposed to be convex with respect to the behaviour* C. *Then for every* $\alpha_k \in \mathbf{R}$, $k = 1, 2, \ldots, n$, *the set* $\alpha_1 Y_1 + \alpha_2 Y_2 + \ldots + \alpha_n Y_n$ *is convex with respect to the behaviour* C.

Example 3.4.1. In $X = \mathbf{R}^n$, for $s \in [0, 1]$, the straight-line segment determined by the points $x \in X$ and $y \in X$ is defined by $\langle x, y\rangle_s = \{t \in \mathbf{R}^n \mid t = \alpha^s x + (1-\alpha)^s y,\ \alpha \in [0, 1]\}$. Let us take $B = \mathbf{R}$, $D = \{1\}$ and let us define the transformation $T: X \times 2^X \to B$ by $T(t, Y) = \chi_Y(t)$, for every $(t, Y) \in X \times 2^X$. Here, χ_Y is the characteristic function of the set Y. The behaviour $C = (D, T)$ is taken into account.

Definition 3.4.2. *A set* $A \subset \mathbf{R}^n$ *is called* s – *convex if for each* $x \in A$, $y \in A$ *and* $\alpha \in [0, 1]$ *the following condition holds:* $\alpha^s x + (1-\alpha)^s y \in A$.

Proposition 3.4.1. *A set* A *is* s – *convex if and only if* A *is convex with respect to the behaviour* C.

Proof. If A is s – convex then for every $x \in A$, $y \in A$ and $\alpha \in [0, 1]$ we have $\alpha^s x + (1-\alpha)^s y \in A$. It means that $\chi_A(\alpha^s x + (1-\alpha)^s y) = T(\alpha^s x + (1-\alpha)^s y) = 1$ and therefore $(\alpha^s x + (1-\alpha)^s y, A) \in C(X \times 2^X)$, i. e. the convexity of A with respect to the behaviour C.

Conversely, if A is convex with respect to the behaviour C then for every $x \in A$, $y \in A$ and $t \in \langle x, y\rangle_s$ the condition $(t, A) \in C(X \times 2^X)$ fulfils. But for every $t \in \langle x, y\rangle_s$ there is a number $\alpha \in [0, 1]$ such that t can be written as $t = \alpha^s x + (1-\alpha)^s y$. From $(t, A) \in C(X \times 2^X)$ it results that $\chi_A(t) = 1$, meaning that $t = \alpha^s x + (1-\alpha)^s y \in A$. This happens for every $t \in \langle x, y\rangle_s$, which is equivalent to every $\alpha \in [0, 1]$ and therefore A is s – convex.

Example 3.4.2. If we take s = 1 in example 3.4.1 we obtain the classical convexity, signifying that the classical convexity is also a convexity with respect to a behaviour.

Example 3.4.3. If the straight-line segment, which appears in definition 3.4.1, is replaced by the geodesic segment defined by N. J. Hicks (1965), the behaviour defined by the characteristic function leads to the geodesic convexity, studied by T. Rapcsák (1989). If the straight-line segment is replaced by the order interval, then the same behaviour leads to the order convexity studied by J. E. Martinez-Legaz and I. Singer (1990, 1991). For the interval determined by two points taken instead of the straight-line segment we obtain the interval convexity of J. R. Calder (1971), using also the behaviour from example 3.4.1.

Example 3.4.4. X is supposed to be a metric space, the distance in X being denoted by d. Taking $B = \mathbf{R}$, $D = [0, \varepsilon]$ for $\varepsilon > 0$ we define the transformation $T: X \times 2^X \to B$ by $T(t, A) = \inf \{d(t, x) \mid x \in A\}$ for $A \subset X$ and $t \in X$. Then the convexity with respect to the behaviour $C = (D, T)$ means that a set $A \subset X$ satisfies the property that for every $x, y \in A$ and $t \in \langle x, y \rangle$ ($\langle x, y \rangle$ is the metric segment) the condition $d(t, A) \le \varepsilon$ holds. This property was studied by G. Cristescu (1996[57], 1997[63], 2000[71]) and called ε - convexity.

The convexity with respect to a behaviour is a non-connected convexity property. The following examples will prove the non-connectivity of sets having a convexity of this type.

Example 3.4.5. If $X = \mathbf{R}^2$, d is the Euclidean distance in the plane, extended to the distance between a point and a set as in example 3.4.4, $A = [0, 1]^2 \cap \mathbf{Q}^2$, then A is convex with respect to the behaviour $C = ([0, \varepsilon], d)$ for every $\varepsilon > 0$. In fact, the set A is dense in $[0, 1]^2$ and is not connected. This particular notion of convexity with respect to a behaviour will be studied in chapter 5.

Example 3.4.6. The discrete convexity defined by J. M. Chassery in 1978 in $\mathbf{Z}^2$ and generalised for general sampling of grid step $h > 0$ over the plane, $\mathbf{Z}^2(h)$, in 1983 can be expressed using the α - convexity. We take here the most general case, published by J. M. Chassery (1983). In the plane $\mathbf{R}^2$ we shall consider the chessboard distance $c(x, y) = \max(|x_1 - y_1|, |x_2 - y_2|)$ for $x = (x_1, x_2) \in \mathbf{R}^2$, $y = (y_1, y_2) \in \mathbf{R}^2$. For the real number $h > 0$ we denote the set of the lattice points of the grid of step h over the plane by $\mathbf{Z}^2(h) = \{(ih, jh) \mid i \in \mathbf{Z}, j \in \mathbf{Z}\}$. The set $\mathbf{Z}^2(h)$ is a non-connected subset of $\mathbf{R}^2$. Therefore, a convexity in this set of lattice points is a non-connected convexity in $\mathbf{R}^2$. But J. M. Chassery defines in $\mathbf{Z}^2(h)$ the connectivity by means of arcs using the distance c and defines the convexity only for connected components in $\mathbf{Z}^2(h)$. But the definition may be formulated without the connectivity condition, as we shall do in what follows.

Definition 3.4.3. *A set* $A \subset \mathbf{Z}^2(h)$ *is* ε *- convex if for every* $x, y \in A$ *and* $t \in]0, 1[$, *there is a number* $\varepsilon \in [h/2, h[$ *and a point* $a \in \mathbf{Z}^2(h)$ *such that* $c(tx + (1 - t)y, a) \le \varepsilon$.

Definition 3.4.4. *A set* $A \subset \mathbf{Z}^2(h)$ *is discrete convex if there is a real number* $\varepsilon \in [h/2, h[$ *such that A is* ε *- convex.*

As we see, the condition c (t x + (1-t) y, a) $\leq \varepsilon$ means that c (t x + (1-t) y, A) $\leq \varepsilon$, signifying that the ε - convexity from definition 3.4.3 is exactly the ε - convexity defined in definition 3.4.1. This take place for the behaviour C (ε) = ([0, ε], T), with T: $\mathbf{Z}^2(h) \times 2^{Z^2(h)} \to \mathbf{R}$, T (a, A) = c (a, A) = inf {c (a, x) | x $\in$ A}. More, definition 3.4.4 of the discrete convexity is equivalent to: a subset A of $\mathbf{Z}^2$ (h) is discrete convex if and only if there is a real number $\varepsilon \in$ [h/2, h[such that A is convex with respect to the behaviour C (ε).

3.5 Convexity space

Let X be an arbitrary set. We recall from the previous chapter the following notions and results.

A pair (X, G), with G $\subseteq 2^X$, is called a convexity space on X if X $\in$ G and $\bigcap_{i \in I} A_i \in$ G, for any arbitrary family $(A_i)_{i\in I}$, with $A_i \in$ G. A subset A $\subseteq$ X is called convex in the convexity space (X, G) if A $\in$ G. A mapping g: $2^X \to 2^X$ is called the convex hull operator on X if: A $\subseteq$ g (A), g (g (A)) = g (A), and g (B) $\subseteq$ g (A), for any A, B $\in 2^X$, B $\subseteq$ A. If g is a convex hull operator defined on X, and G = {A $\subseteq$ X | g (A) = A}, then the pair (X, G) is a convexity space on X, called the convexity space generated by the operator g. If (X, G) is a convexity space on X, then a convex hull operator on X is defined by the mapping g: $2^X \to 2^X$ given by

(3.5.1) $\quad$ g (A) = $\cap$ {C $\in$ G | A $\subseteq$ C}, for all A $\in 2^X$.

This function is called the convex hull operator generated by G. If g is a convex hull operator defined on X then the set <x, y> = g ({x, y}) is said to be the straight line segment determined by x and y, for every two points x $\in$ X and y $\in$ X. The set <xy = {z $\in$ X | y $\in$ <x, z>} is called the half-line containing y and opposite to the point x.

Let us suppose that X is a nonempty set in which the notion of straight-line segment is defined and M is a nonempty set.

Definition 3.5.1. *A behaviour* C $\in$ Comp (M $\times 2^X$, B) *is said to be closed for intersection if the two conditions* (a, Y) $\in$ C (M $\times 2^X$) *and* (a, Z) $\in$ C (M $\times 2^X$) *imply that* (a, Y $\cap$ Z) $\in$ C (M $\times 2^X$).

Proposition 3.5.1. *If* C $\in$ Comp (X $\times 2^X$, B) *and* C *is closed for intersection then the pair* (X, G_C) *is a convexity space on* X.

Proof. Let us suppose that the nonempty subsets Y and Z of X are convex with respect to the behaviour C. Then, for every x, y $\in$ Y $\cap$ Z and t

$\in$ <x, y> the relations (t, Y) $\in$ C (X $\times$ 2^X) and (t, Z) $\in$ C (X $\times$ 2^X). But C is closed for intersection, meaning that (t, Y $\cap$ Z) $\in$ C (X $\times$ 2^X). Therefore, Y $\cap$ Z is convex with respect to the behaviour C.

The condition of closure for intersection is very strong. It is satisfied by the behaviours that define the convexity properties from examples 3.4.1, 3.4.2 and 3.4.3. But there are situations when the behaviour that defines the convexity is not closed for intersection, but the intersection of two convex sets with respect to that behaviour has a convexity property of the same type, if some additional hypotheses are satisfied. The discrete convexity from example 3.4.6 yields to such a result, presented in the following example.

Example 3.5.1. Let us consider the discrete convexity, as it is defined by definitions 3.4.3 and 3.4.4. For the real number h > 0 we consider the set $\mathbf{Z}^2$ (h) = {(ih, jh) | i $\in$ **Z**, j $\in$ **Z**} together with the chessboard distance c. We recall more notions and results that will be used in this example. If a = (ih,jh) $\in$ $\mathbf{Z}^2$ (h) then the 8-neighbourhood of a in $\mathbf{Z}^2$ (h) is the set V_8 (a) = {a, ((i-1)h, jh), ((i+1)h, jh), (ih, (j-1)h), (ih, (j+1)h), ((i-1)h, (j-1)h), ((i-1)h, (j+1)h), ((i+1)h, (j-1)h), ((i+1)h, (j+1)h)}. A succession of points $x_1, x_2, \ldots, x_n$ in $\mathbf{Z}^2$ (h) is said to be a 8-path if $x_{i+1} \in V_8(x_i)$ for $1 \le i < n$. Let be S $\subset$ $\mathbf{Z}^2$ (h). Two points x $\in$ S and y $\in$ S, are said to be 8-connected in S if there is a respectively 8-path $x = x_1, x_2, \ldots, x_n = y$ from x to y such that $x_i \in S$, $1 \le i < n$. The 8-connectivity in S is an equivalence relation. The equivalence classes with respect to the 8-connectivity are called 8-components of S. If S has only one 8-component then S is said to be 8-connected (or simply, connected).

Let p and q be two lattice points of $\mathbf{Z}^2$ (h) and <p, q> be the straight-line segment between p and q in $\mathbf{R}^2$. The straight-line segment <a, b>, with a $\in$ S and b $\in$ S, is said *to lie near* S if for every point t $\in$ <a, b> there is a lattice point s $\in$ S such that c (t, s) < h. If for every a $\in$ S and b $\in$ S the straight-line segment <a, b> lies near S then we say that the set S *has the chord property*. A. Rosenfeld (1974 [240]) proves that for every straight-line segment <a, b>, there is an 8-path ($a = a_1, a_2, \ldots, a_m = b$) that lies near <a, b>. J. M. Chassery (1983) described a method of construction of this 8-path.

First of all, let us remark that if the connected sets A, B $\subset$ $\mathbf{Z}^2$ (h) are discrete convex then A $\cap$ B is not necessary connected. Indeed, if we consider h = 1, A = {(0, 0), (0, 1), (0, 2), (0, 3), (1, 3)} and B = {(0, 0), (1, 0), (1, 1), (1, 2), (1, 3)}, which are discrete convex with $\varepsilon = \frac{3}{4}$, then A $\cap$ B = {(0, 0), (1, 3)}, a non-connected set in $\mathbf{Z}^2$.

Now we are able to prove the intersection property.

Theorem 3.5.1. *Let* X $\subset$ $\mathbf{Z}^2$ (h) *be an* ε_1 – *convex connected component and* Y $\subset$ $\mathbf{Z}^2$ (h) *be an* ε_2 - *convex connected component, with* $\varepsilon_1 \le \varepsilon_2$. *Then if* X $\cap$ Y *is a connected component then it is* ε_2 – *convex.*

Proof. Supposing that $X \cap Y$ is a connected component, let p and q be two points of $X \cap Y$. As consequence of theorem 3.4.1, if the set X is ε_1 – convex and $\varepsilon_1 \le \varepsilon_2$ then X is also ε_2 – convex. Because X is ε_2 – convex, $p \in X$ and $q \in X$ then there is an 8-path $A_x = (p, a_{x0}, a_{x1}, \ldots, a_{xn}, q)$ which lies near <p, q>. Because Y is ε_2 – convex, $p \in Y$ and $q \in Y$ then there is an 8-path $A_y = (p, a_{y0}, a_{y1}, \ldots, a_{yn}, q)$ which lies near <p, q>. The digital arcs A_x and A_y are supposed to contain points situated on the same side of the straight line determined by p and q (which is possible to obtain applying the method from J. M. Chassery (1983), pp.332).

The proof will be divided into two steps:

Step 1) we shall prove that $X \cap Y \cap (A_x \cup A_y)$ is a connected component,

Step 2) we shall prove that $X \cap Y$ is ε_2 – convex.

Proof of the Step 1. We shall consider different cases taking into account the relative positions of the points p and q inside X and Y.

First case. Let us suppose that the points p and q are interior points to X and Y. In this case <p, q> is interior to X and Y and we can choose $A_x = A_y$. Thus, we have $X \cap Y \cap (A_x \cup A_y) = X \cap Y \cap A_x = A_x$ which is connected.

Second case. Let us suppose that the points p and q are interior points to X and boundary points to Y. The digital arc A_y is imposed by the fact that p and q are boundary points for Y, but we can choose A_x and we can take $A_x = A_y$. Then, we have $X \cap Y \cap (A_x \cup A_y) = X \cap Y \cap A_y = A_y$ which is connected.

Third case. The points p and q are boundary points to X and Y such that <p, q> is a side of the two minimum perimeter convex polygons associated with X and Y as defined by H. Freeman and R. Shapira (1975). As we suppose that $X \cap Y$ is a connected component, there is a digital arc E included in X and in Y which connects p to q. As <p, q> is a side of the minimum perimeter convex polygon associated with X, E is one side with respect to the straight line determined by p and q and on the same side as X. A similar reasoning with Y shows that we must have E on the other side. This is a contradiction that proves that this case never occurs when we suppose that $X \cap Y$ is a connected component and X and Y are ε_2 – convex sets.

Proof of the Step 2. Let us suppose that $X \cap Y$ is not ε_2 – convex. Then there are the points p and q in $X \cap Y$ such that there is at least one point s on <p, q> for which there is no point r_{xi} belonging to the sequence A_x satisfying the conditions $s \in B_\infty(r_{xi}, \varepsilon_2)$ and $r_{xi} \in A_y$. That hypothesis can be formulated as: for every point r_{xi} belonging to the sequence A_x such as $s \in B_\infty(r_{xi}, \varepsilon_2)$ we have $r_{xi} \notin A_y$. We must remark that given a point s there may exist 1, 2 or 4 lattice points r_i satisfying $s \in B_\infty(r_{xi}, \varepsilon_2)$. These lattice points r_i are each other neighbours. Now, the point s being chosen, three situations may occur.

First case. We have only one lattice point r such that $s \in B_\infty(r, \varepsilon_2)$. In this case we have $r_{xi} = r$ for the reason that X is ε_2 – convex and $r_{yj} = r$ for the reason that Y is ε_2 – convex. So, r belongs both to X and to Y.

Second case. There are two lattice points r_0 and r_1 satisfying $s \in B_\infty(r_0, \varepsilon_2)$ and $s \in B_\infty(r_1, \varepsilon_2)$. As X is ε_2 – convex, we have, for example, $r_0 \in X$ and $r_0 = r_{xi}$. We must prove that if $r_0 \notin Y$ then a contradiction arise. We suppose that $r_1 \notin X \cap Y$; otherwise we should have chosen r_1 and not r_0. Finally, we have $r_0 \notin X \cap Y$ and $r_1 \notin X \cap Y$; this implies that $X \cap Y \cap (A_x \cup A_y)$ is not connected for the reason that every digital arc from p to q must pass through r_0 or r_1. That result is in contradiction with the result previously proved. So, r_0 belongs to Y.

Third case. There are four points r_0, r_1, r_2, r_3 such that $s \in B_\infty(r_i, \varepsilon_2)$, for i = 0, 1, 2, 3. Since X is ε_2 – convex and $X \cap A_x$ is connected there are among r_0, r_1, r_2, r_3 two lattice points which belong to $X \cap A_x$. We call them r_x and $r_{x'}$. Since Y is ε_2 – convex and $Y \cap A_y$ is connected there are among r_0, r_1, r_2, r_3 two lattice points which belong to $Y \cap A_y$. We call them r_y and $r_{y'}$. If $(r_x, r_{x'}) \cap (r_y, r_{y'}) \neq \varnothing$ then the proof is complete. Otherwise, we are in contradiction with the fact that $X \cap Y \cap (A_x \cup A_y)$ is a connected component.

Corollary 3.5.1. *If* X *and* Y *are two discrete convex connected component and* $X \cap Y$ *is connected then* $X \cap Y$ *is discrete convex.*

3.6 Approximation of the convexity

Let be $X \neq \varnothing$, $B \neq \varnothing$, the sequence of nonempty subsets

(3.6.1) $$\ldots \subseteq D_n \subseteq D_{n-1} \subseteq \ldots \subseteq D_2 \subseteq D_1 \subset B$$

such that

(3.6.2) $$\bigcap_{n=1}^{\infty} D_n = D \subset B.$$

Consider the sequence of behaviours $(C_n)_{n=1,2,\ldots}$, where $C_n = (D_n, T)$, n = 1, 2, ..., with $T: X \to B$. Then, according to the lemma 3.3.2, we have the following relation $\bigcap_{n=1}^{\infty} C_n(X) = C(X)$, where C is the behaviour C = (D, T), the limit of the sequence of behaviours $(C_n)_{n=1,2,...}$, denoted by $C = \lim_{n\to\infty} C_n$.

Let us suppose now that a notion of straight-line segment <x, y>, x ∈ X, y ∈ X, is defined in X.

Theorem 3.6.1. *If the set* $A \subset X$ *has the property that it is convex with respect to every behaviour* C_n, n = 1, 2, ..., *of an ascending sequence of behaviours* $(C_n)_{n=1,2,...}$ *in* $X \times 2^X$ *by means of* B, *than it is convex with respect to the behaviour* $C = \lim_{n\to\infty} C_n$.

Proof. For every x ∈ A, y ∈ A and t ∈ <x, y> the element $(t, A) \in C_n(X \times 2^X)$, n = 1, 2, ... and lemma 3.3.2 leads to $(t, A) \in C(X \times 2^X)$, where $C = \lim_{n\to\infty} C_n$. So, A is convex with respect to the behaviour C.

Definition 3.6.1. *In the conditions of theorem* 3.6.1, *the sequence* $(C_n)_{n=1,2,...}$ *of the properties of convexity with respect to the behaviours* C_n, n = 1, 2, ..., *is said to converge to the property* C *of convexity with respect to the behaviour* C.

Definition 3.6.2. *A convexity property* C_k *is said to be an approximation of another convexity property* C *if it is a term of a sequence of convexities* $(C_n)_{n=1,2,...}$ *convergent to* C.

Chapter 4

Convexity with respect to a set and two behaviours

Another way of obtaining new types of convexity properties for sets by means of the notion of behaviour defined in the chapter 3 consists in the combination of the techniques of superposition of a set on another one, more often a network over its support set, with the description of the relationship between their elements and their subsets using the above presented technique of behaviours. In this chapter we shall show how does the concept of behaviour allow the formulation of the definition of a convexity property of a set, which is general enough to contain, as particular cases, convexities intervening in various domains of mathematics. In the integer programming, the set of points of $\mathbf{Z}^n$ is considered as a network over $\mathbf{R}^n$ and the convexity in $\mathbf{Z}^n$ appears as a restriction of a convex set from $\mathbf{R}^n$ to $\mathbf{Z}^n$. It is the case of the results from L. Lupşa (1980). More general, if the support set is a complete lattice and the network over it is a sub-lattice then we obtain the convexity defined within the geometry of preference spaces by V. B. Kuzmin, S. V. Ovchinnikov (1975), S. V. Ovchinnikov (1980)). L. Lupşa (1986) has unified all these properties into a general theory. We shall also take into account the convexities appearing in image analysis. In this domain, the set of pixels is a network over $\mathbf{R}^2$ and the convexity of figures in the plane is perceived through this network. Various approaches of the convexity from this directions can be found in the papers of J. M. Chassery (1983, 1984), C.E. Kim, A. Rosenfeld (1980), C.E. Kim, J. Sklansky (1982), L. Latecki, A. Rosenfeld, R. Silverman (1955), J. F. Lawrence, W. R. Hare, J. M. Kenelly (1972)). G. Cristescu (1996 [56], [58], [59]) has studied these types of properties in a more general framework. The properties of convexity presented in this chapter do not imply, generally, the connectivity, as it will be shown by means of more examples.

4.1 Convexities with respect to a set and two behaviours. Definition and basic properties

X is assumed to be a nonempty set so that for every two points x, y of X the closed straight-line segment determined by these points, denoted by <x, y>, is defined. Let us suppose that M is a nonempty subset of X, B' and B" two nonempty sets, $D' \subset B'$, $D'' \subset B''$ nonempty subsets, and T ' is a set of transformations, $T': X \times M \to B'$. The behaviour C' = (D', T') (or a strong one s - (D', T '), or a weak one w - (D', T ')) is defined in $X \times M$ by means of B'. For the set T " of transformations, $T'': M \times 2^X \to B''$, we define a behaviour C" = (D", T") (or s - (D', T "), or w - (D', T ")) in $M \times 2^X$ by means of B".

Definition 4.1.1. *i) The set* $Y \subset X$ *is said to be slackly convex with respect to the set* M *and the behaviours* C' *and* C" *if it is empty or if for every two points* x, $y \in Y$ *and for every* $t \in$ <x, y>, *if the element* $(t, a) \in C'(X \times M)$, *with* $a \in M$, *then the element* $(a, Y) \in C''(M \times 2^X)$.

ii) The set $Y \subset X$ *is said to be strongly convex with respect to the set* M *and the behaviours* C' *and* C" *if it is empty or if for every two points* x, $y \in Y$ *and for every* $t \in$ <x, y> *there is an element* $a \in M$ *such that the following conditions take simultaneously place:* $(t, a) \in C'(X \times M)$ *and* $(a, Y) \in C''(M \times 2^X)$.

It is obvious that definition 4.1.1 does not depend on the strictness, slackness or strongness of the behaviours C' and C". If a property depends on the type of the behaviour, then this will be specified in the hypothesis.

The property ii) does not imply i). If a set Y is strongly convex with respect to a set M and two behaviours C' and C", then it is not necessarily slackly convex with respect to a set M and two behaviours C' and C", as it is shown in the following example.

Example 4.1.1. Let $X=\mathbf{R}^2$ together with the Euclidean straight-line segment, $M=\mathbf{Z}^2$, $B'=B''=\mathbf{R}$, $D_1 = \left[0, \frac{4}{3}\right]$, $D_2 = \left[0, \frac{1}{3}\right[$, $D'' = \{0\}$ and the city-block distance $d: \mathbf{R}^2 \times \mathbf{Z}^2 \to \mathbf{R}$, $d(a, b) = |x_a - x_b| + |y_a - y_b|$, where $a = (x_a, y_a) \in \mathbf{R}^2$ and $b = (x_b, y_b) \in \mathbf{Z}^2$. The behaviours $C_1 = \left(\left[0, \frac{4}{3}\right], d\right)$ and $C_2 = \left(\left[0, \frac{1}{3}\right[, d\right)$ are chosen in $\mathbf{R}^2 \times \mathbf{Z}^2$, noticing that $C_1 \le C_2$. By means of this distance it is possible to define the behaviour C" = ({0}, f) in $\mathbf{Z}^2 \times 2^X$, where $f: \mathbf{Z}^2 \times 2^X \to \mathbf{R}$ is defined by $f(a, G) = \inf\{d(a, g) \mid g \in G\}$, for $a \in \mathbf{Z}^2$ and $G \in 2^X$. The set Y = {(0, 0), (0, 1), (1, 1), (2, 1), (3, 1), (0, 2), (1, 2), (2, 2),

(3, 2), (4, 2), (5, 2), (6, 2)} is strongly convex with respect M, C_1 and C", but it is not slackly convex with respect to M, C_1 and C". For example, for the points x = (0, 0) ∈ Y, y = (6, 2) ∈ Y and $t = \left(1, \frac{1}{3}\right) \in < x,y >$, the point a = (1, 0) ∈ C_1 ($\mathbf{R}^2 \times \mathbf{Z}^2$) but a ∉ C" ($\mathbf{Z}^2 \times 2^X$).

Also, if a set Y is slackly convex with respect to a set M and two behaviours C' and C", then it is possible that it is not strongly convex with respect to M and with C' and C", as in the following example.

Example 4.1.2. Let us come back to example 4.1.1. The set Y is slackly convex with respect to $\mathbf{Z}^2$ and the behaviours C_2 and C", but it is not strongly convex with respect to $\mathbf{Z}^2$, C_2 and C". Indeed, for the points x = (0, 0) ∈ Y, y = (6, 2) ∈ Y and $t = \left(1, \frac{1}{3}\right) \in < x,y >$ there is not a point a ∈ $\mathbf{Z}^2$ such that (t, a) ∈ C_2 ($\mathbf{R}^2 \times \mathbf{Z}^2$), therefore definition 4.1.1.ii) does not fulfil.

It is important to study the way in which such a convexity property remains valid when the initial behaviours are replaced by other ones that precede them or that are preceded by them. For this purpose, let us consider the behaviours C_1, C_2 ∈ Comp (X × M, B'), C', C" ∈ Comp (M × 2^X, B"). Let be a set Y ⊆ X.

Proposition 4.1.1. *If* $C_1 \le C_2$ *and if* Y *is strongly convex with respect to the set* M *and the behaviours* C_2 *and* C" *then* Y *is strongly convex with respect to* M *and the behaviours* C_1 *and* C".

Proof. If Y = ∅ then the result is obvious. If Y ≠ ∅ then for every two points x, y ∈ Y and t ∈ <x, y> there is a point a ∈ M such that (t, a) ∈ C_2 (X × M) and (a, Y) ∈ C" (M × 2^X). It is the condition of strong convexity of the set Y with respect to M and the behaviours C_2 and C". Since $C_1 \le C_2$, it follows that C_2 (X × M) ⊆ C_1 (X × M). So, for x, y ∈ Y and t ∈ <x, y> there is a ∈ M such that (t, a) ∈ C_2 (X × M) ⊆ C_1 (X × M) and (a, Y) ∈ C" (M × 2^X), therefore Y is strongly convex with respect to M and the behaviours C_1 and C".

Proposition 4.1.2. *If* $C_1 \le C_2$ *and if* Y *is strongly convex with respect to the set* M *and the behaviours* C_1 *and* C" *then* Y *is slackly convex with respect to* M *and the behaviours* C_2 *and* C".

Proof. If Y = ∅ then the result is obvious. If Y ≠ ∅ then for every two points x, y ∈ Y and t ∈ <x, y>, there is an element a ∈ M such that the following relations take place: (t, a) ∈ C_1 (X × M) and (a, Y) ∈ C" (M × 2^X). But $C_1 \le C_2$, and so C_2 (X × M) ⊆ C_1 (X × M). It is possible either that (t, a) ∈ C_2 (X × M) or that (t, a) ∉ C_2 (X × M). If (t, a) ∈ C_2(X × M), then (a, Y) ∈

C" ($M \times 2^X$) and so Y is slackly convex with respect to M and the behaviours C_2 and C".

Propositions 4.1.1 and 4.1.2 do not remain valid if the strong convexity with respect to the set M and the behaviours taken into account is replaced by the slack convexity. This is shown by the following example:

Example 4.1.3. Let us consider the set Y from example 4.1.1 and the same set M, together with the behaviours C_1 and C". The set Y is slackly convex with respect to $\mathbf{Z}^2$ and the behaviours C_2 and C", but it is not slackly convex with respect to $\mathbf{Z}^2$, C_1 and C". To verify that it is enough to choose x = (0, 0), y = (6, 2), t = (1, 1/3) ∈ <x, y>. Then, for a = (1, 0) ∈ $\mathbf{Z}^2$, d (t, a) = 1/3 < 4/3, but f (a, Y) = 1 ≠ 0 and therefore ((1, 0), Y) ∉ C" ($\mathbf{Z}^2 \times 2^X$). The same result can also be obtained for a = (4, 4/3) ∈ <x, y>.

Proposition 4.1.3. *If* C' ≤ C" *and if* Y *is slackly (or strongly) convex with respect to* M *and the behaviours* C_1 *and* C" *then* Y *is slackly (or strongly) convex with respect to* M *and the behaviours* C_1 *and* C'.

Proof. Let us suppose that Y ≠ ∅ and is slackly convex with respect to M and behaviours C_1 and C". For x, y ∈ Y and t ∈ <x, y>, if there is a ∈ M with the property that (t, a) ∈ C_1 ($X \times M$), then (a, Y) ∈ C" ($M \times 2^X$). But the inequality C' ≤ C" implies C" ($M \times 2^X$) ⊆ C' ($M \times 2^X$) and so (a, Y) ∈ C' ($M \times 2^X$) has as consequence the slack convexity of the set Y with respect to M and the behaviours C_1 and C'.

If Y = ∅ the result is evident both for the slack convexity and for strong convexity.

If Y ≠ ∅ and Y is strongly convex with respect to M and the behaviours C_1 and C", then for every x, y ∈ Y and t ∈ <x, y>, there is a ∈ M such that (t, a) ∈ C_1 ($X \times M$) and (a, Y) ∈ C" ($M \times 2^X$). But C' ≤ C" implies (a, Y) ∈ C' ($M \times 2^X$) and so Y is strongly convex with respect to M and the behaviours C_1 and C'.

Corollary 4.1.1. *If* $C_1 \leq C_2$, C' ≤ C" *and if the set* Y *is slackly (or strongly) convex with respect to* M *and the behaviours* C_2 *and* C", *then* Y *is slackly convex with respect to* M *and the behaviours* C_1 *and* C'.

Proof. According to proposition 4.1.1, if Y is strongly convex with respect to M and the behaviours C_2 and C" then Y is strongly convex with respect to M and the behaviours C_1 and C". Applying now proposition 4.1.3 it follows that Y is strongly convex with respect to M and the behaviours C_1 and C'.

Let us consider now two nonempty subsets of X, M_1 and M_2, such that $M_1 \subseteq M_2$. If C' ∈ Comp ($X \times M_2$, B') and C" ∈ Comp ($M_2 \times 2^X$, B"), then we shall denote by T ' the set of the transformations used to define C' and by

T " the set of the transformations used to define C". Taking into account that from $M_1 \subseteq M_2$ it follows that $X \times M_1 \subseteq X \times M_2$ we shall denote by $T'|_{M1}$ the set of the restrictions of the transformations from T ' to $X \times M_1$. The behaviour defined in $X \times M_1$ using the transformations from $T'|_{M1}$ in the same manner as in the definition of C' in $X \times M_2$ is called *the restriction of the behaviour* C' *to* $X \times M_1$ and will be denoted by $C'|_{M1}$. Also, $T''|_{M1}$ will mean the set of the restrictions of the transformations from T" to $M_1 \times 2^X$. The behaviour defined in $M_1 \times 2^X$ using the transformations from $T''|_{M1}$ in the same manner in which the transformations from T" have been used in the definition of the behaviour C" is said to be *the restriction of the behaviour* C" *to* $M_1 \times 2^X$ and is denoted by $C''|_{M1}$.

Proposition 4.1.4. *If the set* $Y \subset X$ *is strongly convex with respect to* M_1 *and the behaviours* $C'|_{M1}$ *and* $C''|_{M1}$, *then* Y *is strongly convex with respect to* M_2 *and the behaviours* C' *and* C".

Proof. If $x, y \in Y$ and $t \in \langle x, y \rangle$, then there is $a \in M_1$ such that $(t, a) \in C'|_{M1}(X \times M_1)$ and $(a, Y) \in C''|_{M1}(M_1 \times 2^X)$. But $M_1 \subseteq M_2$ implies $C'|_{M1}(X \times M_1) \subseteq C'(X \times M_2)$ and $C''|_{M1}(M_1 \times 2^X) \subseteq C''(M_2 \times 2^X)$. So, for $t \in \langle x, y \rangle$ there is $a \in M_1 \subseteq M_2$ such that $(t, a) \in C'(X \times M_2)$ and $(a, Y) \in C''(M_2 \times 2^X)$, following that Y is strongly convex with respect to M_2, C' and C".

Proposition 4.1.4 does not remain true if the strong convexity is replaced by the slack convexity, as the following example will show.

Example 4.1.4. Let us consider $X = \mathbf{R}^2$ together with the Euclidean straight-line segment, $M_1 = \mathbf{Z}^2$, $M_2 = \mathbf{Z}^2(1/2) = \{(a, b) \mid a = (1/2) h, b = (1/2) k, h \in \mathbf{Z}, k \in \mathbf{Z}\}$. Obviously, we have $M_1 \subseteq M_2$. Let be C' = ([0, 1/2[, d) a behaviour defined in $\mathbf{R}^2 \times \mathbf{Z}^2(1/2)$ by means of the city-block distance d: $\mathbf{R}^2 \times \mathbf{Z}^2(1/2) \to [0, +\infty[$. Also, let be the behaviour C" = ([0, 1/2[, f) defined in $\mathbf{Z}^2(1/2) \times 2^X$ using the function f: $\mathbf{Z}^2(1/2) \times 2^X \to \mathbf{R}$, $f(a, G) = \inf \{d(a, g) \mid g \in G\}$. The set Y = {(0, 0), (0, 1), (1, 1), (2, 1)} is slackly convex with respect to $\mathbf{Z}^2$ and the behaviours $C'|_{\mathbf{Z}^2}$ and $C''|_{\mathbf{Z}^2}$ but it is not slackly convex with respect to $\mathbf{Z}^2(1/2)$ and the behaviours C' and C". Indeed, if x = (0, 0), y = (2, 1), t = $(3/2, 3/4) \in \langle x, y \rangle$ then the point a = $(3/2, 3/2) \in \mathbf{Z}^2(1/2)$ has the property that $(t, a) \in C'(\mathbf{R}^2 \times \mathbf{Z}^2(1/2))$ but the point $(a, Y) \notin C''(\mathbf{Z}^2(1/2) \times 2^X)$.

The converse of proposition 4.1.4 is not true. The following example is relevant.

Example 4.1.5. Let us consider $X = \mathbf{R}^2$ together with the Euclidean straight-line segment, $M_1 = \mathbf{Z}^2$, $M_2 = \mathbf{Z}^2(1/2)$, $M_1 \subseteq M_2$. Let us define the behaviours C' = ([0, 1/2], d) in $\mathbf{R}^2 \times \mathbf{Z}^2(1/2)$ and C" = ([0, 1/2], f) in $\mathbf{Z}^2(1/2)$

$\times 2^X$ as in example 4.1.4 using the city-block distance between the points in $\mathbf{R}^2$. The set Y = {(0, 0), (0, 1/2), (0, 1), (1/2, 0), (1/2, 1/2), (1, 0), (1, 1/2), (3/2, 0), (2, 0), (3, 0)} is strongly convex with respect to $\mathbf{Z}^2$ (1/2) and the behaviours C' and C". But it is not strongly convex with respect to $\mathbf{Z}^2$ and the behaviours C'$|z^2$ and C"$|z^2$. Indeed, if we choose x = (0, 1) ∈ Y and y = (3, 0) ∈ Y we obtain that for the points t_1 = (1/4, 11/12) ∈ <x, y> and t_2 = (3/2, 1/2) ∈ <x, y> there is no point a ∈ $\mathbf{Z}^2$ with the property d (t_1, a) ≤ 1/2 and d (t_2, a) ≤ 1/2.

Theorem 4.1.1. *If the set* A ⊂ X *is convex with respect to the behaviour* C" *then* A *is both slackly and strongly convex with respect to the set* X *and the behaviours* C' *and* C", *with* C' = ({0}, T),

$$T: X^2 \to \{0, 1\}, T(t, a) = \begin{cases} 0, & t = a \\ 1, & t \neq a \end{cases}.$$

Proof. For x ∈ A, y ∈ A and t ∈ <x, y>, the convexity of A with respect to the behaviour C" implies that (t, A) ∈ C" (X × 2^X). If a ∈ X such that (t, a) ∈ C' (X^2) then T (t, a) = 0 implies that a = t ∈ <x, y>. Hence, (a, A) = (t, A) ∈ C" (X × 2^X) and A is both slackly and strongly convex with respect to the set X and the behaviours C' and C".

Theorem 4.1.2. i) *If the set* A ⊂ X *is slackly convex with respect to the set* X *and the behaviours* C' *and* C", *with* C' = ({0}, T),

$$T: X^2 \to \{0, 1\}, T(t, a) = \begin{cases} 0, & t = a \\ 1, & t \neq a \end{cases},$$

then A *is convex with respect to the behaviour* C".

ii) *If the set* A ⊂ X *is strongly convex with respect to the set* X *and the behaviours* C' *and* C", *with* C' = ({0}, T),

$$T: X^2 \to \{0, 1\}, T(t, a) = \begin{cases} 0, & t = a \\ 1, & t \neq a \end{cases},$$

then A *is convex with respect to the behaviour* C".

Proof. Both i) and ii) are a consequence of the fact that if a ∈ X such that (t, a) ∈ C' (X^2) then T (t, a) = 0 implies that a = t ∈ <x, y>. Therefore the condition t ∈ <x, y> replaces (t, a) ∈ C' (X^2).

Theorems 4.1.1 and 4.1.2 show that the convexity with respect to a behaviour is a particular case of slack or strong convexity with respect to a set and two behaviours.

4.2 Properties of sets that are convex with respect to a set and two behaviours in linear spaces

It is of interest, in the case in which X is a set having an algebraic structure, to study the situations when a convexity property, in the sense of definition 4.1.1, is invariant to some transformations (translations, scale changes, linear transformations, etc.). This is important because, many times, it is difficult to study a set, but its properties are studied by the investigation of its image after a transformation: Fourier transform, fast Fourier transform, Karhunen-Loève transform, Walsh-Hadamard transform, Haar transform etc.

Let us consider $(X, +, \cdot)$ a real linear space and the nonempty sets B' and B", together with their nonempty subsets $D' \subset B'$, $D'' \subset B''$ and the nonempty set $M \subset X$. Let us suppose that the straight line segment determined by two points is defined as usual, by means of the convex combinations of the extremities. In this case, the natural straight line in the Cartesian product space of two linear spaces is the Cartesian product of two straight-line segments in the factor linear spaces. This property of decomposability will be important in some proofs in what follows. They might lead to generalisations of the theorems in which they apply to other types of structures, more general than the linear space.

Theorem 4.2.1. *Let the strong (slack, strict) behaviours* $C' = (D', \mathcal{T}')$ *in* $X \times M$ *by means of* B', *with* $\mathcal{T}'$ *being a set of transformations* $T: X \times M \to B'$, *and* $C'' = (D'', \mathcal{T}'')$ *in* $M \times 2^X$ *by means of* B", *where* $\mathcal{T}''$ *is a set of transformations* $T_1: M \times 2^X \to B''$. *If* $s \in X$, *we consider:*

i) *the behaviour* $C'^* = (D', \mathcal{T}'^*)$ *defined in* $X \times (M + s)$ *by means of* B' *and the set* $\mathcal{T}'^*$ *of transformations* $T^*: X \times (M + s) \to B'$ *defined by association with a corresponding transformation* $T \in \mathcal{T}'$ *as it follows:*

(4.2.1) $T^*(x + s, a + s) = T(x, a), x \in X, a \in M;$

ii) *the behaviour* $C''^* = (D'', \mathcal{T}''^*)$ *defined in* $(M + s) \times 2^X$ *by means of* B" *and the set* $\mathcal{T}''^*$ *of transformations associated to the set* $\mathcal{T}''$ *as it follows: if* $T_1 \in \mathcal{T}''$ *then* $T_1^*: (M + s) \times 2^X \to D''$ *is defined by*

(4.2.2) $T_1^*(a + s, A + s) = T_1(a, A), a \in M, A \in 2^X;$

iii) $Y \subset X$ *a strongly (slackly) convex set with respect to the set* M *and the behaviours* C' *and* C".

Then the set $Y + s$ *is strongly (slackly) convex set with respect to the set* $M + s$ *and the behaviours* C'^* *and* C''^*.

Proof. First of all we will prove that:

(4.2.3) $C'(X \times M) + (s, s) = C'^*(X \times (M + s)),$

(4.2.4) $C''(M \times 2^X) + (s, s) = C''^*((M + s) \times 2^X)$.

Let $(y, b) \in C'(X \times M) + (s, s)$. This means that there is $(x, a) \in (X \times M)$ such that $(y, b) = (x, a) + (s, s) = (x + s, a + s)$. Suppose that C' is a strict behaviour, and let $T' = \{T\}$. Then $T(x, a) \in D'$. If T^* is the transformation obtained from T by (4.2.1) then $T^*(y, b) = T^*(x + s, a + s) = T(x, a) \in D'$ and so $(y, b) \in C'^*(X \times (M + s))$, meaning that $C'(X \times M) + (s, s) \subseteq C'^*(X \times (M + s))$.

Let now $(y, b) \in C'^*(X \times (M + s))$, so $(y, b) \in X \times (M + s)$ and $T^*(y, b) \in D'$. Since $X + s = X$ it follows that there is $(x, a) \in X \times M$ such that $(x, a) + (s, s) = (x + s, a + s) = (y, b)$ and then $T(x, a) = T^*(x + s, a + s) = T^*(y, b) \in D'$, so $(x, a) \in C'(X \times M)$. Therefore $(y, b) \in C'((X \times M) + (s, s))$ and then $C'^*(X \times (M + s)) \subseteq C'(X \times M) + (s, s)$ which completes the proof of (4.2.3).

If the behaviour C' is slack or strong the previous reasoning is repeated for every transformation that appears.

The proof of identity (4.2.4) is similar. Consider, for the beginning, $(b, B) \in C''(M \times 2^X) + (s, s)$. This means that there is $(a, A) \in C'(M \times 2^X)$ such that $(b, B) = (a, A) + (s, s) = (a + s, A + s)$. On the other hand, $T_1(a, A) \in D''$, having as consequence, according to (4.2.2), the relation $T_1^*(b, B) = T_1^*(a + s, A + s) = T_1(a, A) \in D''$, therefore $(b, B) \in C''((M + s) \times 2^X)$. This means that $C''(M \times 2^X) + (s, s) \subseteq C''^*((M + s) \times 2^X)$.

Conversely, if $(b, B) \in C''^*((M + s) \times 2^X)$ it means that $(b, B) \in (M + s) \times 2^X$ and, because $X + s = X$, it follows that there is $(a, A) \in M \times 2^X$ with the property that $(b, B) = (a, A) + (s, s) = (a + s, A + s)$. On another hand, $T_1^*(b, B) \in D''$. But this means that $T_1(a, A) = T_1^*(a + s, A + s) = T_1^*(b, B) \in D''$ and therefore $(a, A) \in C''(M \times 2^X)$, implying that $(b, B) \in C''(M \times 2^X) + (s, s)$. So, the converse inclusion is established. If the behaviour is slack or strong the proof is similar for all the behaviour that appear.

In order to prove that the set $Y + s$ is strongly convex with respect to the set $M + s$ and the behaviours C'^* and C''^*, let us consider two arbitrary points x and y from $Y + s$. It follows that there are the points $u, v \in Y$ such that $x = u + s$ and $y = v + s$ and therefore $\langle x, y\rangle = \langle u, v\rangle + s$. Let $t \in \langle x, y\rangle$. Then there is $p \in \langle u, v\rangle$ with the property that $t = p + s$. Because Y is strongly convex with respect to the set M and the behaviours C' and C'', there is $a \in M$ such that $(p, a) \in C'(X \times M)$ and $(a, Y) \in C''(M \times 2^X)$. But because $(p, a) \in C'(X \times M)$, from (4.2.3) it follows that $(t, a + s) \in C'(X \times M) + (s, s) = C'^*(X \times (M + s))$. On another hand, according to (4.2.4), we have $(a + s, Y + s) \in C''(M \times 2^X) + (s, s) = C''^*((M + s) \times 2^X)$. Therefore, for the point $t \in \langle x, y\rangle$ we found the point $(a + s) \in M + s$ which satisfies the condition from definition 4.1.1.i).

Let us suppose now that Y is slackly convex with respect to M, C' and C'' and let us prove that $Y + s$ has the similar property. Let $x, y \in Y + s$ and $t \in$

$\langle x, y\rangle$. Also, let the point $a \in M + s$ such that $(t, a) \in C'^* (X \times (M + s))$. Then there are the points $u, v \in Y$, $p \in \langle u, v\rangle$ and $b \in M$ with the properties $\langle x, y\rangle = \langle u, v\rangle + s$, $t = p + s$ and $a = b + s$. Since $(t, a) \in C'^* (X \times (M + s))$ and, due to (4.2.1) and (4.2.3) it follows that $(p, b) \in C' (X \times M)$. The slack convexity of Y with respect to M, C' and C'' implies that $(b, Y) \in C'' (M \times 2^X)$ and, applying now (4.2.4), we get that $(a, Y + s) \in C''^* ((M + s) \times 2^X)$. Therefore $Y + s$ is slackly convex with respect to the set $M + s$ and the behaviours C'^* and C''^*.

For $Y = \varnothing$ the property is obvious.

Corollary 4.2.1. *Let* $s \in X$ *and* $M \subset X$ *such that* $M + s = M$, C' *a behaviour in* $X \times M$ *by means of* B' *and* C'' *a behaviour in* $M \times 2^X$ *by means of* B''. *If:*

(i) C' *is a translation invariant,*

(ii) C'' *is a translation invariant,*

(iii) $Y \subset X$ *is strongly (slackly) convex with respect to the set M and the behaviours* C' *and* C'',

then $Y + s$ *is strongly (slackly) convex with respect to the* M *and the behaviours* C' *and* C''.

The proof is immediate.

Theorem 4.2.2. *Let* C' *be a behaviour in* $X \times M$ *by means of* B', C'' *a behaviour in* $M \times 2^X$ *by means of* B'', *and* $Y_1, Y_2 \subset X$ *are two strongly convex sets with respect to* M *and the behaviours* C' *and* C''. *The following properties are supposed:*

(i) $M + M = M$;

(ii) C' *is stable with respect to* + *in* $X \times M$;

(iii) C'' *is stable with respect to* + *in* $M \times 2^X$.

Then the set $Y_1 + Y_2$ *is a strongly convex set with respect to the set* M *and the behaviours* C' *and* C''.

Proof. Let Y_1 and Y_2 be strong convex sets with respect to M, C' and C''. If $x, y \in Y_1 + Y_2$ then there are $x_1, y_1 \in Y_1$ and $x_2, y_2 \in Y_2$ such that $x = x_1 + x_2$ and $y = y_1 + y_2$, and therefore $\langle x, y\rangle = \langle x_1, y_1\rangle + \langle x_2, y_2\rangle$. If $t \in \langle x, y\rangle$ then there is $t_1 \in \langle x_1, y_1\rangle$ and $t_2 \in \langle x_2, y_2\rangle$ such that $t = t_1 + t_2$. Since Y_1 is strongly convex with respect to the set M and the behaviours C' and C'' it follows that there is an element $a_1 \in M$ such that $(t_1, a_1) \in C' (X \times M)$ and $(a_1, Y_1) \in C'' (M \times 2^X)$. Since Y_2 is strongly convex with respect to the set M and the behaviours C' and C'' it follows that there is an element $a_2 \in M$ such that $(t_2, a_2) \in C' (X \times M)$ and $(a_2, Y_2) \in C'' (M \times 2^X)$. From $M + M = M$ we have $a_1 + a_2 = a \in M$. From ii) it follows that $(t, a) = (t_1 + t_2, a_1 + a_2) = (t_1, a_1) + (t_2, a_2) \in C'(X \times M)$. Then, due to iii), we get $(a, Y_1 + Y_2) = (a_1 + a_2, Y_1 + Y_2) = (a_1, Y_1) + (a_2, Y_2) \in C''(M \times 2^X)$. Therefore the point $a \in M$ is the corresponding point of t such that definition 4.1.1 fulfils. So, $Y_1 + Y_2$ is strongly convex with respect to M and the behaviours C' and C".

If one or both sets Y_1 and Y_2 are empty, the conclusion is evident.

Theorem 4.2.2 does not remain valid if the slack convexity is supposed instead of the strong convexity, as it is shown in what follows.

Example 4.2.1. In $X = \mathbf{R}^2$ with the Euclidean straight-line segment, consider $M = \mathbf{Z}^2$ and the behaviours C' = ([0, 1/3], c) where c: $\mathbf{R}^2 \times \mathbf{Z}^2 \to \mathbf{R}$ is the distance $c(a, b) = \max(|x_a - x_b|, |y_a - y_b|)$ for $a = (x_a, y_a) \in \mathbf{R}^2$ and $b = (x_b, y_b) \in \mathbf{R}^2$ and C" = ({0}, f) where f: $\mathbf{Z}^2 \times 2^X \to \mathbf{R}$ is defined by the distance from a point to a set, $f(b, A) = \inf\{c(b, x) \mid x \in A\}$. The sets $Y_1 = \{(0, 0), (0, 1), (1, 0), (1, 1), (2, 0), (3, 0)\}$ and $Y_2 = \{(0, 0), (2, 17/12)\}$ are slackly convex with respect to $\mathbf{Z}^2$ and the behaviours C' and C". But the set $Y_1 + Y_2 = \{(0, 0), (0, 1), (1, 0), (1, 1), (2, 0), (3, 0), (2, 17/12), (2, 29/12), (3, 17/12), (3, 29/12), (4, 17/12), (5, 17/12)\}$ is not slackly convex with respect to $\mathbf{Z}^2$, C' and C". Indeed, for $x = (0, 1)$, $y = (2, 29/12)$ and for $t = (1, 41/24) \in$ <x, y>, the point $a = (1, 2) \in \mathbf{Z}^2$ satisfies the condition $c(t,a) = \frac{7}{24} < \frac{1}{3}$ but $f(a, Y_1 + Y_2) = 1 \neq 0$.

Theorem 4.2.3. *Let* C' *be a behaviour in* $X \times M$ *by means of* B', C" *a behaviour in* $M \times 2^X$ *by means of* B" *and* $\alpha \in \mathbf{R}$. *Let be* C' = (D', T ') *(slack, strict, strong) with* T ' *a set of transformations* T: $X \times M \to$ B' *and* C" = (D", T ") *(slack, strict, strong) with* T " *a set of transformations* T_1: $M \times 2^X \to$ B". *Consider:*

(i) B' *a subset of a real linear space* $(V_1, +, \cdot)$;

(ii) *the behaviour* αC' = (α.D', αT '), *where* αT ' *is the set of transformations* αT: $X \times \alpha M \to \alpha$B' *defined by the association with a corresponding transformation* T $\in$ T ', *as it follows:*

(4.2.5) $\quad (\alpha T)(\alpha x, \alpha a) = \alpha \cdot T(x, a)$, *with* $x \in X$ *and* $a \in M$;

(iii) B" *a subset of a real linear space* $(V_2, +, \cdot)$;

(iv) *the behaviour* αC"=(α.D",αT "), *where* αT " *is the set of transformations* αT_1: $\alpha M \times 2^X \to \alpha$B", *everyone defined by the association with a corresponding transformation* $T_1 \in$ T " *as it follows:*

(4.2.6) $\quad (\alpha T_1)(\alpha \cdot a, \alpha \cdot A) = \alpha . T_1(a, A)$, *for* $a \in M$ *and* $A \subseteq X$;

(v) *the strongly (slackly) convex set* $Y \subset X$ *with respect to the set* M *and the behaviours* C' *and* C".

Then the set αY *is strongly (slackly) convex with respect to the set* M *and the behaviours* $\alpha C'$ *and* $\alpha C''$.

Proof. As in the proof of theorem 4.2.1 we shall show first that:

(4.2.7) $\alpha \cdot C' (X \times M) = (\alpha C') (X \times \alpha M)$;

(4.2.8) $\alpha \cdot C'' (M \times 2^X) = (\alpha C'') (\alpha M \times 2^X)$.

Let us suppose, for simplicity, that both C' and C" are strict behaviours. For slack or strong behaviours, the reasoning is similar for each behaviour that is used.

First, let $(x, a) \in \alpha \cdot C' (X \times M)$. Then there are $y \in X$ and $b \in M$ such that $x = \alpha y$, $a = \alpha b$ and $(y, b) \in C' (X \times M)$, therefore $T (y, b) \in D'$. In this case we have:

(4.2.9) $(\alpha T) (x, a) = (\alpha T) (\alpha y, \alpha b) = \alpha \cdot T (y, b) \in \alpha . D'$

and from $\alpha X = X$ it follows that $\alpha \cdot C' (X \times M) \subseteq (\alpha C') (X \times \alpha M)$.

Conversely, if $(x, a) \in (\alpha C') (X \times \alpha M)$ then there are $y \in X$ and $b \in M$ such that $x = \alpha y$, $a = \alpha b$ and calculus (4.2.9) is correct due to definition (4.2.5) of αT. So $T (y, b) \in D'$ and, as consequence $(y, b) \in C' (X \times M)$, therefore $(x, a) = (\alpha y, \alpha b) = \alpha (y, b) \in \alpha C' (X \times M)$. This means that $(\alpha C') (X \times \alpha M) \subseteq \alpha \cdot C' (X \times M)$ and (4.2.7) is proved.

In order to prove (4.2.8), let $(a, A) \in \alpha \cdot C'' (M \times 2^X)$. There are $b \in M$ and $B \in 2^X$ having the property that $a = \alpha b$, $A = \alpha \cdot B$, and $(b, B) \in C'' (M \times 2^X)$, therefore $T_1 (b, B) \in D''$. Then

(4.2.10) $(\alpha T_1) (a, A) = (\alpha T_1) (\alpha b, \alpha B) = \alpha \cdot T_1 (b, B) \in \alpha D''$,

meaning that $\alpha \cdot C'' (M \times 2^X) \subseteq (\alpha C'') (\alpha M \times 2^X)$.

Conversely, if $(a, A) \in (\alpha C'') (\alpha M \times 2^X)$ it means that $(\alpha T_1) (a, A) \in \alpha D''$. So, there are $b \in M$ and $B \subset X$ such that $a = \alpha b$, $A = \alpha \cdot B$ and calculus (4.2.10) is correct, implying that $T_1 (b, B) \in D''$. Therefore $(b, B) \in C'' (M \times 2^X)$ and then $(a, A) = (\alpha b, \alpha B) = \alpha \cdot (b, B) \in \alpha C'' (M \times 2^X)$, which completes the proof of identity (4.2.8).

Let now $x, y \in \alpha Y$ and $t \in <x, y>$. In this case there are the elements $u, v \in Y$ and $p \in <u, v>$ such that $x = \alpha u$, $y = \alpha v$, $<x, y> = \alpha<u, v>$ and $t = \alpha p$. Because Y is strongly convex with respect to the set M and the behaviours C' and C" there is an element $b \in M$ such that $(p, b) \in C' (X \times M)$ and $(b, Y) \in C'' (M \times 2^X)$. Let now $a = \alpha b \in \alpha \cdot M$. We have $(t, a) = (\alpha p, \alpha b) = \alpha (p, b) \in \alpha \cdot C' (X \times M) = (\alpha C') (X \times \alpha M)$ as a consequence of identity (4.2.7). Then, from (4.2.8), it follows that $(a, \alpha Y) = (\alpha b, \alpha Y) = \alpha (b, Y) \in \alpha C' (M \times 2^X) =$

(αC") (αM × 2^X), therefore the element a = αb satisfies definition 4.1.1 ii). So, the set αY is strongly convex with respect to the set αM and the behaviours αC' and αC".

For the slack convexity, let x, y ∈ αY, t ∈ <x, y> and a ∈ αM such that (t, a) ∈ (αC') (x + αM). Then there are u, v ∈ Y, p ∈ <u, v> and b ∈ M such that x = αu and y = αv, <x, y> = α <u, v>, t = αp and a = αb. Now, using (4.2.7) one gets (t, a) ∈ (αC') (X × αM) = $\alpha \cdot$ C'(X × M) which implies that (p, b) ∈ C' (X × M). The slack convexity of Y with respect to M and the behaviours C' and C" has as consequence that (b, Y) ∈ C" (M × 2^X) and then (a, αY) = (αb, αY) = $\alpha \cdot$ (b, Y) ∈ $\alpha \cdot$ C" (M × 2^X) = (αC") (αM × 2^X) the last equality takes place because of identity (4.2.8). This means that the set αY is slackly convex with respect to αM, αC' and αC". If Y ≠ ∅ then the property is evident.

Theorem 4.2.4. *Let* M *be a subspace of* X, (B', +, .) *and* (B", +, .) *real linear spaces,* D' *subspace in* B', D" *subspace in* B", C' *a behaviour in* X × M *by means of* B' *and* C" *a behaviour in* M × 2^X *by means of* B". *Suppose that:*

(i) *the behaviour* C' *is strict or strong, is stable with respect to the addition and every scale change invariant;*

(ii) *the behaviour* C" *is strict or strong, is stable with respect to the addition and every scale change invariant;*

(iii) *the sets* Y_1, $Y_2 \subset$ X *are strongly convex with respect to the set* M *and the behaviours* C' *and* C";

(iv) $\alpha \in \mathbf{R}$, $\beta \in \mathbf{R}$.

Then the set $\alpha Y_1 + \beta Y_2$ *is strongly convex with respect to the set* M *and the behaviours* C' *and* C".

Theorem 4.2.4 is a consequence of theorems 4.2.2 and 4.2.3.

Corollary 4.2.2. *Let* M *be a subspace of* X, C' *a behaviour in* X × M *by means of* B' *and* C" *a behaviour in* M × 2^X *by means of* B". *Suppose that:*

(i) C' *is a strict or strong behaviour which is stable with respect to the addition and every scale change invariant;*

(ii) C" *is a strict or strong behaviour, which is stable with respect to the addition and every scale change invariant;*

(iii) *the sets* $Y_1, Y_2, \ldots, Y_n \subset$ X *are strongly convex with respect to the set* M *and the behaviours* C' *and* C";

(iv) $\alpha_1, \alpha_2, \ldots, \alpha_n \in \mathbf{R}$.

Then the set $\alpha_1 Y_1 + \alpha_2 Y_2 + \ldots + \alpha_n Y_n$ *is strongly convex with respect to the set* M *and the behaviours* C' *and* C".

For the proof theorem 4.2.4 can be used repeatedly.

The following notations will be used:

$G_{st, C', C'', M}$ is the set of all the subsets of X that are strongly convex with respect to the set M and the behaviours C' and C";

$G_{sl, C', C'', M}$ is the set of all the subsets of X that are slackly convex with respect to the set M and the behaviours C' and C".

Theorem 4.2.5. *If* $C' \in \text{Comp}(X \times M, B')$ *and* $C'' \in \text{Comp}(M \times 2^X, B'')$ *and* C" *is closed for intersection then the pairs* $(X, G_{st,C',C'',M})$ *and* $(X, G_{sl,C',C'',M})$ *are convexity spaces on X.*

The next paragraph contains more examples of properties of convexity with respect to a set and two behaviours. The behaviours C" appearing in examples 4.3.3 and 4.3.4 are closed for intersection and, therefore, the convexity properties discussed in these examples define convexity spaces on their support sets. This convexity structure leads to important separation properties with applications in decision making, as it was shown by L. Lupşa (1981, 1986), S. V. Ovchinnikov (1980). This type of properties will be studied in the second part of this book, referring to applications. Other examples of convexity properties of the types defined in this paper have important applications in pattern recognition, see J. M. Chassery (1983) and G. Cristescu (1997 [65]), even if they do not generate a convexity space on X.

4.3 Examples

Many more examples of sets having these kinds of convexity properties can be obtained studying different types of convexity properties used in various domains of applied mathematics.

Example 4.3.1. Let X be a nonempty set endowed with a straight line segment, $M \subset X$ nonempty, B' and B" two nonempty sets and $D' \subset B'$ and $D'' \subset B''$ nonempty subsets. Let T ' be a set of transformations $T': X \times M \to B'$ and T " a set of transformations $T'': M \times 2^X \to B''$. Define the strong, slack or strict behaviours C' = (D', T') and C" = (D", T "). Denote:

$pr_X C'(X \times M) = \{x \in X \mid \exists\, a \in M, (x, a) \in C'(X \times M)\}$,

$pr_M C'(X \times M) = \{a \in M \mid \exists\, x \in X, (x, a) \in C'(X \times M)\}$,

$\text{pr}_M C''(M \times 2^X) = \{a \in M \mid \exists A \in 2^X, (a, A) \in C''(M \times 2^X)\}$,

$\text{pr}_{2^X} C''(M \times 2^X) = \{A \subset X \mid \exists a \in M,\ (a, A) \in C''(M \times 2^X)\}$.

Proposition 4.3.1. *Suppose that* C' *and* C" *satisfy the condition*

(4.3.1) $\quad \text{pr}_M C'(X \times M) \subseteq \text{pr}_M C''(M \times 2^X)$.

In this case:

1) *If for every* $a \in \text{pr}_M C''(M \times 2^X))$ *the condition* $(a, X) \in C''(M \times 2^X)$ *obtains then* X *is slackly convex with respect to the set* M *and the behaviours* C' *and* C";

2) *If* $\text{pr}_X C'(X \times M) = X$ *and for every* $a \in \text{pr}_M C''(M \times 2^X)$ *the condition* $(a, X) \in \text{pr}_M C''(M \times 2^X)$ *obtains then* X *is a strongly convex set with respect to* M *and the behaviours* C' *and* C";

3) *If the set* $A \subset X$ *satisfies the condition*

(4.3.2) $\quad \text{pr}_M C'\left(\bigcup_{x,y \in A} \langle x,y\rangle \times M\right) \subset \{a \in M \mid (a,A) \in C''(M \times 2^X)\}$,

then the set A *is slackly convex with respect to the set* M *and the behaviours* C' *and* C".

Proof. 1) Let $x \in X$, $y \in X$, $t \in \langle x, y\rangle$ and $a \in M$ such that $(t, a) \in C'(X \times M)$. Then $a \in \text{pr}_M C'(X \times M) \subset \text{pr}_M C''(M \times 2^X)$, therefore $(a, X) \in C''(M \times 2^X)$. So, X is slackly convex with respect to the set M and the behaviours C' and C".

2) Let $x \in X$, $y \in X$ and $t \in \langle x, y\rangle$. Because $\text{pr}_X C'(X \times M) = X$ there is a $\in M$ such that $(t, a) \in C'(X \times M)$. Then $a \in \text{pr}_M C'(X \times M) \subseteq \text{pr}_M C''(M \times 2^X)$ and so $(a, X) \in C''(M \times 2^X)$. Therefore X is strongly convex with respect to the set M and the behaviours C' and C".

3) Let $A \subset M$ satisfying condition (4.3.1) and $x, y \in A$, $t \in \langle x, y\rangle$, $a \in M$ such that $(t, a) \in C'(X \times M)$. Then, because A satisfies condition (4.3.1), it follows that $(a, A) \in C''(M \times 2^X)$ therefore the set A is slackly convex with respect to the set M and the behaviours C' and C".

Example 4.3.2. Usually, in a nonempty set X endowed with a notion of straight line segment determined by two points $x, y \in X$, denoted by $\langle x, y\rangle$, a set A is classically convex whether together with two points a, b belonging to A the whole $\langle a, b\rangle \subset A$. It is obvious that a set $A \subset X$ is convex if and only if it is strongly convex with respect to X (or to itself) and the behaviours $C' = (\{1\}, f)$ and $C'' = (\{1\}, g)$ where $f: X \times X \to \mathbf{R}$, $g : X \times 2^X \to \mathbf{R}$ defined by

$$(4.3.3) \quad f(a, b) = \begin{cases} 1 & \text{if } a = b \\ 0 & \text{if } a \neq b \end{cases} ; \quad g(a, A) = \begin{cases} 1 & \text{if } a \in A \\ 0 & \text{if } a \notin A \end{cases},$$

where $(a, b) \in X \times X$ and $(a, A) \in X \times 2^X$.

The behaviours C' and C" satisfy the conditions of proposition 4.3.1. 3), therefore A is also slackly convex with respect to the set X and the behaviours C' and C".

If X is a real linear space endowed with the classical straight-line segment then one notices that both behaviours C' and C" satisfy the conditions of theorems 4.2.1, 4.2.2, 4.2.3 and 4.2.4 and also of corollary 4.2.1. Therefore, the classical properties of the convex sets are retrieved.

Example 4.3.3. Liana Lupşa has proved in her thesis (1981) that the sets of dominated points of a set with respect to a preference relation can have two types of convexity properties, studied in chapter 2 of this book. We remember here the definition of the convexities taken into account in our example. In what follows, a real linear space (V, +, .) will be considered, $M \subset V$, $A \subset V$ and the following definition is quoted from L. Lupşa (1982, 1986).

Definition 4.3.1. i) *The set* A *is said to be slackly convex with respect to* M *if* A *is empty or if for every* $x, y \in A$ *and* $t \in \,]0, 1[$, *if* $t\,x + (1 - t)\,y \in M$ *then* $t\,x + (1 - t)\,y \in A$.

ii) *The set* A *is said to be strongly convex with respect to* M *if it is empty or* $\operatorname{conv}(A) \cap M \subseteq A$.

Here, conv (A) means the union of all the straight-line segments determined by pairs of points of A.

Proposition 4.3.2. *The set* A *is slackly convex with respect to* M *if and only if* A *is slackly convex with respect to* M, *to a behaviour* C' *in* $V \times M$ *and a behaviour* C" *in* $M \times 2^V$, *both defined by means of* 2^V.

Proof. If $f: V \times M \to 2^V$, $f(a, b) = \{a\} \cap \{b\}$, for $a \in V$ and $b \in M$, and $D' = \{\{m\} \mid m \in M\}$ then C' = (D', f) is a behaviour in $V \times M$ by means of 2^V. To define C", a function $g: M \times 2^V \to 2^V$, $g(b, A) = \{b\} \cap A$, for $b \in M$ and $A \subset V$, is considered and then C" = (D", g), with $D'' = \{\{a\} \mid a \in A\}$. Let now $x, y \in A$ and $t \in \,]0, 1[$ such that $t\,x + (1 - t)\,y \in M$. This is equivalent to the following statement: there is a point $a \in M$ such that

$$(4.3.4) \quad f(t\,x + (1 - t)\,y, a) = \{t\,x + (1 - t)\,y\} \cap \{a\} = \{a\}$$

and hence $(t\,x + (1 - t)\,y, a) \in C'(V \times M)$.

If A is assumed to be slackly convex with respect to M then it follows that $t\,x + (1 - t)\,y \in A$. But this is equivalent, by (4.3.4), to

$$(4.3.5) \quad g\,(a, A) = \{a\} \cap A = \{a\},$$

so that $(a, A) \in C''\,(M \times 2^V)$. The equivalence of the two slack convexities is now obvious.

Proposition 4.3.3. *The set* A *is strongly convex with respect to* M *if and only if* A *is slackly convex with respect to* M, *to a behaviour* C' *in* $V \times M$ *and a behaviour* C" *in* $M \times 2^V$, *both defined like in* (4.3.3).

Proof. Let $x, y \in A$ and $t \in]0, 1[$. The element $t\,x + (1 - t)\,y \in \text{conv}\,(A)$. Consider the function $f\colon V \times M \to \mathbf{R}$, defined like in (4.3.3) for every $(a, b) \in V \times M$ and the behaviour $C' = (\{1\}, f)$. It is obvious that $t\,x + (1 - t)\,y \in \text{conv}\,(A) \cap M$ is equivalent to $(t\,x + (1 - t)\,y, a) \in C'\,(V \times M)$ for an element $a \in M$ and hence $t\,x + (1 - t)\,y = a$. From the strong convexity of A with respect to M it follows that $t\,x + (1 - t)\,y \in A$. Considering now the function $g\colon M \times 2^V \to \mathbf{R}$ defined as in (4.3.3) for every $(a, A) \in M \times 2^V$ and the behaviour $C'' = (\{1\}, g)$, the equivalent result $(t\,x + (1 - t)\,y, A) \in C''\,(M \times 2^V)$ is obtained. Therefore, the slack convexity of A with respect to M, C', C" is equivalent to its strong convexity with respect to the set M.

Example 4.3.4. Let L be a distributive lattice and M be an arbitrary subset of L. If $x \in L$ and $y \in L$ then a point $z \in L$ is said to lie between x and y, see S. V. Ovchinnikov (1980), if and only if

$$(4.3.6) \quad x \wedge y \leq z \leq x \vee y$$

where $\wedge, \vee, \leq$ mean the infimum, the supremum and the order relation in L. Then, if $a \in M$ and $b \in M$

$$(4.3.7) \quad I\,(M, [a, b]) = \{x \in M \mid a \wedge b \leq x \leq a \vee b\}$$

is said to be the M-interval determined by a and b. The assertion $x \in I\,(L, [a, b])$ means that x lies between a and b and, using the terminology of L. M. Blumenthal, K. Menger (1970) and S. V. Ovchinnikov (1980), the betweenness relation $B(x, a, b)$ is true. According to V. B. Kuzmin, S. V. Ovchinnikov (1975), *a set* $A \subseteq M$ *is said to be 2-convex with respect to* M *if and only if the truth of* B (x, a, b) *implies* $x \in A$ *for all* $x \in M$ *and all the pairs* $(a, b) \in A \times A$. This notion of convexity was introduced by V. B. Kuzmin, S. V. Ovchinnikov (1975) as n - convexity, generalised in S. V. Ovchinnikov (1980), and used by S. V. Ovchinnikov (1980) to study Pareto's principle in the theory of group choice.

Proposition 4.3.4. *Let* L *be a lattice and* $M \subseteq L$. *The set* $A \subseteq M$ *is 2-convex with respect to* M *if and only if it is slackly convex with respect to* M,

a behaviour C' *in* L × M *by means of* **R** *and a behaviour* C" *in* M × 2^L *by means of* **R** *defined like in* (4.3.3).

Proof. If a ∈ L and b ∈ L then the straight-line segment is considered to be the set

(4.3.8) <a, b> = I (L, [a, b])

of those points of L that lie between a and b. A possibility of defining the behaviours C' and C" is: C' = ({1}, T') where T': L × M → **R**, is defined as in (4.3.3) for every (a, b) from L × M, and C" = ({1}, T") where T": M × 2^L → **R** is defined in the same manner like in (4.3.3) for (a, A) from M × 2^L. Let now A ⊆ M be a 2-convex set with respect to M. Then for every p ∈ A and q ∈ A and for every x ∈ M such that B(x, p, q) is true it follows that x ∈ A. Let now p ∈ A, q ∈ A, z ∈ <p, q> be three points such that B (z, p, q) is true. Then two situations are possible:

i) There is an element y ∈ M such that (z, y) ∈ C'(L × M). Then z = y ∈ M and the 2-convexity property of A with respect to M involves that z ∈ A. That means that (z, A) ∈ C" (M × 2^L) and therefore A is slackly convex with respect to M, C' and C".

ii) There does not exist a point y ∈ M such that (z, y) ∈ C' (L × M). The definitions of both convexity properties the proposition 4.3.4 deals with do not refer to the situation that z ∉ M.

Now, if A ⊆ M is supposed to be slackly convex with respect to M, C' and C", the remark that (t, z) ∈ C' (L × M) means that t ∈ M. The condition (t, A) ∈ C" (M × 2^L) means that t ∈ A. Now the 2-convexity of A with respect to M is obvious.

Example 4.3.5. In order to describe the convexity property of discrete regions of $\mathbf{Z}^2$, where **Z** is the set of the integer numbers, J.M. Chassery (1978) has formulated the following definition with applications to pattern recognition. In this example by c (x, y) the chessboard distance between two points x = (a', b') and y = (a", b") in the real plane, c (x, y) = max (|a' - a"|, |b' - b"|), will be denoted.

Definition 4.3.2. *A set* A ⊂ $\mathbf{Z}^2$ *is said to be discrete convex if there is a number* ε ∈ [1/2, 1[*such that for each* x, y ∈ A *and* t ∈]0, 1[, *there is a point* a ∈ $\mathbf{Z}^2$ *such that* c (t x + (1 - t) y, a) ≤ ε.

Now, taking X = $\mathbf{R}^2$, together with the usual straight-line segment from the Euclidean geometry, and M = $\mathbf{Z}^2$ it is possible to prove that the discrete convexity defined above is a particular case of the convexities studied in the previous paragraph. The following property is now evident.

Proposition 4.3.5. *Let* B' = B" = **R**, D' = [0, ε] *where* ε ∈ [1/2, 1], D" = {0}, T': X×M → **R**, T' (a, b) = c (a, b) *for every* (a, b) ∈ X × M. *let us define* *the behaviours* C' = ([0, ε], T') *and* C" = ({0}, T"), *with* T": $M \times 2^X \to R$, T"(x, Y) = c (x, Y) = inf {c (x, y) | y ∈ Y} *for* (x, Y) ∈ $M \times 2^X$. *A set* $A \subset \mathbf{Z}^2$ *is discrete convex if and only if it is strongly convex with respect to* M *and the behaviours* C' *and* C".

Proof. If the set A is discrete convex then there is ε ∈ [1/2, 1] such that if x∈ A, y ∈ A and t ∈ <x, y> there is a ∈ A satisfying c (t x + (1 - t) y, a) ≤ ε. But a ∈ A is equivalent to (a, A) ∈ C" (M × 2^X), and c (t x + (1 - t) y, a) ≤ ε is equivalent to (t x + (1 - t) y, a) ∈ C' (X × M). These conditions are simultaneously valid and therefore A is strongly convex with respect to M and the behaviours C' and C".

The discrete convexity defined by definitions 3.4.3 and 3.4.4 in $\mathbf{Z}^2$(h) might be characterised by a necessary and sufficient condition similar to proposition 4.3.5. Hence, the convexity defined in definition 4.3.2 is a particular case of the property presented in example 3.4.6, taking h = 1.

From these examples it follows both the consistence of definition 4.1.1 and the usefulness of the above-discussed notions. There are also in the literature other convexities, quoted in the monograph of P. M. Gruber and J. M. Willis (1993), that are of the same type. A classification of these convexities having as criteria certain properties of the behaviours used for their definition might be interesting.

4.4 Approximation of the classical convexity property

In this paragraph we study those convexity properties that might have a set that is strongly (slackly) convex with respect to a sequence (M_n, C_n, $C^{(n)}$)$_{n=1,2,...}$, where $(M_n)_{n=1,2,...}$ is a sequence of sets and $(C_n)_{n=1,2,...}$ and $(C^{(n)})_{n=1,2,...}$ are monotone sequences of behaviours.

Let X be a nonempty set and the sequence $(M_n)_{n=1,2,...}$ of nonempty subsets of X, having the property that

$$(4.4.1) \quad M_1 \subset M_2 \subset \ldots \subset M_n \subset \ldots \subset \bigcup_{n=1}^{\infty} M_n = X .$$

This has as consequence that

$$(4.4.2) \quad X \times M_1 \subset X \times M_2 \subset \ldots \subset X \times M_n \subset \ldots \subset \bigcup_{n=1}^{\infty} X \times M_n = X \times 2^X .$$

Let B be a nonempty set and the function f: X × X → B.

Suppose that there is an element $0 \in B$ such that

(4.4.3) $f(x, a) = 0 \Leftrightarrow x = a$.

Consider the descending sequence of nonempty subsets of B, $(D_n)_{n=1,2,\ldots}$, with the property

$$(4.4.4) \quad D = \{0\} = \bigcap_{n=1}^{\infty} D_n \subset \ldots \subset D_n \subset \ldots \subset D_2 \subset D_1$$

and define the sequence of behaviours $C_n = (D_n, f)$, n = 1, 2,..., which has the limit $C = (\{0\}, f)$.

From (4.4.1) it also follows that

$$(4.4.5) \quad M_1 \times 2^X \subset M_2 \times 2^X \subset \ldots \subset M_n \times 2^X \subset$$
$$\ldots \subset \lim_{n \to \infty} M_n \times 2^X = X \times 2^X.$$

Another nonempty set, denoted by B', is considered together with the function $T: X \times 2^X \to B'$. For them, suppose that there is an element $O' \in B'$ with the property that:

(4.4.6) $T(a, A) = O' \Leftrightarrow a \in A$.

Also, the descending sequence of nonempty subsets of B', $(D^{(n)})_{n=1,2,\ldots}$, is considered, having the property

$$(4.4.7) \quad D' = \{O'\} = \bigcap_{n=1}^{\infty} D^{(n)} \subset \ldots \subset D^{(n)} \subset \ldots \subset D^{(2)} \subset D^{(1)}.$$

and the sequence of behaviours $C^{(n)}=(D^{(n)},T)$, n = 1,2 ..., is defined. Its limit is the behaviour $C'=(\{O'\},T)$.

Suppose that X is endowed with a notion of straight-line segment determined by pairs of its points.

From condition (4.4.1) immediately follows that for each point $x \in X$ the condition is satisfied:

(4.4.8) $\exists N \in \mathbf{N}^*$ *such that* $x \in M_n$ *for each* $n \geq N$.

Definition 4.4.1. *If*

1. $(M_n)_{n=1,2,\ldots}$ *is an ascending sequence of sets such that* $\bigcup_{n=1}^{\infty} M_n = X$;
2. $(C_n)_{n=1,2,\ldots}$ *is a monotone sequence of behaviours in* $X \times M_n$, *which has as a limit the behaviour* $C = (\{0\}, f)$, *with f satisfying* (4.4.3);

3. $(C^{(n)})_{n=1,2,...}$ *is a monotone sequence of behaviours in* $M_n \times 2^X$ *which has as a limit the behaviour* $C' = (\{O'\}, T)$, *with* T *satisfying* (4.4.6),

and if every nonempty set $A \subset X$ *which is strongly (slackly) convex with respect to* $(M_n, C_n, C^{(n)})$, *for every* n = 1, 2,..., *has the property of classical convexity then we say that the sequence of strong (slack) convexities with respect to* $(M_n, C_n, C^{(n)})$, n = 1, 2, ..., *converges to the classical convexity.*

Definition 4.4.1 is consistent, as it will be shown in what follows.

Theorem 4.4.1. *If the nonempty set* $A \subset X$ *has the property that it is strongly convex with respect to the set* M_n *and the behaviours* C_n *and* $C^{(n)}$ *for every* n = 1, 2, ..., *then* A *is classically convex in* X.

Proof. A is supposed to be strongly convex with respect to the set M_n and with the behaviours C_n and $C^{(n)}$ for every n = 1, 2,.... Then for every points $x \in A$ and $y \in A$ and for every $t \in <x, y>$ there is the point $a_n \in M_n$ such that $(t, a_n) \in C_n (X \times M_n)$ and $(a_n, A) \in C^{(n)} (M_n \times 2^X)$, for every n = 1, 2, ... But $(t, a_n) \in C_n (X \times M_n)$ means that $f(t, a_n) \in D_n$. Then (4.4.4) implies the existence of a point $a \in X = \bigcup_{n=1}^{\infty} M_n$ having the property that $f(t, a) \in \{0\} = \bigcap_{n=1}^{\infty} D_n$. But (4.4.4) means the existence of a number $N \neq 0$, such that for every $n \geq N$, $a \in M_n$ takes place and then, it follows that $f(t, a) = 0 \in D_n$ for every $n \geq N$. Condition (4.4.3) implies that a = t. Now, $t \in A$ must be proved. Since $f(t, t) \in D_n$, equivalent to $(t, t) \in C_n (X \times M_n)$ for $n \geq N$, the property that A is strongly convex with respect to the set M_n and the behaviours C_n and $C^{(n)}$ for every n = 1, 2, ..., yields that $(t, A) \in C^{(n)} (M_n \times 2^X)$, for $n \geq N$. This means that $T(t, A) \in D^{(n)}$ for $n \geq N$ and, from (4.4.7), one obtains $T(t, A) = O'$. But, according to property (4.4.6), it takes place if and only if $t \in A$. Therefore, the set A is classically convex.

The proof of theorem 4.4.1 does not remain valid if condition (4.4.8) is eliminated. This is the example of the plane case $X = \mathbf{R}^2$ and $M_n = \{(x, y) \mid x = k \cdot \frac{1}{2^n}, y = h \cdot \frac{1}{2^n}, k \in \mathbf{Z}, h \in \mathbf{Z}\}$ where $\lim_{n \to \infty} M_n = \mathbf{R}^2$, but for point $(\sqrt{2}, \sqrt{2})$ there is not $N \in \mathbf{N}^*$ such that $(\sqrt{2}, \sqrt{2}) \in M_n$ holds for $n \geq N$.

It is not difficult to prove that sometimes condition (4.4.8) might be replaced, in a topological space, by density and closure properties.

Let us suppose that X is a topological space, together with a notion of straight-line segment such that it becomes a topological convexity space. Let us consider the sequences of behaviours previously defined. Then we are able to prove the following result.

Theorem 4.4.2. *If the closed set* $A \subset X$, $A \neq \varnothing$, *is strongly convex with respect to the set* M_n *and the behaviours* C_n *and* $C^{(n)}$ *for every* $n = 1, 2, \ldots$, *then* A *is classically convex in* X.

The proof follows the same stages as the previous one and will be omitted.

Definition 4.4.2. *A convexity property with respect to a set and two behaviours is said to be an approximation of the classical convexity if it is a term of a sequence of convexities convergent to the classical convexity.*

In the case of some particular behaviours defined in $\mathbf{R}^2 \times M_n$ it is possible to obtain the discrete convexity properties defined by J.M. Chassery (1978, 1983) and used in pattern recognition by J.M. Chassery (1984), J.M. Chassery, C. Garbay, P. Cinquin (1984), J.M. Chassery, C. Garbay (1984). The problem of the convergence of some discrete convexities to the classical convexity appears for the first time in J.M. Chassery (1983). There, condition (4.4.8) is replaced by a property concerning the covering of a straight-line segment by minimum radius balls centred in the knots of a network having the step h applied over $\mathbf{R}^2$. That convergence property will be discussed and generalised in the next paragraph.

Here we shall deal with another interesting example. Even if theorem 4.4.1 does not remain valid if the slack convexity replaces the strong convexity of the set A with respect to the set M_n and the behaviours C_n and $C^{(n)}$, in the case of the convexities defined in Liana Lupşa (1980, 1982) and presented by us in example 4.3.3, the convergence properties towards the classical convexity are valid.

In what follows $(X, +, \cdot)$ is supposed to be a real linear space and the sequence $(M_n)_{n=1, 2, \ldots}$ of nonempty subsets of X satisfy (4.4.1) and, therefore, (4.4.8). It was proved, in propositions 4.3.2 and 4.3.3, that if a set $A \subset X$ is slackly (strongly) convex with respect to a given set M, in the sense of definition 4.3.1, then it has a property of slack convexity with respect to M and two behaviours. The converse is also true. It is the reason for which the theorem 4.4.1 does not apply in this case, but it is easy to obtain other convergence properties.

Theorem 4.4.3. *If the set* $A \subseteq X$ *is slackly convex with respect to* M_n *for every* $n = 1, 2, \ldots$, *then* A *is classically convex in* X.

Proof. Let x and y be two points of A and $t \in]0, 1[$. From the slack convexity of the set A with respect to the set M_n it follows that if $t x + (1 - t) y \in M_n$ then $t x + (1 - t) y \in A$. But (4.4.8) implies the existence of a natural number N_a such that $t x + (1 - t) y \in M_n$ for every $n \geq N_a$ and so $t x + (1 - t) y \in A$. Therefore, A is classically convex in X.

Theorem 4.4.4. *If the set* $A \subseteq X$ *is strongly convex with respect to* M_n *for every* $n = 1, 2, ...$, *then* A *is classically convex in* X.

Proof. If the set $A \subset X$ is strongly convex with respect to M_n, $n = 1, 2, \ldots$, then for every $n = 1, 2, \ldots$, the inclusion $\text{conv}(A) \cap M_n \subseteq A$ takes place. This means that

$$\bigcap_{n=1}^{\infty}(\text{conv}(A) \cap M_n) = \text{conv}(A) \cap \left(\bigcap_{n=1}^{\infty} M_n\right) = \text{conv}(A) \cap X \subseteq A,$$

according to condition (4.4.1). But then $\text{conv}(A) \subseteq A$, which implies the classical convexity of the set A.

The proofs of theorems 4.4.3 and 4.4.4 are easy to be formulated in terms of behaviours, according to propositions 4.3.2 and 4.3.3.

As a consequence of the above discussed convergence theorems and according to definition 4.4.2 the convexity properties involved in them are said to approximate the classical convexity.

It is interesting to find necessary and sufficient conditions for a sequence of convexities to be convergent to the classical convexity.

4.5 Weak cases of convergence to the classical convexity

There are situations when a process of convergence of some sequences of convexity properties to the classical convexity in weaker conditions can be proved. Some typical situations will be discussed in this section.

If X is supposed to be a topological space, together with a notion of straight line segment such that it becomes a topological convexity space, if the sequences of behaviours defined in section 4.4 are considered, then a weaker definition then definition 4.4.1 can be formulated.

Definition 4.5.1. *If for every nonempty set* $A \subset X$ *that is strongly (slackly) convex with respect to* $(M_n, C_n, C^{(n)})$, *for every* $n = 1, 2, ...,$*where :*

I) $(M_n)_{n=1,2,...}$ *is an ascending sequence of sets such that* $\bigcup_{n=1}^{\infty} M_n$ *is dense in* X*;*

II) $(C_n)_{n=1,2,...}$ *is an ascending sequence of behaviours in* $X \times M_n$, *having as a limit* C=({o}, f), f *satisfying* (4.4.3);

III) $(C^{(n)})_{n=1,2,...}$ *is an ascending sequence of behaviours in* $M_n \times 2^X$, *having as a limit* C'=({o'}, T), T *satisfying* (4.4.6);

it follows that the closure of A *is classically convex, then the sequence of strong (slack) behaviours with respect to* $(M_n, C_n, C^{(n)})$, n = 1, 2, ..., *is said to weakly converge to the classical convexity.*

Theorem 4.5.1 *If the set* $A \subset X$, $A \neq \emptyset$, *is strongly convex with respect to the set* M_n *and the behaviours* C_n *and* $C^{(n)}$ *for every* n = 1, 2, ..., *then the closure of* A *is classically convex in* X.

Proof. A is supposed to be strongly convex with respect to the set M_n and with the behaviours C_n and $C^{(n)}$ for every n = 1, 2,.... Then for every points $x \in A$ and $y \in A$ and for every $t \in \langle x, y\rangle$ there is the point $a_n \in M_n$ such that $(t, a_n) \in C_n (X \times M_n)$ and $(a_n, A) \in C^{(n)}(M_n \times 2^X)$, for every n = 1, 2, ... But $(t, a_n) \in C_n (X \times M_n)$ means that $f(t, a_n) \in D_n$. Then I) implies the existence of a point $a \in X = \overline{\bigcup_{n=1}^{\infty} M_n}$ having the property that $f(t, a) \in \{0\} = \bigcap_{n=1}^{\infty} D_n$. The density condition I) also implies the existence of a number $N \neq 0$, such that for every $n \geq N$, $a \in M_n$ takes place and then, it follows that $f(t, a) = 0 \in D_n$ for every $n \geq N$. Condition (4.4.3) implies that a = t. Now, $t \in \overline{A}$ must be proved. Since $f(t, t) \in D_n$, equivalent to $(t, t) \in C_n (X \times M_n)$ for $n \geq N$, the property that A is strongly convex with respect to the set M_n and the behaviours C_n and $C^{(n)}$ for every n = 1, 2, ..., yields that (t, A) $\in C^{(n)} (M_n \times 2^X)$, for $n \geq N$. This means that $T(t, A) \in D^{(n)}$ for $n \geq N$ and, from (4.4.7), one obtains T (t, A) = O'. But, according to property (4.4.6), it takes place if and only if $t \in A$. Therefore, closure of the set A is classically convex.

Chapter 5

Convexities defined by means of distance functions

It was proved in J. M. Chassery (1983) that if a set $A \subset \mathbf{Z}^2(h)$ is discrete convex (in the sense of definition 4.3.2) then there is a convex set in $\mathbf{R}^2$ such that A is its digitisation. In this chapter we intend to study the sets that are the solution of the converse problem: to find the properties of the sets that become, after digitisation, the same discrete convex set. The properties of convexity with respect to a set and two behaviours, studied in the previous chapter, will allow us to solve the above-mentioned problem. The methods of digitisation are not studied in this chapter. They are discussed in the thesis of M. L. P. van Lierop (1987) and further in M. Maes (1990). In chapter 10 we shall choose a particular digitisation method that leads to all the results of this chapter, but it is not unique, the results being easy to retrieve if the digitisation method is changed.

The sets that are solutions of the above formulated problem satisfy a condition of convexity, which is a particular case of the convexities defined by means of behaviours in chapters 3 and 4.

5.1 α - convex sets in a metric space

Let (X, d) be a metric space. For every two points x and y of X, a notion of straight-line segment is defined, denoting it by $\langle x, y \rangle$. It can be different from the metric straight-line segment. Let us consider the nonnegative number α.

Definition 5.1.1. *The set* $A \subset X$ *is said to be* α*-convex if it is empty or if for every two points* $x \in A$ *and* $y \in A$ *every point* $t \in \langle x, y \rangle$ *satisfies the condition* $d(t, A) \leq \alpha$..

This convexity property is a generalisation of the classical convexity that is obtained as a particular case of a set that is 0-convex. It is also a particular case of the convexity with respect to a behaviour, taking $T: X \times 2^X \rightarrow \mathbf{R}$ defined by $T(t, A) = d(t, A) = \sup\{d(t, x) \mid x \in A\}$, and the behaviour $C = (\{0\}, T)$ in $X \times 2^X$.

Example 5.1.1. If $X \subset \mathbf{R}^2$, with the Euclidean distance, such that the set $A = ([0, 1] \setminus \mathbf{Q}) \times ([0, 1] \setminus \mathbf{Q}) \subset X$, where $\mathbf{Q}$ is the set of all the rational numbers, then the set A is α - convex, $\alpha \geq 0$.

Example 5.1.2. Let $A = G^2$, where $G \subset [0, 1]$ is the Cantor set in the space $X = \mathbf{R}^2$ defined classically, as G. Cantor (1872) did, by the elimination of the middle thirds of [0, 1]. The set A is ½ - convex. Using principles of construction similar to those described in B. Mandelbrot (1975), similar sets to G, denoted by G', are obtained. The set G'^2, called *the carpet of Sierpinski* and quoted by B. Mandelbrot (1975), have properties of α - convexity for $\alpha \geq 1/2$. Also, entering three dimensions, the sets of the type G'^3, are objects generically called *the sponge of Sierpinski*, B. Mandelbrot (1975). They have a zero area, the total perimeter of their holes is infinite and they are α-convex, for $\alpha \geq 1/2$. B. Mandelbrot (1975) described the method of generation of this type of three-dimensional fractals together with a graphical three-dimensional representation.

Example 5.1.3. For $m \geq 2$, and X including the square $[-m, m]^2$, let us consider the astroide together with its inside,

$$A = \{(x, y) \in X \mid -(m^{2/3} - x^{2/3})^{3/2} \leq y \leq (m^{2/3} - x^{2/3})^{3/2}, -m \leq x \leq m\}.$$

The set A is α - convex for $\alpha > \dfrac{m\left(-1+\sqrt{2}\right)}{2\sqrt{2}}$.

In order to present more results, by $B(a, \alpha)$ we denote the closed ball in X centred in a and having the radius α.

Theorem 5.1.1. *The set* $A \subset \mathbf{R}^2$ *is* α *- convex if and only if for every* x *and* y *from* A *and for every* $a \in \langle x, y\rangle$ *the set* $B(a, \alpha) \cap A$ *is not empty.*

Proof. The property of α - convexity is supposed to be satisfied by A. Then, if x and y are two points of A and $t \in \langle x, y\rangle$, the validity of the condition $d(t, A) \leq \alpha$ means that $B(t, \alpha) \cap A \neq \Phi$.

Now, the converse property is supposed. For $x, y \in A$ and $a \in \langle x, y\rangle$, it follows that $B(a, \alpha) \cap A \neq \Phi$. It means that there is a point $t \in A$ such that the inequality $d(t, a) \leq \alpha$ holds. Therefore, the definition of the property of α - convexity of A is fulfilled.

According to theorems 4.1.1 and 4.1.2 we can conclude that the property of α - convexity for sets is a particular case of slack or strong convexity with respect to a set and two behaviours.

5.2 (α, δ) - convexity with respect to a network

Let (X, d) be a totally bounded metric space and for $\varepsilon > 0$ let E' be a finite ε-network over X, and by E a finite 2ε - network over E' is denoted.

For a set $A \subset X$ and a point $x \in X$, the distance between x and A is defined as usual by the number $d(x, A) = \inf \{d(x, A) \mid a \in A\}$. The closed ball centred in x having the radius $\alpha > 0$ will be denoted, as in the previous chapters, by $B(x, \alpha) = \{m \in X \mid d(m, x) \leq \alpha\}$, and in E, the corresponding closed ball is denoted by $B(x, \alpha; E) = \{m \in E \mid d(m, x) \leq \alpha\}$.

In X it is supposed that, for every two points $x \in X$ and $y \in Y$ a notion of straight-line segment $<x, y>$ is defined. This notion of straight-line segment might be the classical one in metric spaces, see V. Bryant (1994), defined by means of the distance and then all the properties following from this situation, see P. M. Gruber, J. M. Willis (1993), V. P. Soltan (1984), R. Webster (1994), will be used if necessary. But the notion of straight-line segment might be different than that determined by d. It is a situation appearing frequently in image analysis when the Euclidean straight-line segment is used together with distances that are integer approximations of the Euclidean distance: see J. M. Chassery (1978, 1984), G. Cristescu (1996 [57], [58], [59]), A. Rosenfeld, (1969, 1987), A. Rosenfeld and A. Kak (1976). In this paragraph we need no special conditions on the notion of straight-line segment.

Let the real numbers $\alpha \geq 0$ and $\delta \geq 0$, and let E be a network chosen as above.

Definition 5.2.1. *i) The set* $A \subset X$ *is said to be slackly* (α, δ) - *convex with respect to* E *if it is empty or if for every pair of points* $x, y \in A$ *and for every* $t \in <x, y>$ *if the element* $a \in E$ *has the property that* $d(t, a) \leq \alpha$ *then the inequality* $d(a, A) \leq \delta$ *also holds.*

ii) The set $A \subset X$ *is said to be strongly* (α, δ)*-convex with respect to* E *if it is empty or for every pair of points* $x, y \in A$ *and for every* $t \in <x, y>$ *there is an element* $a \in E$ *such that* $d(t, a) \leq \alpha$ *and* $d(a, A) \leq \delta$.

These notions are particular cases of the convexities with respect to a given set and two behaviours introduced by definition 4.1.1. To prove it we shall use the behaviours defined in examples 3.1.4 and 3.1.5. Let B' = B'' =

R, $D_1 = [0, \alpha]$, T_1: $X \times E \to$ **R**, $T_1 (x, a) = d (x, a)$ for $(x, a) \in X \times E$ and let us consider the behaviour $C_1 = (D_1, T_1)$. Let now $D_2 = [0, \delta]$, T_2: $E \times 2^X \to$ **R**, $T_2 (a, M) = d(a, M)$ for $(a, A) \in E \times 2^X$ and define the behaviour $C_2 = (D_2, T_2)$. The following property is evident.

Property 5.2.1. *i) Set* $A \subset X$ *is slackly* (α, δ) *- convex with respect to the network* E *if and only if* A *is slackly convex with respect to the set* E *and the behaviours* C_1 *and* C_2.

ii) Set $A \subset X$ *is strongly* (α, δ) *- convex with respect to the network* E *if and only if* A *is strongly convex with respect to the set* E *and the behaviours* C_1 *and* C_2.

Proof. If A is strongly (α, δ)-convex with respect to E then for every x and y from A and every $t \in \langle x, y\rangle$ there is $a \in E$ such that $d (t, a) \le \alpha$ and $d (a, A) \le \delta$. But $d (t, a) \le \alpha$ is equivalent to $T_1 (t, a) \in D_1$ and $d (a, A) \le \delta$ is equivalent to $T_2 (a, A) \in D_2$. Therefore the (α, δ) - convexity of the set A with respect to E is equivalent to the property: for every x and y from A and every $t \in \langle x, y\rangle$ there is $a \in E$ such that $(t, a) \in C_1 (X \times E)$ and $(a, A) \in C_2 (E \times 2^X)$, which means that A is strongly convex with respect to the set E and the behaviours C_1 and C_2. For the slack property the proof is similar.

As consequence of property 5.2.1, many properties of the strong (slack) (α, δ) - convex sets with respect to a network can be deduced from the properties proved in the previous chapter.

Theorem 5.2.1. *If the set* $A \subset X$ *is strongly* (α, δ) *- convex with respect to* E, *for* $\delta \ge 2\varepsilon$, *then it is slackly* (α, δ) *- convex with respect to* E.

Proof. If $x, y \in A$ and $t \in \langle x, y\rangle$ then there is a point $a \in E$ such that the following conditions take simultaneously place: $d (t, a) \le \alpha$ and $d (a, A) \le \delta$. But if $B = \{b \in X \mid b \notin \text{conv}(A)\}$ then $\min \{d (b, \text{conv}(A)) \mid b \in B\} \le 2\varepsilon$ and for every point $z \in \text{conv}(A) \cap E$ there is a point $u \in B$ such that $d (z, u) = 2\varepsilon$. Hence, for every $t \in \langle x, y\rangle$ and the point $c \in E$ such that $d (t, c) \le \alpha$, relation $d (c, A) \le 2\varepsilon \le \delta$ is true and the slackly (α, δ) - convexity of A with respect to E is proved.

Theorem 5.2.1 is not true for $\delta < 2\varepsilon$, as in the following example.

Example 5.2.1. Let $X \subseteq \mathbf{R}^2$ be a rectangle, $E = X \cap \mathbf{Z}^2$ is an 1-network for $X \cap \mathbf{Z}^2$ and an $\frac{1}{2}$ - network for (X, c), with distance $c (a, b) = \max \{|x_a - x_b|, |y_a - y_b|\}$, for $a = (x_a, y_a) \in \mathbf{R}^2$ and $b = (x_b, y_b) \in \mathbf{R}^2$. The notion of straight-line segment we use in $\mathbf{R}^2$ is the Euclidean one. Let $A = \{(0, 0), (0, 1), (1, 0)\}$. The set A is strongly $\left(\frac{1}{2}, \frac{3}{4}\right)$-convex with respect to E, but is not slackly $\left(\frac{1}{2},\frac{3}{4}\right)$-convex with respect to E. Indeed, for the points x = (0, 1), y = (1, 0) and $t = \left(\frac{1}{2},\frac{1}{2}\right) \in \langle x, y\rangle$ the point a = (1, 1) has the property that $c(a, t) = \frac{1}{2}$ but $c(t, A) = 1 > \frac{3}{4}$.

Theorem 5.2.2. *If set* $A \subset X$ *is slackly* (α, δ) *- convex with respect to* E, *with* $\alpha \in [\varepsilon, 2\varepsilon[$ *and* $\delta \geq 0$, *then it is strongly* (α, δ) *- convex with respect to* E.

Proof. Let $x, y \in A$ and $t \in \langle x, y\rangle$. Then there is a point $a \in E$ such that $d(t, a) \leq \varepsilon < \alpha$. Using the property of (α, δ) - convexity of A with respect to E it follows that $d(a, A) \leq \delta$. But this property takes place for every $t \in \langle x, y\rangle$, hence A is strongly (α, δ)-convex with respect to E.

In order to compare more convexity properties with respect to E let us consider the nonnegative real numbers $\alpha, \alpha', \alpha'', \delta, \delta', \delta''$ such that $\alpha' \leq \alpha''$ and $\delta' \leq \delta''$. The following properties are consequences of the propositions 4.1.1, 4.1.2 and 4.1.3, but their direct proofs are also easy.

Proposition 5.2.1. *If set* $A \subset X$ *is strongly* (α', δ)*-convex with respect to* E *then it is strongly* (α'', δ)*-convex with respect to* E.

Proposition 5.2.2. *If set* $A \subset X$ *is slackly (strongly)* (α, δ') *- convex with respect to* E *then the set* A *is slackly (strongly)* (α, δ'')*-convex with respect to* E.

Corollary 5.2.1. *If set* $A \subset X$ *is strongly* (α', δ') *- convex with respect to* E *then it is strongly* (α'', δ'') *- convex with respect to* E.

5.3 Particular plane case. Examples.

In this paragraph X is supposed to be a rectangular surface in $\mathbf{R}^2$ endowed with the Euclidean straight-line segment and the distance c. The set $E = X \cap \mathbf{Z}^2(h)$ is a h/2-network over X and a h-network over itself, for $h \geq 0$ and $\mathbf{Z}^2(h) = \{(ih, jh) \mid i \in \mathbf{Z}, j \in \mathbf{Z}\}$. First of all we shall give some examples of sets having various types of convexities defined in the previous paragraph. They will show that these convexities are distinct from the classical convexity and also that the slack property and the strong property are different.

Example 5.3.1. Let us suppose that $0 \in X$ and $1 \in X$ and let us consider the set $A = ([0, 1] - \mathbf{Q})^2$, where **Q** is the set of all the rational numbers. Set A is strongly (α, δ) - convex with respect to $X \cap \mathbf{Z}^2$ for every $\alpha \geq 1/2$ and $\delta > 0$. A is not strongly (α, δ)-convex with respect to $X \cap \mathbf{Z}^2$ for $\alpha \geq 1/2$, but is slackly (α, δ) - convex with respect to $X \cap \mathbf{Z}^2$, for every $\delta > 0$. This set is strongly (α, δ) - convex with respect to $X \cap \mathbf{Z}^2(1/2)$ for $\alpha \geq 1/4$ and $\delta > 0$, bit if $\alpha < 1/4$ and $\delta > 0$ then it is only slackly (α, δ) - convex with respect to $X \cap \mathbf{Z}^2(1/2)$.

Example 5.3.2. Let $A = G^2$, where $G \subset [0, 1]$ is the Cantor set (see G. Cantor (1872)), in space X defined as in example 5.3.1. Then A is strongly (1/2, 0) - convex with respect to $X \cap \mathbf{Z}^2$. Also, A is strongly (1/4, 1/6) - convex with respect to $X \cap \mathbf{Z}^2(1/2)$.

Using principles of construction similar to those described by G. Cantor (1872) (generalised by B. Mandelbrot (1975)), similar sets to G, denoted by G', are obtained. The set G'^2, called *the carpet of Sierpinski* and quoted by B. Mandelbrot (1975), has properties of strong (α, δ) - convexity with respect to $X \cap \mathbf{Z}^2$ for $\alpha \geq 1/2$ and $\delta \in [0, 1[$. Also, passing in three dimensions, the sets of the type G'^3, are objects generically called *the sponge of Sierpinski*, see L. M. Blumenthal, K. Menger (1953). They are strongly (α, δ)-convex with respect to $\mathbf{Z}^3$, for $\alpha \geq 1/2$ and $\delta \in [0, 1[$.

Example 5.3.3. For $m \geq 2$, and X including the square $[-m, m]^2$, let us consider the astroide together with its inside,

$$A = \{(x, y) \in X \mid -(m^{2/3} - x^{2/3})^{3/2} \leq y \leq (m^{2/3} - x^{2/3})^{3/2}, -m \leq x \leq m\}.$$

Set A is strongly (α, δ)-convex with respect to $X \cap \mathbf{Z}^2$ for each $\alpha \geq \frac{m}{4}$ and $\delta > \frac{m\left(-1+\sqrt{2}\right)}{2\sqrt{2}}$.

Example 5.3.4. The set $A = \{(0, 0), (0, 1), (1, 0), (2, 0)\}$, in a rectangle X which contains it, is strongly $(\frac{2}{3}, \frac{1}{2})$-convex with respect to $X \cap \mathbf{Z}^2$, but it is not slackly $(\frac{2}{3}, \frac{1}{2})$ - convex with respect to $X \cap \mathbf{Z}^2$. Indeed, choosing three points $x = (0, 1)$, $y = (2, 0)$, $t = (1, \frac{1}{2}) \in \langle x, y \rangle$ and taking $a = (1, 1)$, we have $c(t, a) = \frac{1}{2}$, but $c(a, A) = 1 > \frac{1}{2}$.

Hence, generally if A is strongly (α, δ) -convex with respect to $\mathbf{Z}^2(h)$ then the conclusion that it is also slackly (α, δ) - convex with respect to $\mathbf{Z}^2(h)$ is not true, as it was shown in example 5.3.4. Example 5.3.1 contains the converse situation: if A is slackly (α, δ) - convex with respect to $\mathbf{Z}^2(h)$ then it is not necessarily strongly (α,δ) - convex with respect to $\mathbf{Z}^2(h)$.

On the other hand, from example 5.3.1 it follows that theorem 5.2.2 is not valid for $\alpha < \varepsilon$. Also, from example 5.3.4 it follows that the converse of theorem 5.2.2 is not true.

The consistence of definition 5.1.1, hence of definition 4.1.1, is once again proved by the following results.

Theorem 5.3.1. *For every bounded set* $A \subset X$ *there are three real numbers* $h > 0$, $\alpha > 0$, $\delta \geq 0$, *such that* A *is strongly* (α, δ) - *convex with respect to* $E = X \cap \mathbf{Z}^2(h)$.

Proof. 1) If $A = \{a\}$ then for every $h > 0$, $\alpha \geq h/2$ and $\delta \geq h$ the set A is strongly (α, δ) - convex with respect to E;

2) If A is not singleton then let be $h \geq \mathrm{diam}(A) = \sup\{c(x, y) \mid x \in A, y \in A\}$, $\alpha \geq h/2$, $\delta = h$. In this case four situations are possible:

a) There are the numbers $(i, j) \in \mathbf{Z}^2$ such that $A \subseteq [ih, (i+1)h] \times [jh, (j+1)h]$. If

(5.3.1) $\quad S = \{(ih, jh), (ih, (j+1)h), ((i+1)h, jh), ((i+1)h, (j+1)h\}$

then for each point $x \in A$ there is a point $y \in S$ such that $c(x, y) \leq h/2 \leq \alpha$. This property is true in particular in the case when $x \in \langle a, b\rangle$, where $a \in A$ and $b \in A$. More, for every point $y \in S$ it is true that $c(y, A) \leq h$ and therefore A is strongly (α, h) - convex with respect to E.

b) There are $(i, j) \in \mathbf{Z}^2$ such that $A \subseteq [(i-1)h, (i+1)h] \times [jh, (j+1)h]$ (a similar situation is when $A \subseteq [(ih, (i+1)h] \times [(j-1)h, (j+1)h]$). Then the reasoning from case a) remains valid if S is replaced by the set

(5.3.2) $\quad T = S \cup \{((i-1)h, jh), ((i-1)h, (j+1)h)\}$.

c) There are $(i, j) \in \mathbf{Z}^2$ such that $A \subseteq \{[(i-1)h, (i+1)h] \times [jh, (j+1)h]\} \cup \{[ih, (i+1)h] \times [(j-1)h, jh]\}$ (or all the situations in which the set is situated in three squares similarly situated). Then the reasoning from a) remains valid if S is replaced by

(5.3.3) $\quad V = T \cup \{(ih, (j-1)h), ((i+1)h, (j-1)h)\}$

and $h/2 < \alpha \leq h$.

d) If there is a point $y \in E$ such that $y \in \text{int}(\text{conv}(A))$ then $A \subseteq B(y, h) = \{z \in X \mid c(y, z) \le h\}$. Hence, for each $x \in A$, there is a point $y \in B(y, h) \cap \mathbf{Z}^2(h)$ such that $c(x, y) \le h/2 \le \alpha$ and the conclusion is obtained using the previous reasoning.

Theorem 5.3.2. *For every bounded set* $A \subset X$ *there are three real numbers* $h > 0$, $\alpha > 0$, $\delta \ge 0$ *such that* A *is slackly* (α, δ) - *convex with respect to* $E = X \cap \mathbf{Z}^2(h)$.

Proof. 1) If $A = \{a\}$ then let be $h > 0$, $\alpha \le c(a, E)$, $\delta \ge \alpha$ and the conclusion is evident.

2) If A is not singleton, then we choose $h \ge \text{diam}(A)$. The proof is similar to that of the previous theorem for $0 \le \alpha \le h$ and $\delta \ge \sup\{c(x, A) \mid x \in E\}$.

5.4 Properties of the (α, δ) - convex sets with respect to a network

In this paragraph we shall suppose that X is a hyper-parallelepiped from $\mathbf{R}^n$ and $E = X \cap \mathbf{Z}^n(h)$ with $h > 0$ is a h/2 - network over X and a h-network over itself, if the c, d_1 or the Euclidean distance is used in X. We shall denote by d the distance in X, fixed from the beginning. The straight-line segment that will be considered is the Euclidean one.

Let now consider the set $A \subset X$. Let be the sets $B' = B'' = \mathbf{R}$, $D' = [0, \alpha]$, $D'' = [0, \delta]$, and the functions $T': X \times E \to \mathbf{R}$, $T'(a, b) = d(a, b)$ for $(a, b) \in X \times E$, $T'': E \times 2^X \to \mathbf{R}$, $T''(b, Y) = d(b, Y)$ for $(b, Y) \in 2^X$. Consider the behaviours $C' = (D', T')$ and $C'' = (D'', T'')$.

Lemma 5.4.1. *a) If* $(x, a) \in C'(X \times E)$ *and* $(y, b) \in C'(X \times E)$ *then* $(x, a) + (y, b) \in C(X \times E)$, *where* $C = ([0, 2\alpha], T')$.

b) If $(\alpha, A) \in C''(E \times 2^X)$ *and* $(b, B) \in C''(E \times 2^X)$ *then* $(a, A) + (b, B) \in C_1(E \times 2^X)$, *where* $C_1 = ([0, 2\delta], T'')$.

Proof. a) If $(x, a) \in C'(X \times E)$ then $d(x, a) \le \alpha$. If $(y, b) \in C'(X \times E)$ then $d(y, b) \le \alpha$. Consider the element $(x, a) + (y, b) = (x + y, a + b)$. Then $T'((x, a) + (y, b)) = T'(x + y, a + b) = d(x + y, a + b) \le d(x, a) + d(y, b) \le \alpha + \alpha = 2\alpha$, therefore $(x, a) + (y, b) \in C(X \times E)$.

b) Similarly, we have $T''((a, A) + (b, B)) = T''(a + b, A + B) = d(a + b, A + B) = \inf\{d(a + b, x + y) \mid x \in A, y \in B\} \le \inf\{d(a, x) + d(b, y) \mid x \in A, y \in B\} \le 2\delta$, therefore $(a, A) + (b, B) \in C_1(E \times 2^X)$.

Corollary 5.4.1. *a)* If $(x, a) \in C'(X \times E)$ *and* $s \in E$ *then* $(x, a) + (s, s) \in C'(X \times E)$.

b) If $(a, A) \in C''(E \times 2^X)$ *and* $s \in E$ *then* $(a, A) + (s, \{s\}) \in C''(E \times 2^X)$.

Proof. For every $x, y, s \in \mathbf{R}$, $|x + s - (y + s)| = |x - y|$ takes place and hence $T((x, a) + (s, s)) = d(x + s, a + s) = d(x, a) \leq \alpha$ and $T''((a, A) + (s, \{s\})) = d(a + s, A + s) = d(a, A) \leq \delta$. Therefore the invariance of the behaviours C' and C" to translations with vectors of (s, s) and $(s, \{s\})$ types, for $s \in E$.

Lemma 5.4.2. *a) If* $(x, a) \in C'(X \times E)$ *and* $\lambda \in \mathbf{R}_+$ *then one has* $\lambda(x, a) \in \lambda C'(\lambda X \times \lambda E)$.

b) If $(a, A) \in C'(E \times 2^X)$ *and* $\lambda \in \mathbf{R}_+$ *then* $\lambda(a, A) \in \lambda C'(\lambda E \times 2^{\lambda X})$.

Proof. The property follows from the positive homogeneity of the distance and from (4.2.5) and (4.2.6). □

These properties of behaviours that are used in the definition of (α, δ) - convexities with respect to E permit the study of the manner in which these convexities change if the sets that have these properties are changed by geometric transformations. This is an application of the general study made in section 4.1 for these particular cases of convexity, but lemmas 5.4.1 and 5.4.2 give more information in these situations.

Theorem 5.4.1. *If the sets* $A \subset X$ *and* $B \subset X$ *are strongly (slackly)* (α, δ) *- convex with respect to* E *then the set* $(A + B) \cap X$ *is strongly (slackly)* $(2\alpha, 2\delta)$ *- convex with respect to* E.

Proof. If $x, y \in (A + B) \cap X$, $t \in \langle x, y\rangle$ then there are $x_1 \in A$, $y_1 \in A$, $x_2 \in B$, $y_2 \in B$, $t_1 \in \langle x, y\rangle$, $t_2 \in \langle x_2, y_2\rangle$ such that $x = x_1 + x_2$, $y = y_1 + y_2$ and $t = t_1 + t_2$. If A is supposed to be strongly (α, δ) - convex with respect to E then there is $a_1 \in E$ such that $d(t_1, a_1) \leq \alpha$ and $d(a_1, A) \leq \delta$. Also, supposing that B is strongly (α, δ) - convex with respect to E, then it follows that there is $a_2 \in E$ such that $d(t_2, a_2) \leq \alpha$ and $d(a_2, B) \leq \delta$. Taking $a = a_1 + a_2$ and using lemma 4.3.1 we get that $d(t, a) \leq 2\alpha$ and $d(a, (A + B) \cap X) \leq 2\delta$, meaning that $(A + B) \cap X$ is strongly $(2\alpha, 2\delta)$ - convex with respect to E.

For the slack convexity the proof is similar.□

Corollary 5.4.2. *If the sets* $A \subset X$ *and* $B \subset X$ *are strongly (slackly)* (α, δ) *- convex with respect to* E *for* $\alpha \geq h/2$ *and* $\delta \geq 2\alpha$ *then* $(A + B) \cap X$ *is strongly (slackly)* (α, δ)*-convex with respect to* E.

Proof. The conclusion follows from the fact that E is a h/2-network over X, therefore X might be covered with balls centred in points of E and having the radius h/2. For this reason, if $\alpha \geq h/2$ then the inequalities from the previous proof are still valid as d (t, a) $\leq$ h/2 $\leq \alpha$ and d (a, (A + B) $\cap$ X) $\leq$ h $\leq 2\alpha \leq \delta$.□

Corollary 5.4.2 is not valid for the slack (α, δ)-convexity with respect to E without conditions: $\alpha \geq$ h/2 and $\delta \geq 2\alpha$. This may be proved by the example 4.2.1.

Theorem 5.4.2. *If the set* A $\subset$ X *is strongly (slackly)* (α, δ) - *convex with respect to* E *and* s $\in$ E *then the set* (A + s) $\cap$ X *is strongly (slackly)* (α, δ) - *convex with respect to E.*

Proof. The conclusion is an immediate consequence of theorem 4.2.1 and of corollary 5.4.1. □

As consequence, the property of strong (slack) (α, δ)-convexity with respect to E is an invariant to the translation of amplitude s $\in$ E. If s $\notin$ E then the translation of the set A from the hypothesis of theorem 5.4.2 will have a convexity property but, generally, with other parameters then α and δ, as it results from the next example.

Example 5.4.1. Considering the distance d = c, X = [-1, 4] $\times$ [-1, 4] and E = $\mathbf{Z}^2 \cap$ X, the set A = {(0, 0), (0,1/2), (0, 1), (1/2, 0), (1, 0), (3/2, 0), (2, 0)} is strongly (1/2, 0) - convex with respect to E. If s = (0, 1/4) $\notin$ E then A + s = {(0, 1/4), (0, 3/4), (0, 5/4), (1/2, 1/4), (1, 1/4), (3/2, 1/4), (2, 1/4)} and this set is strongly (1/2, 1/4) - convex with respect to E.

Regarding the manner in which the strong (slack) (α, δ) - convexity of a set changes when the scale changes, theorem 4.2.3 together with lemma 5.4.2 immediately lead to the following result.

Theorem 5.4.3. *If the set* A *is strongly (slackly)* (α, δ) - *convex with respect to* E *and if* $\lambda \in \mathbf{R}_+$ *then the set* (λA) $\cap$ X *is strongly (slackly)* ($\lambda\alpha$, $\lambda\delta$) - *convex with respect to the set* λE.

Also, in this case the set (λA) $\cap$ X has a property of convexity with respect to E, but the parameters are, generally, different from α, δ or $\lambda\alpha$, $\lambda\delta$. The following example will show that.

Example 5.4.2. Again, using the distance c and the sets X, E and A from example 5.4.1, the set 2A = {(0, 0), (0, 1), (0, 2), (1, 0), (2, 0), (3, 0), (4, 0)} is strongly (1/2, 1) - convex with respect to E. According to theorem 5.4.3, the set 2A = {(0, 0), (0, 2), (0, 4), (2, 0), (2, 2), (2, 4), (4, 0), (4, 2), (4, 4)} is strongly (1, 0) - convex with respect to 2E.

Referring to the preservation of the strong or slack (α, δ) - convexity with respect to E in case of the union or intersection of sets having this property, one can notice that the union of two strong (slack) (α, δ) - convex has not the same property. This is easy to see in the case of the convexity with respect to a given set defined by Liana Lupşa in her thesis, L. Lupşa (1981), and presented in example 4.3.3. One can notice that for $\alpha = \delta = 0$ definition 5.1.1 reduces to definition 4.3.1. Hence, the same convexity notion might be obtained using different behaviours. L. Lupşa (1981) has proved that the intersection of two sets that are strongly (slackly)-convex with respect to a given set is strongly (slackly)-convex with respect to that set, but the property is not valid for the union.

In the general case, $\alpha \geq 0$ and $\delta \geq 0$, the intersection of two strongly (slackly) (α, δ) - convex sets with respect to E is not strongly (slackly) (α, δ) - convex with respect to E, as it will be proved by some examples.

Example 5.4.3. For d = c the sets A = {(0, 0), (0, 1), (1, 0), (2, 0)} and B = {(0, 1), (1, 1), (2, 1), (2, 0)} are strongly (1/2, 0) - convex with respect to $\mathbf{Z}^2$. But $A \cap B$ = {(0, 1), (2, 0)} is not strongly (1/2, 0) - convex with respect to $\mathbf{Z}^2$. Indeed, for t = (1, 1/2) $\in$ <(0,1), (2,0)>, the points a = (1, 0) and b = (1, 1) have the property that c (t, a) = c (t, b) = 1/2, but c (a, $A \cap B$) = c (b, $A \cap B$) = 1 > 0.

Example 5.4.4. For d = c the sets A = {(0, 1/3), (1/3, 1), (1/2, 0), (2/3, 0), (1, 2/3)} and B = {(1/2, 0), (1, 2/3), (0, 1/3), (4/3, 0)} are slackly (1/2, 1/3) - convex with respect to $\mathbf{Z}^2$. But $A \cap B$ = {(0, 1/3), (1/2, 0), (1, 2/3)} is not slackly (1/2, 1/3) - convex with respect to $\mathbf{Z}^2$ because if t = (1/2, 0) $\in$ <(1/2, 0), (1, 2/3)> and a = (1, 0) are considered then c (t, a) = 1/2 but c (a, $A \cap B$) = ½ > 1/3.

In the case of discrete sets $X \subset \mathbf{Z}^2(h)$, X being all the points having the x and y coordinates the multiples of h from a rectangle ir the plane, J.M.Chassery (1979, 1983) has proved theorems containing c nditions in which a convexity property is preserved by intersection. In what ͻllows we shall quote these results, without insisting on their proofs.

The following convexity property of a set $A \subset \mathbf{Z}^2$ is defined b‸ J. M. Chassery (1979).

Definition 5.4.1. *A connected component* A *is convex in* $\mathbf{Z}^2$ *if for every pair of points* $P \in A$ *and* $Q \in A$ *and for every real* $t \in]0, 1[$ *there is a point* $R \in A$ *such that*

(5.4.1) $\quad tP + (1 - t)Q \in B_\infty(R, 1/2)$.

Here, by connectivity is meant the 8-connectivity and the notation $B_\infty(R, h/2) = \{S \in \mathbf{Z}^2(h) \mid c(R, S) \leq h/2\}$ is used.

Property 5.4.1. *If* A *and* B *are two convex connected components in* $\mathbf{Z}^2$ *then* $A \cap B$ *is convex.*

It is obvious that the convexity from definition 5.4.1 is, in fact, the strong (h/2, 0)-convexity of a set in $\mathbf{Z}^2(h)$ with respect to $\mathbf{Z}^2(h)$. It is preserved by the intersection.

Four years later, in 1983, the same author generalises his results as follows:

Definition 5.4.2. *A connected component* $A \subset \mathbf{Z}^2(h)$ *is* α *- convex with* $\alpha \in [h/2, h[$ *if for every* $P \in A$, $Q \in A$ *and* $t \in [0, 1]$, *there is a point* $R \in A$ *satisfying*

(5.4.2) $\quad tP + (1 - t)Q \in B_\infty(R, \alpha)$.

It is easy to see that this convexity is a strong $(\alpha, 0)$ - convexity with respect to $\mathbf{Z}^2(h)$, for $\alpha \in [h/2, h[$. The result concerning the preservation of this result by intersection is the following.

Theorem 5.4.4. *Let* $A \subset \mathbf{Z}^2(h)$ *be a connected component that is* α_1 *- convex and* $B \subset \mathbf{Z}^2(h)$ *a connected component that is* α_2 *- convex, with* $\alpha_1 \leq \alpha_2$. *Then, if* $A \cap B$ *is a connected component then it is* α_2 *- convex.*

In the same paper J. M. Chassery (1983) defined the notion of discrete convexity of a connected set in $\mathbf{Z}^2(h)$. We have quoted it in example 4.3.5 for the case $h = 1$. Generally, a connected component $A \subset \mathbf{Z}^2(h)$ is discrete convex if there is a number $\alpha \in [h/2, h[$ such that A is α - convex. Theorem 5.4.4 implies the fact that if the intersection of two connected components discretely convex is connected then it is discretely convex.

From examples 5.4.3, 5.4.4, property 5.4.1 and theorem 5.4.4 it follows that, generally, in the case of (α, δ) - convexities with respect to a network it is not possible to define the strong (slack) (α, δ) - convex hull with respect to the network. Hence, this type of convexity does not allow the organisation of the space X as a convexity space in the sense discussed by V. W. Bryant, R. J. Webster (1972, 1973, 1974), W. Prenowitz (1961), V. Soltan (1984) and studied by us, G. Cristescu (1980).

5.5 The geometrical characterisation

In this paragraph we intend to discuss the significance of number δ which appears in the definition of the properties of (α, δ) - convexity with respect to a given set. For this purpose the geometrical significance of these convexity properties will be studied. As consequence, the context in which the number δ reveals its importance and its applications will arise. The results from this paragraph are valid for every distance d that might be used.

For every two points x and y from $\mathbf{R}^2$ and for every number $\alpha > 0$, we denote by $\mathrm{Seg}(\alpha; x, y) = \mathbf{Z}^2(h) \cap \mathrm{conv}(B(x, \alpha) \cup B(y, \alpha))$.

Lemma 5.5.1. *For each* $x, y \in \mathbf{R}^2$ *and for every* $t \in <x, y>$, *the inclusion* $B(t, \alpha) \cap \mathbf{Z}^2(h) \subseteq \mathrm{Seg}(\alpha; x, y)$ *holds for every* $\alpha > 0$.

Proof. Let be $a \in B(t, \alpha) \cap \mathbf{Z}^2(h)$. Then $d(a, t) \leq \alpha$. Let $m = d(a, t)$ and $B(<x, y>, m) = \{z \in \mathbf{R}^2 \mid d(z, t) \leq m, t \in <x, y>\}$. Then $a \in B(<x, y>, m) \subseteq \mathrm{conv}(B(x, m) \cup B(y, m)) \subseteq \mathrm{conv}(B(x, \alpha) \cup B(y, \alpha))$, therefore $a \in \mathrm{Seg}(\alpha; x, y)$.

Theorem 5.5.1. *The set A is slackly* (α, δ) *- convex with respect to* $\mathbf{Z}^2(h)$ *if and only if for every two points* x *and* y *from* A *and for every point* $a \in \mathrm{Seg}(\alpha; x, y)$, *the set* $B(\alpha, \delta) \cap A$ *is nonempty.*

Proof. Let us suppose that set A is slackly (α, δ) - convex with respect to $\mathbf{Z}^2(h)$. Then, if $x, y \in A$ and $t \in <x,y>$, if the point $a \in \mathbf{Z}^2(h)$ satisfies $d(a, t) \leq \alpha$ then it follows that $d(a, A) \leq \delta$. But $d(a, t) \leq \alpha$ is equivalent to $a \in B(t, \alpha) \subseteq \mathrm{Seg}(\alpha; x, y)$ according to lemma 5.5.1, which implies, due to the slack (α, δ) - convexity of A with respect to $\mathbf{Z}^2(h)$, that $d(a, A) \leq \delta$, therefore $B(\alpha, \delta) \cap A \neq \varnothing$.

Now, let us prove the converse property. If we choose $x \in A$, $y \in A$ and $a \in \mathrm{Seg}(\alpha; x, y)$ then $d(a, <x, y>) \leq \alpha$ and so $B(a, \alpha) \cap <x, y> \neq \varnothing$. Let us consider the set $V(\alpha; x, y) = \cup\{B(a, \alpha) \cap <x, y> \mid a \in \mathrm{Seg}(\alpha; x, y)\}$. Then for each point $t \in V(\alpha; x, y)$ there is $a \in \mathbf{Z}^2(h)$ such that $d(a, t) \leq \alpha$ and for these points condition $B(\alpha, \delta) \cap A \neq \varnothing$ means that $d(a, A) \leq \delta$. If $t \notin V(\alpha; x, y)$ then for every $a \in \mathbf{Z}^2(h)$ the inequality $d(t, a) > \alpha$ takes place. So, set A is slackly (α, δ) - convex with respect to $\mathbf{Z}^2(h)$.

Theorem 5.5.1 shows that a set A of points in the plane is slackly (α, δ) - convex with respect to $\mathbf{Z}^2(h)$ if for every two points x and y of A, every point from $\mathbf{Z}^2(h)$ situated in an α-neighbourhood of the straight-line segment $<x, y>$, which is closed in $\mathbf{R}^2$, has the property that its δ- neighbourhood closed in $\mathbf{R}^2$ contains points from A. A similar result for the case of the strong property is contained in the following theorem.

Theorem 5.5.2. *The set* A *is both strongly and slackly* (α, δ) *- convex with respect to* $\mathbf{Z}^2(h)$ *if and only if for every* x *and* y *from* A *the following conditions are satisfied:*

(1) $<x, y> \subseteq \cup\{B(a, \alpha) \mid a \in \mathrm{Seg}(\alpha; x, y)\}$;

(2) $a \in \mathrm{Seg}(\alpha; x, y) \Rightarrow B(\alpha, \delta) \cap A \neq \varnothing$.

Proof. Condition (1) means that for every pair of points $x, y \in A$ the straight-line segment $<x, y>$ is covered by closed α-neighbourhoods in $\mathbf{R}^2$, centred in the grid knots. Hence, for every t ∈ <x, y>, there is a ∈ $\mathbf{Z}^2$ (h) such that a ∈ Seg (α; x, y). But since A is both slackly and strongly (α, δ) - convex with respect to $\mathbf{Z}^2$ (h), applying theorem 5.5.1 we get that B (α, δ) ∩ A ≠ ∅. The converse property is also the consequence of theorem 5.5.1.

Corollary 5.5.1. *If set* A *is neither strongly nor slackly* (α, δ) - *convex with respect to* $\mathbf{Z}^2$ (h) *then there are two points* x ∈ A *and* y ∈ A *and there is* t ∈ <x, y> *such that, for every* a ∈ B (t, α) ∩ $\mathbf{Z}^2$ (h), *the set* B (α, δ) ∩ A *is empty.*

Proof. Suppose that A is neither strongly nor slackly (α, δ) - convex with respect to $\mathbf{Z}^2$ (h). Then there are x ∈ A, y ∈ A and t ∈ <x, y> such that, for every a ∈ $\mathbf{Z}^2$ (h), with d (t, a) ≤ α the inequality d (a, A) > δ takes place. Hence, for every a ∈ B (t, α) ∩ $\mathbf{Z}^2$ (h), we have B (α, δ) ∩ A = ∅.

This means that, working with the c distance, in the case of a set A which is neither strongly nor slackly (α, δ) - convex with respect to $\mathbf{Z}^2$ (h) there are two points x and y in A having the following property: on the straight-line segment determined by them there is a point that introduces between x and y a square having the side δ centred in a grid knot of $\mathbf{Z}^2$ (h) which does not contain points from A. Therefore, if this set A is conveniently digitised (see chapter 9 of this book) then, if α ≥ h/2 and δ ≥ h/2, there is a point of the complement of the digitisation of A lying between two points of the digitisation of A. As consequence, a set that is not discrete convex, in the sense of J. M. Chassery, was obtained.

Similar theorems might also be obtained in the three-dimension case. These types of properties have been forecast in some comments of A. Rosenfeld (1985) in the case of the use of the Euclidean distance. This type of theorems has led to interesting applications connected to the way of "measuring the concavities" of a set that has not a property of strong (α, δ) - convexity with respect to $\mathbf{Z}^2$ (h). For this purpose we will adjust the definition of the concavity of a set, formulated by A. Rosenfeld (1985), extending it to more general sets than the bounded and simply connected sets considered by A. Rosenfeld (1985).

5.6 Particular approximations of the classical convexity

In this paragraph the convergence properties of some sequences of convexities will be discussed without condition (4.4.8). Our aim is to prove that the discrete convex sets, in the sense of J.M. Chassery (1983), are digital approximations of some sets having various convexity properties, sometimes classical convexity.

In addition, there are cases in which, some classes of sets A satisfy the property from definition 4.4.1 even if conditions (4.4.1) and (4.4.8) are not satisfied. This is the case when X is a topological space and $(M_n)_{n=1,2,...}$ is an ascending sequence of sets having their union dense in X. In these conditions we will find situations when there are closed sets A that are strongly and/or slackly convex with respect to a sequence of sets and two sequences of behaviours and, as consequence, they are classically convex.

Let be $X = \mathbf{R}^2$, d one of the distances d_1, c or the Euclidean distance, the real number $h > 0$, and $M_n = \mathbf{Z}^2\left(\frac{h}{2^n}\right)$, $n \in \mathbf{N}^*$. Here, $X = \overline{\bigcup_{n \in N'} M_n}$ replaces condition (4.4.1). Let be $B = \mathbf{R}$ and the sequence $(\alpha_n)_{n=1,2,...}$ of real numbers with the property

(5.6.1) $\alpha > \min \{r \mid [0, h/n]^2 \subset B((0, 0), r) \cup B((0, h/n), r) \cup$

$\cup B((h/n, 0), r) \cup B((h/n, h/n), r)\}$, for $n = 1, 2, \ldots$.

Evidently, the sequence $(\alpha_n)_{n=1,2,...}$ is strictly decreasing and convergent towards zero. Also, consider the decreasing sequence of real numbers $(\delta_n)_{n=1,2,...}$, convergent to zero.

Let $A \subset \mathbf{R}^2$, $A \neq \emptyset$, be a closed set.

Theorem 5.6.1. *If the set* A *is strongly* (α_n, δ_n) *- convex with respect to the set* $\mathbf{Z}^2\left(\frac{h}{2^n}\right)$, *for every* $n \in \mathbf{N}^*$, *then* A *is classically convex in* $\mathbf{R}^2$.

Proof. If A is strongly (α_n, δ_n) - convex with respect to the set $\mathbf{Z}^2\left(\frac{h}{2^n}\right)$, for every $n \in \mathbf{N}^*$, then, for every $n \in \mathbf{N}^*$, $x \in A$, $y \in A$ and $t \in \langle x, y\rangle$, there is an element $a_n \in \mathbf{Z}^2\left(\frac{h}{2^n}\right)$, such that

(5.6.2) $d(t, a_n) \leq \alpha_n$ and $d(a_n, A) \leq \delta_n$.

Let us fix the arbitrary points x, y and t. Condition (5.6.1) implies the existence of at least a point a_n that satisfies (5.6.2) for these points x, y and t. But, from the continuity of the distance we work with, the conditions

$$d(t, a_n) \leq \alpha_n \text{ and } d(a_n, A) \leq \delta_n, \text{ for every } n \in \mathbf{N}^*,$$

imply the convergence to zero of the sequences $(d(t, a_n))_{n=1,2,...}$ and $(d(a_n, A))_{n=1,2,...}$. This means that, the inequality $d(t, A) \leq d(t, a_n) + d(a_n, A)$, taking place for every $n \in \mathbf{N}^*$, implies $d(t, A) = 0$. Since A is closed it follows that $t \in A$ and therefore A is classically convex.

Theorem 5.6.2. *If the set* A *is slackly* (α_n,δ_n) *- convex with respect to the set* $\mathbf{Z}^2\left(\frac{h}{2^n}\right)$*, for every* $n \in \mathbf{N}^*$*, then* A *is classically convex in* $\mathbf{R}^2$.

Proof. Condition (5.6.1) implies that for every $n \in \mathbf{N}^*$, and the points $x \in A$, $y \in A$ and $t \in \langle x, y\rangle$, there is $a_n \in \mathbf{Z}^2\left(\frac{h}{2^n}\right)$ such that $d(t, a_n) \le \alpha_n$. Then, since A is slackly (α_n,δ_n) - convex with respect to the set $\mathbf{Z}^2\left(\frac{h}{2^n}\right)$, it follows that $d(a_n, A) \le \delta_n$, for every $n \in \mathbf{N}^*$. In this case, the reasoning from the proof of theorem 5.6.1 is repeated.

One can notice that the closure of set A replaces condition (4.4.8) in the previous proofs.

Hence, it was proved that, for the closed sets, the properties of the strong (slack) (α, δ) - convexity with respect to set $\mathbf{Z}^2(h)$ are approximations, in the sense of definition 5.6.1 of the classical convexity. In this case, the problem of the evaluation of the error of this approximation must be discussed.

Now, we will take $X = \mathbf{R}^2$ and $\varepsilon_n = h_n$, with $\lim_{n\to\infty} h_n = 0$. Then $M_n = \mathbf{Z}^2(h_n)$. If A is strongly $(h_n/2, \delta_n)$ - convex with respect to $\mathbf{Z}^2(h_n)$ and B is a strongly $(h_m/2, \delta_m)$ - convex with respect to $\mathbf{Z}^2(h_m)$, with $m > n$, then, taking $\Delta(A) = \sup\{\delta \mid \exists\, x, y \in A, \exists\, t \in \langle x, y\rangle, d(t, A) > \delta\}$, we can deduce that $\Delta(A) \le \Delta(B)$.

Let us suppose that $(A_n)_{n=1,2,\ldots}$ is a sequence of sets in $\mathbf{R}^2$ such that :

(5.6.3) $A_1 \subset A_2 \subset \ldots \subset A_n \subset \ldots \subset \bigcup_{n\in\mathbf{N}^*} A_n$,

(5.6.4) A_n is strongly $(h_n/2,\delta_n)$-convex with respect to $\mathbf{Z}^2(h_n)$.

Then, obviously, sequence $(\Delta(A_n))_{n=1,2,\ldots}$ is a decreasing sequence of nonnegative numbers. The properties of number $\Delta(A)$ will be studied in chapter 9 of this book.

Theorem 5.6.3. *If sequence* $(A_n)_{n=1,2,\ldots}$ *satisfies conditions* (5.6.3) *and* (5.6.4) *then:*

1. $\overline{\bigcup_{n\in\mathbf{N}^*} A_n}$ *is classically convex;*

2. $\lim_{n\to\infty} \Delta(A_n) = 0$.

The proof is similar to that of theorem 5.6.2.

Number $\Delta(A)$ will appear, in chapter 9, as a measure of the error committed when the classical convexity is replaced by a strongly $(h_n/2, \delta_n)$-convex with respect to $\mathbf{Z}^2(h_n)$.

Sometimes condition (5.6.8) is replaced by other properties of sets M_n that play a similar role in the proof. For example, in the case of the digital convexity, J. M. Chassery (1983) uses the possibility of a convenient choice of the set of grid points.

Lemma 5.6.1. *For every two points* p *and* q *from* $\mathbf{R}^2$, *there is a number* h *such that* p *and* q *are knots of the network* $\mathbf{Z}^2(h)$.

Proof. Let us consider the distance d = c in $\mathbf{R}^2$ and the distinct points p and q belonging to $\mathbf{R}^2$. We are interested to build the rectangular triangle with the straight-line segment determined by p and q as hypotenuse. Therefore, it is important to find the integer numbers k, k' and j satisfying the following relations: $s = \frac{kc(p,q)}{j}$ and $s' = \frac{k'c(p,q)}{j}$, where s and s' are the lengths of the two orthogonal sides of the desired triangle. The Pythagorean theorem gives

$$\left(\frac{kc(p,q}{j}\right)^2 + \left(\frac{k'c(p,q)}{j}\right)^2 = c(p,q)^2 .$$

From the identity $(x^2 - y^2)^2 + (2xy)^2 = (x^2 + y^2)^2$, taking the numbers

$$k = x^2 - y^2,\ k' = 2xy,\ j = x^2 + y^2,$$

we obtain the step size of the network that contains the points p and q as lattice points. This equals to $h = \frac{c(p,q)}{x^2 + y^2}$.

For the case of the discrete convexity, J.M. Chassery has proved in 1983 the following approximation theorem.

Theorem 5.6.4. *Let* A *be a 8-connected set in* $\mathbf{R}^2$. *If for every step* h, *the digitisation* A_h *of set* A *in* $\mathbf{Z}^2$ (h) *is discrete convex then* A *is classically convex in* $\mathbf{R}^2$.

Proof. Let p and p be two points of A and $\alpha \in [0, 1[$. The relation $\alpha p + (1-\alpha) q \in A$ is to be proved. According to the previous lemma, there is a real number h, which is the step of a tessellation that contains p and q as grid points. The digitisation A_h of A in $\mathbf{Z}^2$ (h) is discretely convex, according to the hypothesis.

Let us assume that A is not convex. This assumption implies the existence of two points a and b of A and the existence of a point $t \in \langle a, b \rangle$ that does not belong to A. Therefore, there is a number $h_0 > 0$ such that for every $h < h_0$ there is no point $s \in A_h$ satisfying $s \in B_\infty(t, h)$.

Using lemma 5.6.1 we can find sampling steps of size $k = \frac{c(p,q)}{x^2 + y^2}$ for which p and q are lattice points. Changing x by jx and y by jy, with $j \in \mathbf{N}$ we obtain the new sampling step $h_1 = \frac{k}{j^2}$. It makes possible to find $j \in \mathbf{N}$ such that $\frac{k}{j^2} < h_0$. Hence, for that step of size $h_1 = \frac{k}{j^2} < h_0$ there is no point $t \in A_{h_1}$ and satisfying $s \in B_\infty(t, h_1)$. Because for every sampling of step size h the set A_h is discrete convex, we obtain that the relations $p \in A_h$ and $q \in A_h$ imply that for every point $t \in \langle p, q \rangle$ there is a point $s \in A_{h_1}$ such that $s \in B_\infty(t, h_1)$. This is a contradiction with the hypothesis of non-convexity of A. Therefore A is classically convex in $\mathbf{R}^2$.

It is easy to see that theorems 5.2.1, 5.2.2, 5.2.3, 5.3.1, 5.3.2 are less restrictive that the result of J.M. Chassery. They replace the condition of discrete convexity for *every step* with a sequence of convexities with respect to some *particular networks* over $\mathbf{R}^2$.

5.7 Weak particular cases of convergence to the classical convexity

In what follows, some interesting particular cases of weak approximation of the classical convexity are presented. The weak approximation of a convexity property means the approximation by means of the weak convergence.

Let (X, d) be a totally bounded metric space, that is complete in the topology induced by the metric d, let $(\varepsilon_n)_{n=1,2,...}$ be a decreasing sequence of positive numbers converging to zero, such that there is a sequence of finite $2\varepsilon_n$ - networks over finite ε_n - networks over X, $(M_n)_{n=1,2,...}$, satisfying

$$(5.7.1) \qquad M_1 \subset M_2 \subset \ldots \subset M_n \subset \ldots \subset \bigcup_{n=1}^{\infty} M_n$$

and

(5.7.2) $$\overline{\bigcup_{n\in N'} M_n} = X.$$

Let A be a set in X, $(\alpha_n)_{n=1,2,...}$ and $(\delta_n)_{n=1,2,...}$ two decreasing sequences of positive real numbers such that

(5.7.3) $$\alpha_n > \max_{x\in M_n} \min\{r \mid B(x, \varepsilon_n) \subset \bigcup_{d(x,y)\leq 2\varepsilon_n} B(y,r)\},\ n=1,2,\ldots .$$

Here, by B(x,r) the closed ball centred in x and having the radius r is denoted.

The continuity of the metric d is supposed in what follows.

Theorem 5.7.1. *If the set* $A \subset X$ *is strongly* (α_n, δ_n) *- convex with respect to the set* M_n *for every* n = 1, 2, ... , *then the closure of* A *is classically convex in* X.

Proof. If A is strongly (α_n, δ_n)-convex with respect to M_n, for every n = 1, 2, ... then , for every $n \in \mathbf{N}^*$, $x \in A$, $y \in A$ and $t \in \langle x, y\rangle$, there is an element $a_n \in M_n$, such that

(5.7.4) $d(t, a_n) \leq \alpha_n$ and $d(a_n, A) \leq \delta_n$.

Condition (5.7.3) implies that the first inequality of (5.7.4) takes place for at least one point a_n. The continuity of the metric d implies the convergence to zero of sequences $(d(t, a_n))_{n=1,2,...}$ and $(d(a_n, A))_{n=1,2,...}$. From inequality

(5.7.5) $d(t, A) \leq d(t, a_n) + d(a_n, A)$, for every n = 1, 2, ...,

it follows that d (t, A) = 0. Therefore t belongs to the closure of A, so the closure of A is classically convex in X. If $x \in \overline{A}$ and/or $y \in \overline{A}$ then the proof is similar.

Theorem 5.7.2. *If the set* $A \subset X$ *is slackly* (α_n, δ_n) *- convex with respect to the set* M_n *for each* n = 1, 2, ..., *then the closure of* A *is classically convex in* X.

Proof. Condition (5.7.3) implies that for each $n \in \mathbf{N}^*$ and two points $x \in A$, $y \in A$ and for each $t \in \langle x, y\rangle$, there is $a_n \in M_n$ such that $d(t, a_n) \leq \alpha_n$. Then, the slack (α_n, δ_n) - convexity of A with respect to M_n implies that $d(a_n, A) \leq \delta_n$, for every $n \in \mathbf{N}^*$. In this case, the reasoning of the proof of the previous theorem is repeated.

Example 5.7.1. Consider $X = \mathbf{R}^2$ endowed with the chessboard metric and $A = \{[0, 1] \cap \mathbf{Q}\}^2$. If $\alpha_n \in \left[\frac{1}{2^{n+1}}, \frac{1}{2^n}\right[$ then the set A is strongly $(\alpha_n, 0)$-

convex with respect to $\mathbf{Z}^2\left(\frac{1}{2^n}\right)$, for every $n \in \mathbf{N}^*$. But A is not classically convex. The closure of A has this property.

Example 5.7.2. Consider $X = \mathbf{R}^2$ endowed with the chessboard metric and $A = \{[0, 1] \cap \mathbf{Q}\}^2$. If $\alpha_n \in \left[\frac{1}{2^{n+1}}, \frac{1}{2^n}\right[$ then set A is strongly $\left(\alpha_n, \frac{1}{2^{n+1}}\right)$- convex with respect to $\mathbf{Z}^2\left(\frac{1}{2^n}\right)$, for every $n \in \mathbf{N}^*$. But A is not classically convex. The closure of A has also this property.

Therefore, for a closed set $A \subset X$, if $\lim_{n\to\infty} \alpha_n = 0$ and $\lim_{n\to\infty} \delta_n = 0$, then the sequence of strong (slack) (α_n, δ_n) - convexity properties with respect to sequence of sets $(M_n)_{n=1,2,...}$ converges to the classical convexity. If set A is not closed, then this convergence is retrieved for its closure.

For $\delta_n = 0$, the properties of strong (α_n, δ_n) - convexity with respect to $\mathbf{Z}^2(\varepsilon_n)$ become the discrete convexity of sets in $\mathbf{Z}^2(\varepsilon_n)$, in the sense of J. M. Chassery (1983). By means of a digitisation process of a set $A \subset \mathbf{R}^2$ in $\mathbf{Z}^2(\varepsilon_n)$, $n \in \mathbf{N}^*$, a stronger property than the result of J. M. Chassery is obtained.

Therefore, all the above-discussed convexities are approximations, in strong or weak sense, of the classical convexity.

Chapter 6

Induced convexity

Everyday life sets frequently in front of us situations in which an object from a set A is replaced by another object a' from a set B, a' being obtained from a as its image by means of a transformation T: $A \to B$, therefore a' = T(a). The human perception provides us with the most known example: every object is transformed, through the senses, into a set of impulses. The synthesis of the impulses generates an "image" of that object at the level of the brain. An important problem arises in these situations: for every property p of the object a, a property p' of the object a' is sought for such that for each object b from A having the property p, the object T(b) has the property p'. In this case, the property p' is said to be the image by T of the property p. The problem of the description (identification) of the property p' is called recognition problem.

But there are situations in which the elements of the image have a property that is not directly noticeable in the case of the elements of the set A. In this type of cases the converse procedure is followed, identifying the property p" of the elements of A satisfying the condition that for every object b of the image set B, which has the property p', the element $T^{-1}(b)$ has the property p". In this situation the property p" will be called induced property by the transformation T, and the problem of the description of the property p" will be called the converse recognition problem. Both the recognition problem and its converse have been formulated in G. Cristescu (2000 [71]).

In this context, in this chapter we give the notion of "the induced best approximation point" (i. b. a. p.) of an element x^0 from a set X by elements from a nonempty subset A of X. For this purpose, a function f: $X \to Y$, $(Y,+,\cdot,\| \|)$ being an H-normed space, is used. After that, the properties of the induced best approximation points of x^0 by elements of A are studied in the case in which A is an induced seg-convex set (see L. Blaga and L. Lupşa (1998)). This type of approximation accompanies both the recognition and the converse recognition problems, and in a totally different and more general context was also formulated and partially solved in G. Cristescu (2000 [71]).

6.1 Induced convexity

Let X, Y be two arbitrary sets, let M be a nonempty subset of Y and let us consider the set valued mapping s: $2^Y \to 2^Y$. Let k be a natural number, k ≥ 2.

Problem 6.1.1. Given a mapping f: $X \to Y$, how does one endow X with a convexity structure, using the structure of Y and, when required, s and/or set M and/or number k?

Let a, b$\in$X. Using the points a, b we define the set

$$<a, b>_{s,f,M} = f^{-1}(s(\{f(a), f(b)\}) \cap M).$$

Clearly

$$<a, b>_{s,f,M} = \{z \in X \mid f(z) \in s(\{f(a), f(b)\}) \cap M\}.$$

Definition 6.1.1. *We call a subset* A *of* X

a) *induced* k*-strongly convex with respect to* s, f *and* M *if*

$$s(\{f(a_1), ..., f(a_k)\}) \cap M \subseteq f(A), \textit{for any } a_1, ..., a_k \in A;$$

b) *induced strongly-convex with respect to* s, f *and* M *if*

$$s(f(A)) \cap M \subseteq f(A).$$

c) *induced strictly strong-convex with respect to* s, f *and* M *if*

$$s(f(B)) \cap M \subseteq f(B), \textit{for any } B \subseteq A.$$

d) *induced seg-convex with respect to* s, f *and* M *if*

$$<a, b>_{s,f,M} \subseteq A, \textit{for any } a, b \in A.$$

In what follows we use the notations:

$G_{st,k,s,f,M}$ denotes the set of all subsets of X which are induced k-strongly convex with respect to s, f and M.

$G_{st,s,f,M}$ denotes the set of all subsets of X which are induced strongly-convex with respect to s, f and M.

$G_{sst,s,f,M}$ denotes the set of all subsets of X which are induced strictly strong-convex with respect to s, f and M.

$G_{seg,s,f,M}$ denotes the set of all subsets of X which are induced seg-convex with respect to s, f and M.

In general neither does $A \in G_{st,2,s,f,M}$ imply $A \in G_{seg,s,f,M}$ nor does $A \in G_{seg,s,f,M}$ imply $A \in G_{st,2,s,f,M}$.

Example 6.1.1. Let X = {-1, 1, 2}, Y={-1, 0, 1, 2} and let f: X → Y given by f (-1) = 1, f (1) = 1, f (2) = 2. Let s: $2^Y \to 2^Y$ given by

$$s(\varnothing) = \varnothing,\ s(\{-1\}) = \{-1\},\ s(\{0\}) = \{0\},\ s(\{1\}) = \{0, 1\},$$
$$s(\{2\}) = \{0, 1, 2\},\ s(\{-1, 0\}) = \{0\},\ s(\{-1, 1\}) = \{0, 1\},$$
$$s(\{-1, 2\}) = \{0, 1, 2\},\ s(\{0, 1\}) = \{0, 1\},\ s(\{0, 2\}) = \{0, 1, 2\},$$
$$s(\{1, 2\}) = \{1, 2\},\ s(\{-1, 0, 1\}) = \{0, 1\},\ s(\{-1, 0, 2\}) = \{0, 2\},$$
$$s(\{-1, 1, 2\}) = \{1, 2\},\ s(\{0, 1, 2\}) = \{0, 1, 2\},$$
$$s(\{-1, 0, 1, 2\}) = \{-1, 0, 1, 2\}.$$

Let M = Y and A = {-1, 1}. Because

$$s(\{f(-1), f(1)\}) = \{0, 1\} \not\subseteq \{1\} = f(\{-1, 1\}) = f(A),$$

it results that A is not strongly 2-convex with respect to s, f and M. Since

$$<-1, 1>_{s,f,M} = \{z \in X \mid f(z) \in \{0, 1\}\} = \{-1, 1\} \subseteq A$$

the set A is seg-convex with respect to s, f and M.

Let B = {1, 2}. Because s ({f (1), f (2)}) = s ({1, 2}) = f (A), the set B is strongly 2-convex with respect to s, f and M. Since

$$<1, 2>_{s,f,M} = \{z \in X \mid f(z) \in \{1, 2\}\} = \{-1, 1, 2\} \not\subseteq \{1, 2\}$$

the set B is not seg-convex with respect to s, f and M.

Theorem 6.1.1. *The following statements are true*:

a) *If* f *is injective, then* $G_{st,2,s,f,M} \subseteq G_{seg,s,f,M}$.

b) *If* f *is surjective, then* $G_{seg,s,f,M} \subseteq G_{st,2,s,f,M}$.

c) *If* f *is bijective, then* $G_{seg,s,f,M} = G_{st,2,s,f,M}$.

Proof. a) Suppose that f is injective. Let $A \in G_{st,2,s,f,M}$, and let a, b ∈ A. As A is induced 2-strongly convex with respect to s, f and M, we have

(6.1.1) $\quad s(\{f(a), f(b)\}) \cap M \subseteq f(A)$.

Let be $c \in <a, b>_{s,f,M} = f^{-1}(s(\{f(a), f(b)\}) \cap M)$. Then

$$f(c) \in s(\{f(a), f(b)\}) \cap M,$$

and, from (6.1.1), we get $f(c) \in f(A)$. As f is injective, it follows that $c \in A$. But c is arbitrarily chosen in $<a, b>_{s,f,M}$ Therefore, $<a, b>_{s,f,M} \subseteq A$. This implies that $A \in G_{seg,s,f,M}$.

b). Suppose now that f is a surjective function. Let $A \in G_{seg,s,f,M}$, and let $a, b \in A$. As A is induced seg-convex with respect to s, f and M, we have

$$(6.1.2) \quad <a, b>_{s,f,M} = f^{-1}(s(\{f(a), f(b)\}) \cap M) \subseteq A.$$

If we suppose that $A \notin G_{st,2,s,f,M}$, then there are two point $a, b \in A$ such that $s(\{f(a), f(b)\}) \cap M \not\subseteq f(A)$. Let $c \in (s(\{f(a), f(b)\}) \cap M) \setminus f(A)$. It follows $c \in Y \setminus f(A)$, and because f is surjective, there is $\alpha \in X \setminus A$, such that $f(\alpha) = c$. Then $\alpha \in f^{-1}(s(\{f(a), f(b)\}) \cap M)$, and from (6.1.2) we get

$$\alpha \in <a, b>_{s,f,M} \subseteq A.$$

This contradicts the fact that $\alpha \in X \setminus A$. Therefore the hypothesis that $A \notin G_{st,2,s,f,M}$, is false. Hence $A \in G_{st,2,s,f,M}$.

c) It results immediately from a) and b).

It is known that if $(V, +, \cdot)$ is a linear space and A is a convex subset of V, convex in classical sense, then any convex combination of elements of A is an element of A. In the case of induced convexity this is not true. Moreover, if h and k are two non null natural numbers with $h < k$, we don't have $G_{st,h,s,f,M} = G_{st,k,s,f,M}$.

Example 6.1.2. Let $X = Y = \mathbf{R}^2$, $M = \mathbf{Z}^2$, $f = 1_{R^n}$, and let $s: 2^Y \to 2^Y$, $s(B) = \text{conv}(B)$, for each $B \in 2^Y$. If we take $A = \{(1, 3), (0, 2), (2, 1)\}$, then A is induced 2 – strongly convex with respect to M, but M is not induced 3-strongly convex with respect to M. We have

$$(1, 2) \in \text{conv}\{(1, 3), (0, 2), (2, 1)\} \text{ and } (1, 2) \notin A.$$

Remark 6.1.1. a) Because in the definition of induced k-strongly convex with respect to s, f and M is not necessary that the points $a_1, ..., a_k$ to be distinct, it follows that if h and k are two non null natural number with $h < k$, we have $G_{st,h,s,f,M} \supset G_{st,k,s,f,M}$. Therefore, if the set $A \subseteq X$ is not induced k-strongly convex with respect to s, f and M, then A is not induced s-strongly convex with respect to s, f and M, for all natural numbers s, $s \geq k$.

b) Obviously we have $G_{sst,s,f,M} \subset G_{st,s,f,M}$.

c) If a set $A \subseteq X$ is induced strongly convex with respect to M, then, in general, it is not a natural number k, $k \geq 2$, such that A is induced k-strongly convex with respect to (s, f, M).

Example 6.1.3. Let X = Y = {-1, 0, 1, 2}, f = 1_X, M = X, and let s:$2^Y \to 2^Y$, given by: s (∅) = ∅, s ({-1}) = {1}, s ({0}) = 0, s ({1}) = {1}, s({2}) = 2, s ({-1, 0}) = {1}, s ({-1, 1}) = {1}, s ({-1, 2}) = {2}, s ({1, 2}) = {1, 2}, s ({-1, 0 ,1}) = {0, 1}, s ({-1, 0, 2}) = {0, 2}, s ({-1, 1, 2}) = {1, 2}, s({0, 1, 2}) = {0, 1, 2}, s ({-1, 0, 1, 2}) = {0, 1, 2}. Because

$$s(f(\{-1, 0, 1\})) \cap M = \{0, 1\} \subseteq \{-1, 0, 1\} = f(\{-1, 0, 1\}),$$

it follows that set A = {-1, 0, 1} is induced strongly convex with respect to M. The set A is not induced 1-strongly convex with respect to M because –1 ∈ A and s ({f(-1}) ∩ M = 1 ∉ {-1} = f (-1). In view of remark 6.1.1. a), it follows that A is not induced k-strongly convex with respect to (s, f, M).

Also, from example 6.1.3, it results that if card A = k and A is induced strongly convex with respect to (s, f, M), it is not necessary that A is induced (k+1) – strongly convex with respect to (s, f, M).

Remark 6.1.2. 1) In the case of the classical convexity, the empty set is, by definition, a convex set. In the induced strong convexity, the condition that s (∅) ∩ M = ∅ is necessary for the induced strong convexity of ∅. A sufficient condition for s (∅) ∩ M = ∅ is s (∅) = ∅.

2) In the case of the classical convexity each singleton set is convex. To keep this property in the case of the notions defined in this section it is necessary that s ({a}) ∩ M ⊆ {a}, for each a ∈ A. A sufficient condition for this is s ({a}) = {a}, for each a ∈ A.

In what follows we shall give necessary, sufficient or necessary and sufficient conditions for keeping in the framework of the induced convexities, the most important properties of the classical convexity. It is well known that any straight-line segment is a convex set and the intersection of two convex sets is also convex. This does not remain valid in the case of the induced convexity.

Theorem 6.1.2. *If* f: X → Y *and* s: $2^Y \to 2^Y$ *are functions, s satisfied the conditions that*

(6.1.3) $\quad s(A) \subseteq s(B)$ *for any* A, B ∈ 2^Y *with* A ⊆ B

and

(6.1.4) $\quad s(s(B) \cap M) \subseteq s(B) \cap M$, *for each* B ∈ 2^Y,

then for any two points a *and* b *of* X *the set* $<a, b>_{s,f,M}$ *is induced seg-convex with respect to* (s, f, M).

Proof. Let a, b in X. Let c, d ∈ $<a, b>_{s,f,M} = f^{-1}(s(\{f(a), f(b)\}) \cap M)$. We have

$$f(c) \in s(\{f(a), f(b)\}) \cap M, \text{ and } f(d) \in s(\{f(a), f(b)\}) \cap M.$$

It results that s ({f (c), f (d)}) $\subseteq$ s (s ({f (a), f (b)}) $\cap$ M), and, in view of (6.1.4), it follows s ({f (c), f (d)}) $\cap$ M $\subseteq$ s ({f (a), f (b)}) $\cap$ M. This implies that

$$<c, d>_{s,f,M} = f^{-1}(s(\{f(c), f(d)\}) \cap M) \subseteq$$
$$\subseteq f^{-1}(s(\{f(a), f(b)\}) \cap M) = <a, b>_{s.f.M}.$$

Hence $<a, b>_{s,f,M}$ is induced seg-convex with respect to (s, f, M).

From theorem 6.1.1 and 6.1.2 we get

Corollary 6.1.1. *If* f: X → Y *is surjective and* s: $2^Y \to 2^Y$ *satisfies condition* (6.1.4), *then for any two points* a *and* b *of* X *the set* $<a, b>_{s,f,M}$ *is induced* 2-*strongly convex with respect to* (s, f, M).

Theorem 6.1.3. *If* X, Y, M *are nonempty sets, and* f: X → Y, s: $2^Y \to 2^Y$ *are functions, then the set* $G_{seg,s.f.M}$ *is a convexity space.*

Proof. It is easy to see that the sets ∅ and X are induced seg-convex with respect to (s, f, M). Let be the family $(A_j)_{j\in J}$, with $A_j \in G_{seg,s.f.M}$, and let a, b elements of $\bigcap_{j\in J} A_j$. Then a, b $\in A_j$, for all j ∈ J. As sets A_j, j ∈ J, are induced seg-convex with respect to (s, f, M), we have $<a, b>_{s,f,M} \subseteq A_j$, for each j ∈ J. Then $<a, b>_{s,f,M} \subseteq \bigcap_{j\in J} A_j$. Therefore $\bigcap_{j\in J} A_j$ is induced seg-convex with respect to (s, f, M).

Theorem 6.1.4. *If* X, Y, M *are nonempty sets, if* s: $2^Y \to 2^Y$ *is a function satisfying* (6.1.3), f: X → Y *is an injective function and* k *is a natural number,* k ≥ 2, *then the following sentences are true:*

i) *if* $(A_j)_{j\in J}$ *is a family of induced k-strongly convex sets, then*

$$\bigcap_{j\in J} A_j \in G_{st,k,s.f.M};$$

ii) *if* $(A_j)_{j\in J}$ *is a family of induced strongly convex sets, then*

$$\bigcap_{j\in J} A_j \in G_{st,s.f.M};$$

iii) *if* $(A_j)_{j\in J}$ *is a family of induced strict strongly convex sets, then*

$$\bigcap_{j\in J} A_j \in G_{sst,s.f.M}.$$

Proof. We prove sentence ii). By analogy, one can prove the other sentences. From hypothesis we have

(6.1.5) $s(f(A_j)) \cap M \subseteq f(A_j)$, for all $j \in J$.

As $\bigcap_{j\in J} A_j \subseteq A_k$, for all $k \in J$, we get

(6.1.6) $s(f(\bigcap_{j\in J} A_j)) \cap M \subseteq s(f(A_k)) \cap M$, for all $k \in J$.

From (6.1.5) and (6.1.6) it results

(6.1.7) $s(f(\bigcap_{j\in J} A_j)) \cap M \subseteq f(A_k)$ for each $k \in J$.

As f is an injective function, we have

(6.1.8) $f(\bigcap_{j\in J} A_j) = \bigcap_{j\in J} f(A_j)$.

From (6.1.7) and (6.1.8) it results $s(f(\bigcap_{j\in J} A_j) \cap M \subseteq f(\bigcap_{j\in J} A)$.

Therefore $\bigcap_{j\in J} A_j \in G_{st,s.f.M}$.

6.2 The element of (f, Y)-induced best approximation

In this paragraph we say that a subset A of X is induced seg-convex with respect to f if A is induced seg-convex with respect to s, f and M, where $M = Y$ and $s: 2^Y \to 2^Y$ is given by $s(C) = \text{conv}(C)$, for all $C \in 2^Y$. Obviously, set $A \subseteq X$ is seg-convex with respect to f if and only if for any $a, b \in A$ we have

$$<a, b> = \{x \in X \mid f(x) \in \{(1-t)\cdot f(a) + t\cdot f(b) \mid t \in [0, 1]\}\} \subseteq A.$$

Then, it is easy to see that $a \in <a, b>$, and $b \in <a, b>$, for any $a, b \in X$.

Let X be an arbitrarily set, $(Y,+,\cdot,\|\ \|)$, be an H-normed linear space and $f:X \to Y$ be an injective function. Let be $x^0 \in X$, and $A \subseteq X$, $A \neq \varnothing$.

Definition 6.2.1. *A point* $a^0 \in A$ *is said to be an element of* (f, Y)-*induced best approximation of* x^0 *by elements of* A *if*

(6.2.1) $\|f(a^0) - f(x^0)\| \leq \|f(a) - f(x^0)\|$, *for all* $a \in A$.

We recall that in the classical case, i.e. $X = \mathbf{R}^n$, if $A \subseteq \mathbf{R}^n$ is a convex set and if y^0 is a given point of $\mathbf{R}^n$, then there is at most one element of the best approximation (in classical sense) of y^0, by elements of A. In what follows, we show that this property of convex sets remains true under additional hypothesis, if the set A is not convex, but it is induced seg-convex.

Theorem 6.2.1. *If* x^0 *is a given point of* X, *and* A *is a nonempty induced seg-convex set with respect to* f *such that* f(A) *is a convex set in* Y, *then there exists at most one element of* (f, Y)*-induced best approximation of* x^0 *by elements of* A.

Proof. Two cases may arise:

Case 1: $x^0 \in A$. Then, from the injectivity of f, it follows that x^0 is the single (f, Y)-induced best approximation point of x^0, by elements of A.

Case 2: $x^0 \notin A$. We suppose that there are at least two elements a^0 and a of (f, Y)-induced best approximation of x^0, by elements of A. Then we have

$$(6.2.2) \qquad \|f(a) - f(x^0)\| = \|f(a^0) - f(x^0)\| \neq 0.$$

Because f is an injective function, $f(a) \neq f(a^0)$. Then

$$(6.2.3) \qquad \{f(a), f(a^0)\} \subset \langle f(a), f(a^0)\rangle \subseteq f(A).$$

It follows that there is $\gamma \in f(A) \setminus \{f(a), f(a^0)\}$, such that

$$(6.2.4) \qquad \gamma = \frac{1}{2} f(a) + \frac{1}{2} f(a^0).$$

Since the set A is induced seg-convex with respect to f, we get

$$(6.2.5) \qquad f^{-1}(\gamma) \in A.$$

Then there is $c \in A$, such that

$$(6.2.6) \qquad \gamma = f(c) = \frac{1}{2} f(a^0) + \frac{1}{2} f(a).$$

From (6.2.4) and (6.2.5) we have

$$\|f(c) - f(x^0)\|^2 = \frac{1}{4}\|(f(a^0) - f(x^0)) + (f(a) - f(x^0))\|^2 <$$

$$< \frac{1}{4}(\|(f(a^0) - f(x^0)) + (f(a) - f(x^0))\|^2 +$$

$$+ \|(f(a^0) - f(x^0)) - (f(a) - f(x^0))\|^2$$

Applying the parallelogram equality for the H-norm we get

$$\|(f(a^0) - f(x^0)) + (f(a) - f(x^0))\|^2 + \|(f(a^0) - f(x^0)) - (f(a) - f(x^0))\|^2 =$$

$$= 2\,\|f(a^0) - f(x^0)\|^2 + 2\|f(a) - f(x^0)\|^2.$$

Then $||f(c) - f(x^0)|| < ||f(a^0) - f(x^0)||$. This contradicts the fact that a^0 is an element of the (f, Y)-induced best approximation of x^0, by elements of A.

Remark 6.2.1. If f (A) is a not convex set in Y, then the conclusion of theorem 6.2.1 is not always true.

Example 6.2.1. Let be $X = \{(0, 1), (0, 0), (1, 0)\}$, $Y = \mathbf{R}^2$, $x^0 = (0, 0)$, $A = \{(0, 1), (1, 0)\}$. If we take $f: X \to Y$, $f(x_1, x_2) = (x_1, x_2)$, for each $(x_1, x_2) \in X$, then it is easy to see that set A is induced seg-convex with respect to f. But we have that (0, 1), and (1, 0), are, in the same time, elements of the (f,Y)-induced best approximation of x^0 by elements of A.

Lemma 6.2.1. *If the function* $f: X \to Y$ *is injective, if the set* f (X) *is convex (in classical sense) and if set* A *is induced seg-convex with respect to* f, *then set* f (A) *is also convex (in classical sense).*

Proof. Let be $u', u'' \in f(A)$. There exist $a', a'' \in A$, such that $u' = f(a')$, and $u'' = f(a'')$. Let $t \in]0, 1[$. Because set f (X) is convex, we have

$$u = t u' + (1-t) u'' \in f(X).$$

From the induced convexity of A, it follows that

$$f^{-1}(u) \in \langle a', a'' \rangle \subseteq A.$$

Then there is $a \in A$, such that $f(a) = u$. We get that

$$u = t f(a') + (1 - t) f(a'') \in f(A).$$

Because t is arbitrarily chosen in]0, 1[, it results that f (A) is a convex set in Y.

From theorem 6.2.1 and lemma 6.2.1 we get:

Corollary 6.2.1. *If the function* $f: X \to Y$ *is injective,* f (X) *is a convex set,* x^0 *is a given point of* X, *and* A *is a nonempty induced seg-convex set with respect to* f, *then there is at most one element of the* (f, Y)*-induced best approximation of* x^0 *by elements of* A.

Theorem 6.2.2. *If the function* $f: X \to Y$ *is injective, if* x^0 *is a given point of* X, *and* A *is a nonempty induced seg-convex set with respect to* f, *then the set* $A(x^0)$ *of all elements of the* (f, Y)*-induced best approximation of* x^0 *by elements of* A *is also induced seg-convex with respect to* f.

Proof. Two cases are possible:

Case 1. $\operatorname{card}(A(x^0)) \in \{0, 1\}$; obviously $A(x^0)$ is an induced seg-convex set.

Case 2. $\operatorname{card}(A(x^0)) > 1$. Let be $a', a'' \in A(x^0)$, and

$$\lambda = ||f(x^0) - f(a')|| = ||f(x^0) - f(a'')||.$$

Let be $a \in f^{-1}(<f(a'), f(a'')>)$. Then there is $t \in [0, 1]$ such that $f(a) = tf(a') + (1 - t)f(a'')$. We have

$$\|f(x^0) - f(a)\| = \|t(f(x^0) - f(a')) + (1 - t)(f(x^0) - f(a''))\| \leq$$
$$\leq t\|f(x^0) - f(a')\| + (1 - t)\|f(x^0) - f(a'')\| = \lambda.$$

On the other hand, we have

$$\|f(x^0) - f(a)\| \geq \lambda.$$

It follows that $\|f(x^0) - f(a)\| = \lambda$. Then $a \in A(x^0)$. Hence the set $A(x^0)$ is induced seg-convex with respect to f.

Chapter 7

Convexity defined by means of given function

The last few years' mathematical literature contains more definitions of new convexity properties of sets of points in $\mathbf{R}^n$. Some of them are in connection with applications in discrete sets, in metric spaces, or in various types of Cartesian products of sets, related to multiple criteria programming problems. We recall some of them because they will be referred to in our study.

7.1 (φ, ψ) – convex sets

An interesting generalization of the convex sets is given by E.A. Youness (1999). It is the starting point of the results from this chapter.

Definition 7.1.1. *A set* $A \subset \mathbf{R}^n$ *is said to be* E-*convex iff there is a map* $E: \mathbf{R}^n \to \mathbf{R}^n$ *such that* $(1-\lambda)\cdot E(x) + \lambda\cdot E(y) \in A$, *for each* $x, y \in A$ *and* $0 \le \lambda \le 1$.

We remark that this notion is very large because every nonempty set A, $A \subseteq \mathbf{R}^n$, is E-convex if we take $E: \mathbf{R}^n \to \mathbf{R}^n$, $E(x) = a$, for each $x \in A$, where a is a given element of A. Interesting is the case when E is a given, not constant function. For this reason, in what follows, we suppose that the function $E: \mathbf{R}^n \to \mathbf{R}^n$ is given and we give another definition for E – convex sets.

Definition 7.1.1'. *The set* $A \subseteq \mathbf{R}^n$ *is called* E – *convex if for each* $x, y \in A$ *and* $0 \le \lambda \le 1$, *the relation* $(1-\lambda)\cdot E(x) + \lambda\cdot E(y) \in A$ *occurs*.

D. Duca, E. Duca, L. Lupşa and L. R. Blaga (2000), gave other four types of convexity using two given functions and investigating some properties of these classes of sets. They might have applications in vectorial programming, their applications being the subject of further research.

Let $\varphi: \mathbf{R}^n \to \mathbf{R}^n$ and $\psi: \mathbf{R}^n \to \mathbf{R}^n$ two given functions.

Definition 7.1.2. *A subset* A *of* $\mathbf{R}^n$ *is said to be* (φ, ψ) – *convex if we have* $(1-t)\cdot\varphi(x) + t\cdot\psi(y) \in A$, *for each* $x, y \in A$ *and each* $t \in [0, 1]$.

Let m and n be two natural non-null numbers and let $\varphi: X \subseteq \mathbf{R}^m \to \mathbf{R}^m$, and $\psi: Y \subseteq \mathbf{R}^n \to \mathbf{R}^n$, two given non-constant functions.

Definition 7.1.3. i) *A set* $A \subset X \times Y$ *is said to be* bi-(φ, ψ) - *convex, if for all* (x, y), (x, v), (u, y) *elements of* A *and for each* $t \in [0, 1]$ *we have both*

$$(\varphi(x), (1-t)\,\psi(y) + t\,\psi(v)) \in A$$

and

$$((1-t)\,\varphi(x) + t\,\varphi(u), \psi(y)) \in A.$$

ii) *A set* $A \subset X \times Y$ *is said to be* bi-$(\varphi, 1)$ - *convex if for each* $y \in Y$, x' *and* $x'' \in X$ *such that* $(x', y) \in A$ *and* $(x'', y) \in A$ *and each* $t \in [0, 1]$

$$((1-t)\,\varphi(x') + t\,\varphi(x''), y) \in A.$$

iii) *A set* $A \subset X \times Y$ *is said to be* bi-$(1, \psi)$ - *convex if for each* $x \in X$, y' *and* $y'' \in Y$, *such that* $(x, y') \in A$ *and* $(x, y'') \in A$, *and each* $t \in [0, 1]$ *we have*

$$(x, (1-t)\,\psi(y') + t\,\psi(y'')) \in A.$$

Remark 7.1.1. i) If $A \subseteq X \times Y$ is bi-(φ, ψ) – convex, then for each $(x, y) \in A$ we have $(\varphi(x), \psi(y)) \in A$.

ii) If $A \subseteq X \times Y$ is bi-$(1, \psi)$ – convex and bi-$(\varphi, 1)$ – convex, then set A is bi-(φ, ψ) – convex. Generally, the converse is not true. But if $A = A_1 \times A_2 \subseteq X \times Y$ and set A is bi-(φ, ψ) – convex, then A is bi-$(1, \psi)$ – convex and bi-$(\varphi, 1)$ – convex.

iii) If $A \subseteq X \times Y$ is an E-convex set, where $E = (E_1, E_2): \mathbf{R}^m \times \mathbf{R}^n \to \mathbf{R}^m \times \mathbf{R}^n$, is a given function, set A is bi-(φ, ψ) - convex, with $\varphi = E_1$, and $\psi = E_2$. The converse is not true in general, but if $A = A_1 \times A_2 \subseteq X \times Y$ is bi-(φ, ψ)-convex, then it is E-convex, where $E: \mathbf{R}^m \times \mathbf{R}^n \to \mathbf{R}^m \times \mathbf{R}^n$, is given by

$$E(x, y) = \begin{cases} (\varphi(x), \psi(y)), & (x, y) \in X \times Y \\ (0_m, 0_n), & (x, y) \in R^{m+n} \setminus (X \times Y) \end{cases}$$

iv) Each classically convex subset $A \subseteq \mathbf{R}^{m+n}$ is bi-$(1_{R^m}, 1_{R^n})$ - convex.

7.2 (k, g, h, M) – convex sets

In this paragraph a new type of convexity, namely (k, g, h, M) - convexity, is introduced. The classical notion of convex set (see for example W.W. Breckner (1974), H.C. Eggleston (1959), J. Stoer and C. Witzgall (1970), F. Valentine (1964)), the notion of convex set with respect to a given set (see L. Blaga and L. Lupşa (1998), L. Lupşa (1986) [163]), the notion of E-convex set (see E.A. Youness (1999)) and the notions of (φ, ψ) – convexity and bi -(φ, ψ) – convexity (see L. Lupşa, D. Duca, E. Duca, and L. Blaga (2000)) result (all this notions) as particular cases of the new type of convexity. The set of (k, g, h, M) - convex sets strictly includes the above mentioned sets. This will be proved by means of examples. Some properties of the class of (k, g, h, M) – convex sets are studied.

Notation. Let n be a natural number and $B = \{0, 1\}$. If $h = (h_1, ..., h_n) \in B^n$, we put $|h| = h_1 +...+h_n$. Let be the function $\Lambda = (\Lambda_1, ..., \Lambda_n)$: $\mathbf{R}^n \to \mathbf{R}^n$, given, by

$$\Lambda_i(x) = \begin{cases} x_{n-|h|+\sum_{k=1}^{i} h_k}, & \text{if } h_i = 1 \\ x_1, & \text{if } h_i = 0 \text{ and } i = 1 \\ x_{i-\sum_{k=1}^{i-1} h_k}, & \text{if } h_i = 0 \text{ and } i > 1 \end{cases}$$

for each $i \in \{1, ..., n\}$ and for all $x = (x_1, ..., x_n) \in \mathbf{R}^n$.

Let $A \subseteq \mathbf{R}^n$ and $h = (h_1,...,h_n) \in B^n$. We denote

$$pr(h,A) = \begin{cases} A, \text{if } |h| = 0 \\ \{y \in R^{n-|h|} \mid (\exists) z \in R^{|h|} \text{ with } \Lambda(y,z) \in A\}, \text{if } 0 < |h| < n \\ \varnothing, \text{if } |h| = n \end{cases}$$

and, if $|h| \neq n$, for each $y \in pr(h, A)$, we put

$$s(y, h, A) = \begin{cases} \varnothing, \text{if } |h| = 0 \\ \{z \in R^{|h|} \mid \Lambda(y,z) \in A\}, \text{if } 0 < |h| < n. \end{cases}$$

Example 7.2.1. Let n = 3, $h = (1, 0, 1) \in B^3$, and

$$A = \{(x_1, x_2, x_3) \in \mathbf{R}^3 \mid x_1^2 + x_2^2 \le x_3^2, \quad 0 \le x_3 \le 1\}.$$

Then $|h| = 2$ and the function Λ: $\mathbf{R}^3 \to \mathbf{R}^3$, is given by

$$\Lambda(x_1, x_2, x_3) = (x_2, x_1, x_3), \text{ for all } (x_1, x_2, x_3) \in \mathbf{R}^3.$$

We have

$$\mathrm{pr}(h, A) = \{y \in \mathbf{R} \mid (\exists)\, z = (z_1, z_2) \in \mathbf{R}^2 \text{ such that } (z_1, y, z_2) \in A\} =$$
$$= [-1, 1],$$

and, for each $y \in [-1, 1]$,

$$s(y, h, A) = \{z = (z_1, z_2) \in \mathbf{R}^2 \mid z_1^2 + z_2^2 \geq y^2,\ 0 \leq z_2 \leq 1\}.$$

Remark 7.2.1. i) For any $h \in B^n$, the function $\Lambda: \mathbf{R}^n \to \mathbf{R}^n$ is bijective.

ii) If $h = 0_n$, or $h = e_n$, then $\Lambda = 1_{\mathbf{R}^n}$.

iii) For any $h \in B^n$, if $A \subseteq \mathbf{R}^n$ and $D \subseteq A$, then we have

(7.2.2) $\quad \mathrm{pr}(h, D) \subseteq \mathrm{pr}(h, A)$,

and, for each $y \in \mathrm{pr}(h, D)$,

(7.2.3) $\quad s(y, h, D) \subseteq s(y, h, A)$.

iv) For any $h \in B^n$, if A and B are nonempty subsets of $\mathbf{R}^n$, with $A \cap B \neq \emptyset$, then, for each $y \in \mathrm{pr}\,(h, A \cap B)$, we have

(7.2.4) $\quad s(y, h, A \cap B) = s(y, h, A) \cap s(y, h, B)$.

If n, and k are natural numbers and $C \subseteq \mathbf{R}^n$, by conv (k, C) we denote the set of all convex combinations of k elements of C, i.e.

$$\mathrm{conv}\,(k, C) = \{\sum_{i=1}^{k} t_i c^i \mid t_i \geq 0, (\forall)\, i \in \{1,\ldots,k\}, \sum_{i=1}^{k} t_i = 1\}.$$

Let n and k be natural numbers, $h \in B^n \setminus 0_n$, $g: \mathbf{R}^{|h|} \to \mathbf{R}^{|h|}$ a function and let $M \subseteq \mathbf{R}^n$, $M \neq \emptyset$.

Definition 7.2.1. *A set* $A \subseteq R^n$ *is said to be* (k, g, h, M) – *convex if at least one of the following conditions is satisfied*:

i) $\mathrm{pr}\,(h, A) \neq \emptyset$, *and*

$$M \cap \{\Lambda\,(y, z) \mid z \in \mathrm{conv}\,(k, g\,(s\,(y, h, A)))\} \subseteq \{\Lambda\,(y, z) \mid z \in s\,(y, h, A)\},$$

for all $y \in \mathrm{pr}\,(h, A)$;

ii) $\mathrm{pr}\,(h, A) = \emptyset$, *and* $M \cap \mathrm{conv}\,(k, g\,(A)) \subseteq A$.

Remark 7.2.2. Let n and k be natural numbers, $A \subseteq \mathbf{R}^n$, let M be a nonempty subset of $\mathbf{R}^n$, and let E: $\mathbf{R}^n \to \mathbf{R}^n$, be a function.

i) The set A is convex in the classical sense if and only if it is $(2, 1_{\mathbf{R}^n}, e_{\mathbf{R}^n}$ $\mathbf{R}^n)$ – convex;

ii) The set A is E-convex (in sense of definition 7.1.1') if and only if it is (2, E, e_{R^n} $\mathbf{R}^n$) – convex.

iii) The set A is k-convex with respect to M (see L. Blaga, L. Lupşa (1998)), if and only if it is (k, 1_{R^n} , e_{R^n} , M) - convex.

iv) If r and s are non-null natural numbers, with r + s = n, and a function $\varphi: \mathbf{R}^s \to \mathbf{R}^s$, then set A is ($\varphi$, 1) - convex (see definition 7.1.3, ii)) if and only if it is (2, g, e_{R^n} $\mathbf{R}^n$) – convex, where g = (φ, 1_{R^s}).

vi) If r and s are non-null natural numbers, with r + s = n, and $\psi: \mathbf{R}^r \to \mathbf{R}^r$, then set A is (1, ψ) - convex (see definition 7.1.3, iii)) if and only if it is (2, g, e_{R^n} $\mathbf{R}^n$) – convex, where g = (1_{R^r} ,ψ).

We remark that the set of (k, g, h, M) - convex sets strictly includes the set of classically convex sets, the set of convex sets with respect to a given set and the set of E-convex sets.

Example 7.2.2. Let A = [0, 1] × [0, 1] ∪ [1, 2] × [2, 3], and h = (1, 0) ∈ B^2. We have

$$\text{pr}\,((1,0), A) = [0, 1] \cup [2, 3], \text{ and } s\,(y, (1,0), A) = \begin{cases} [0,1], & \text{if } y \in [0,1] \\ [1,2], & \text{if } y \in [2,3] \end{cases}.$$

Because for each y ∈ pr ((1,0), A) the set s (y, (1, 0), A) is convex, then, if we take g = 1_{R^2} , we get

$$\mathbf{R}^2 \cap \text{conv}\,(2, g\,(s\,(y, (1,0), A))) = s\,(y, (1,0), A) \subseteq A.$$

Hence A is (2, 1_{R^2} ,(1, 0), $\mathbf{R}^2$) – convex, but it is not a convex set (in the classical sense) in $\mathbf{R}^2$. We remark that the set A is not a connected set. The set A is not (2,1_{R^2} , (0,1), $\mathbf{R}^2$) – convex (we have

$$\text{pr}\,((0,1), A) = [0,2],\ 1 \in [0, 2],\ s\,(1, (0,1), A) = [0, 1] \cup [2, 3];$$

hence

$$(1, \frac{3}{2}) \in M \cap \{(1, z) \mid z \in \text{conv}\,(2, g\,(s\,(1, (0,1), A)))\},$$

but $(1, \frac{3}{2}) \notin A$).

Example 7.2.3. Let A = [-2, 1] × [-2, -1] ∩ [1, 2] × [1, 2]. It is easy to see that set A is (2, 1_{R^2} , (1, 0), $\mathbf{R}^2$) – convex and (2, 1_{R^2} , (0, 1), $\mathbf{R}^2$) - convex. It is not convex in the classical sense. More, there is no linear function $E:\mathbf{R}^2 \to \mathbf{R}^2$, such that set A is $(2, E, (1, 1), \mathbf{R}^2)$ – convex. We prove this. Suppose that set A is $(2, E, (1, 1), \mathbf{R}^2)$ – convex. Then, in view of remark 7.2.1, we have

$$\mathbf{R}^2 \cap \text{conv}\,(2, E\,(A)) = \text{conv}\,(E\,(A)) \subseteq A.$$

As the set conv (E (A)) is convex, we get that

$$(7.2.5) \quad E\,(A) \subseteq \text{conv}\,(E\,(A)) \subseteq [1, 2] \times [1, 2],$$

or

$$(7.2.6) \quad E\,(A) \subseteq \text{conv}\,(E\,(A)) \subseteq [-2, -1] \times [-2, -1].$$

Let

$$(7.2.7)\; E\,(x_1, x_2) = (E_1\,(x_1, x_2), E_2\,(x_1, x_2)) = (ax_1 + bx_2, cx_1 + dx_2),$$

for all $x = (x_1, x_2) \in \mathbf{R}^2$. If we have (7.2.3), then

$$1 \le E_1\,(-2, -2) \le 2 \text{ and } 1 \le E_2\,(2, 2) \le 2.$$

But there are no real numbers a and b such that

$$1 \le -2a - 2b \le 2 \text{ and } 1 \le 2a + 2b \le 2.$$

A similar situation is in the second case.

Example 7.2.4. Set $A = \{(0, 0), (2, 2)\} \subseteq \mathbf{R}^2$ is not 2-convex with respect to $\mathbf{Z}^2$, but it is $(2, 1_{\mathbf{R}}, (1, 0), \mathbf{Z}^2)$ - convex. We have

$$\text{pr}\,((1, 0), A) = \{0, 2\},$$
$$g\,(s\,(0, (1, 0), A)) = 1_{\mathbf{R}}(\{0\}) = \{0\},$$
$$g\,(s\,(2, (1, 0), A)) = 1_{\mathbf{R}}(\{2\}) = \{2\};$$

because

$$\mathbf{Z}^2 \cap \text{conv}\,(2, g\,(s\,(0, (1, 0), A)) = \{0\} \subseteq A,$$

and

$$\mathbf{Z}^2 \cap \text{conv}\,(2, g\,(s\,(2, (1, 0), A)) = \{2\} \subseteq A,$$

it follows that set A is $(2, 1_{\mathbf{R}}, (1,0), \mathbf{Z}^2)$ - convex. Set A is not $(2, 1_{\mathbf{R}}, (1, 1), \mathbf{Z}^2)$ - convex. We have $\text{pr}\,((1, 1), A) = \varnothing$, then set A is $(2, 1_{\mathbf{R}}, (1, 1), \mathbf{Z}^2)$ – convex if and only if $\mathbf{Z}^2 \cap \text{conv}\,(2, g(A)) \subseteq A$. But

$$\mathbf{Z}^2 \cap \text{conv}\,(2, g\,(A)) = \mathbf{Z}^2 \cap \{(1-t)\cdot(0, 0) + t\cdot(2, 2) \mid t \in [0, 1]\} =$$
$$= \{(0, 0), (1, 1), (2, 2)\} \not\subseteq A.$$

Therefore, set A is not $(2, 1_{\mathbf{R}}, (1, 1), \mathbf{Z}^2)$ - convex.

Let n, k and m be natural numbers and u, v $\in B^n$. Let us take two functions g: $\mathbf{R}^{|h|} \to \mathbf{R}^{|h|}$, and f: $\mathbf{R}^{|h|} \to \mathbf{R}^{|h|}$ and let M, and L be nonempty subsets of $\mathbf{R}^n$.

Theorem 7.2.1. *If* A *is a nonempty subset of* $\mathbf{R}^n$, $k < m$ *and* $L \subseteq M$, *then the following propositions are true*:

i) *If* A *is* (m, g, u, M) – *convex, then* A *is also* (k, g, u, M) – *convex.*

ii) *If set* A *is* (k, g, u, M) – *convex and* (k, f, u, $\mathbf{R}^n$) – *convex, then set* A *is* (k, $g \circ f$, u, M) – *convex.*

iii) *If set* A *is* (m, g, u, M) – convex, *then* A *is also* (k, g, u, L) – *convex.*

Proof. i) Because $k < m$, for each set $D \subseteq \mathbf{R}^n$, we have conv (k, D) $\subseteq$ conv (m, D); in view of remark 7.2.1 it follows that the set A is (k, g, u, M)–convex.

ii) Two cases may arise:

Case 1: $|u| = n$. Because A is (k, f, u, $\mathbf{R}^n$) – convex, we have conv (k, f(A)) $\subseteq$ A. This implies that $M \cap \text{conv}(k, (g \circ f), A) \subseteq M \cap \text{conv}(k, g(A)) \subseteq A$.

Case 2: $0 < |u| < n$. Let $y \in \text{pr}(h, A)$. Since

$$f(s(y, u, A)) \subseteq \text{conv}(k, f(s(y, u, A))) \subseteq s(y, u, A),$$

we get

$$M \cap \Lambda(y, \text{conv}(k, (g \circ f)(s(y, u, A))) \subseteq$$
$$\subseteq M \cap \Lambda(y, \text{conv}(k, g(s(y, u, A))) \subseteq A.$$

iii) As $L \cap U \subseteq M \cap U$, from any subset U of $\mathbf{R}^n$, it follows that the conclusion of proposition iii) is true.

In general, the converse of propositions from theorem 7.2.1 is not true.

Theorem 7.2.2. *If* $h \in B^n \setminus \{1_n, e_n\}$, *then the set* $G_{(k,\, g,\, h,\, M)}$, *of all subsets of* $\mathbf{R}^n$ *which are* (k, g, h, M) – *convex, is a convexity space.*

Proof. As pr (h, ∅) = ∅ and $M \cap \text{conv}(k, g, h, \varnothing) \subseteq \varnothing$, it follows that $\varnothing \in G_{(k,\, g,\, h,\, M)}$. Since pr (h, $\mathbf{R}^n$) = $\mathbf{R}^{n-|h|}$, and s (y, h, $\mathbf{R}^n$) = $\mathbf{R}^{|h|}$ then for each point $y \in \mathbf{R}^{n-|h|}$, we get

$$M \cap \{\Lambda(y, z) \mid z \in \text{conv}(k, g(s(y, h, \mathbf{R}^n)))\} \subseteq$$
$$\subseteq \{\Lambda(y, z) \mid z \in s(y, h, \mathbf{R}^n)\},$$

for all $y \in \text{pr}(h, \mathbf{R}^n)$. Then $\mathbf{R}^n \in G_{(k,\, g,\, h,\, M)}$.

Let now A and B be two nonempty sets, elements of $G_{(k,\, g,\, h,\, M)}$. Two cases may arise:

$A \cap B = \varnothing$; then obviously we have $A \cap B \in G_{(k,\, g,\, h,\, M)}$.

$A \cap B \neq \varnothing$; then $pr\,(h, A \cap B) \neq \varnothing$.

Let

(7.2.8) $\quad y \in pr\,(h, A \cap B)$.

We can prove that

$$M \cap \{\Lambda\,(y, z) \mid z \in conv\,(k, g\,(s\,(y, h, A \cap B)))\} \subseteq$$
$$\subseteq \{\Lambda\,(y, z) \mid z \in s\,(y, h, A \cap B)\}.$$

Let $w \in M \cap \{\Lambda\,(y, z) \mid z \in conv\,(k, g\,(s\,(y, h, A \cap B)))\}$. Then

(7.2.9) $\quad w \in M$

and then there are k non negative real numbers $t_1, ..., t_k$, with $t_1 + ... + t_k = 1$, and k elements $z^1, ..., z^k$ of $s\,(y, h, A \cap B)$ such that

(7.2.10) $\quad w = \Lambda\,(y, \sum_{j=1}^{k} t_j g(z^j))$.

In view of remark 7.2.1 we get $z^j \in s\,(y, h, A)$ and $z^j \in s\,(y, h, B)$, for each $j \in \{1, ..., k\}$. Then $w \in M \cap \{\Lambda\,(y, z) \mid z \in conv\,(k, g\,(s\,(y, h, A)))\}$ and $w \in M \cap \{\Lambda\,(y, z) \mid z \in conv\,(k, g\,(s\,(y, h, B)))\}$. As $A \in G_{(k,\, g,\, h,\, M)}$, and $B \in G_{(k,\, g,\, h,\, M)}$, from (7.2.9) and (7.2.10), it follows that there are the points $u \in s\,(y, h, A)$ and $v \in s\,(y, h, B)$ such that

(7.2.11) $\quad w = \Lambda\,(y, u)$ and $w = \Lambda\,(y, v)$.

Since Λ is a bijective function, (7.2.11) implies

(7.2.12) $\quad u = v$.

On the other hand we have:

(7.2.13) $\quad (y, u) \in A$ and $(y, v) \in B$.

From (7.2.12) and (7.2.13), it follows that

$$(y, u) = (y, v) \in A \cap B.$$

Hence, taking into account (7.2.8), we obtain

$$w = \Lambda\,(y, u) \in \{\Lambda\,(y, z) \mid z \in s\,(y, h, A \cap B)\}.$$

In view of definition 2.8.1, we conclude that set $G_{(k,\, g,\, h,\, M)}$, of all subsets of $\mathbf{R}^n$ which are (k, g, h, M) – convex, is a convexity space. .

7.3 (g, h, M) – convex sets

Let n be a non-null natural number, $h \in B^n \setminus \{0_n\}$, let g: $\mathbf{R}^{|h|} \to \mathbf{R}^{|h|}$, be a given function and let $M \subseteq \mathbf{R}^n$, $M \neq \emptyset$.

Definition 7.3.1. *A set* $A \subseteq \mathbf{R}^n$ *is said to be* (g, h, M) – *convex if at least one of the following conditions is satisfied:*

i) $pr(h,A) \neq \emptyset$ *and*

$$M \cap \{\Lambda(y, z) \mid z \in conv(g(s(y, h, A)))\} \subseteq$$
$$\{\Lambda(y, z) \mid z \in s(y, h, A)\}, \textit{for all } y \in pr(h,A);$$

or

ii) $pr(h, A) = \emptyset$, *and* $M \cap conv(g(A)) \subseteq A$.

Remark 7.3.1. Let n be a natural number, $A \subseteq \mathbf{R}^n$, M a nonempty subset of $\mathbf{R}^n$ and let E: $\mathbf{R}^n \to \mathbf{R}^n$ be a function. The following propositions are true:

i) set A is convex in the classical sense if and only if it is $(1_{R^n}, e_{R^n}, \mathbf{R}^n)$ – convex.

ii) set A is E-convex, in the sense of definition 7.1.1', if and only if it is $(E, e_{R^n}, \mathbf{R}^n)$ – convex.

iii) set A is strongly convex with respect to M if and only if it is $(1_{R^n}, e_{R^n}, M)$ – convex.

We remark that a set $A \subseteq \mathbf{R}^n$ is $(g, e_{R^n}, \mathbf{R}^n)$ convex if and only if $A = \emptyset$, or $A \neq \emptyset$, and, in this case we have $conv(g(A)) \subseteq A$, i.e. A is E-convex in the sense of definition 7.1.1', if we take E = g.

Some interesting properties of (g, h, M) convex sets are spotlight in the following propositions

Theorem 7.3.1. *If* $A \subseteq \mathbf{R}^n$ *is a nonempty* $(g, e_{R^n}, \mathbf{R}^n)$ – *convex set, and if* $x^0 \in g(A)$ *is an element of the best approximation of* $y^0 \in \mathbf{R}^n$ *by elements of* A, *then* x^0 *is the unique element of the best approximation of* y^0 *by elements of* g (A).

Theorem 7.3.2. *If* $A \subseteq \mathbf{R}^n$ *is a* (g, e_{R^n}, M) – *convex set and if* $x^0 \in A$ $(g(A) \cap int\, M)$ *is an element of the best approximation of* $y^0 \in (M \setminus A)$ *by elements of* A, *then* x^0 *is the single element of the best approximation of* y^0 *by elements of* g (A).

Theorem 7.3.3. *If* $A \subseteq \mathbf{R}^n$ *is a* (g, h, M) – *convex set and* $x^0 \in \mathbf{R}^n$, *then the set* $A(x^0)$ *of all the elements of the best approximation of* x^0 *by elements of* A *is also a* (g, h, M) – *convex set.*

In order to prove these theorems see D. I. Duca and L. Lupşa (2001).

Chapter 8

Classification of the convexity properties

During the last ten years a tendency of "organising" the wide range of convexity properties for sets and for functions appeared in the mathematical literature. For example, the great collection of P. M. Gruber and J. M. Willis (1993) refers to 21 types of convexity for various types of sets. I. Singer (1997) writes about 26 types of convexity for sets, trying to obtain a classification of them according to the principle of definition and of some properties. The thesis of J. M. Chassery (1984 [43]) refers to 6 types of discrete convexity properties. Our own book, G. Cristescu (2000 [71]), contains 19 notions of convexity for sets, trying to build a unifying theory of most of them. Other references and studies on more types of convexities or concavities for sets are in the books published or coordinated by S. Schaible (1981), M. Avriel et al. (1988) and V.P. Soltan (1984).

This is the reason why we made a collection of 100 convexities for sets and we made a thorough study of their definitions in order to identify the hidden reason of this diversity of properties of convexity for sets. We obtained a method of constructing definitions of convexity for sets that has led both to a general unifying approach of this domain and to a classification of the convexities. Our definition, itself, will provide us with the classification criteria. On the other hand, this general notion of convexity property for sets provides the reader with a method of obtaining new convexities, with new properties, enlarging the domain in which the tools associated with convexity might be used. Of course, this chapter opens the problem of improving our classification and also the problem of completing the collection with missing pieces. A better classification is always welcome.

First of all we shall describe the main elements identified by us to be key points in a definition of the notion of convexity for sets. At the same time, we shall present the conventions established in order to elaborate this very general theory of convexity. After that, the main definitions will be formulated and the classification of the 100 convexity properties will be presented, by means of the elements of each class.

8.1 The main elements and language conventions.

The key elements that are involved in the process of the elaboration of a convexity theory for sets are:

1. A nonempty set X, which is the support set.
2. A nonempty set Y.
3. A nonempty subset $S \subseteq 2^X$.
4. A family S of functions s: $S \to 2^Y$.
5. A family R of functions r: $2^X \to 2^Y$.

The elements of the subset S of the set of the parts of X will replace the pairs of points which appear in the definition of the classical convexity. Function s: $S \to 2^Y$ will replace the straight-line segment determined by a pair of elements, which appear in the definition of the classical convexity. Function r: $2^X \to 2^Y$ will be used to rewrite the condition that every straight-line segment determined by pairs of elements of the set under consideration belong to this set. In the case of the classical convexity the family R is singleton and its unique element is the identity on 2^X.

It is known that there are convexity properties that, apparently, do not use the straight-line segment in their definition (for example, the H-convexity). But all the convexity properties from our collection may be modelled using the above-described elements.

The study of all the convexity properties from our collection showed that all of them may be obtained if these elements appear in the general definitions in the order (S, S, R), together with the following terminology generated by the manner in which they are used:

1. $\exists\, C \in S$ will determine a *partial* property of convexity.

2. $\forall\, C \in S$ will determine a *strong* property of convexity, named simply, *convexity*.

3. $\exists\, s \in \mathsf{S}$ will be denoted by the letter e (from exist) on the first position of a pair of symbols, (e, ...).

4. ∀ s ∈ S will be denoted by the letter a (from all) on the first position of a pair of symbols, (a, ...).

5. ∃ r ∈ R will be denoted by the letter e (from exist) on the second position of a pair of symbols, (..., e).

6. ∀ r ∈ R will be denoted by the letter a (from all) on the second position of a pair of symbols, (..., a).

Condition s (C) ⊆ r (A) is necessary to be satisfied in every definition of a convexity property.

Table 8.1.1 contains all the distinct combinations of these elements, together with the terminology generated by them.

C ∈ S	s ∈ S	r ∈ R	Condition	Name
∃	∃	∀	s(C) ⊆ r(A)	partial (e,a) - ((S,S), R) convexity
∃	∃	∃	s(C) ⊆ r(A)	partial (e,e) - ((S,S), R) convexity
∃	∀	∃	s(C) ⊆ r(A)	partial (a,e) - ((S,S), R) convexity
∃	∀	∀	s(C) ⊆ r(A)	partial (a,a) - ((S,S), R) convexity
∀	∃	∀	s(C) ⊆ r(A)	(e,a) - ((S,S), R) convexity
∀	∃	∃	s(C) ⊆ r(A)	(e,e) - ((S,S), R) convexity
∀	∀	∃	s(C) ⊆ r(A)	(a,e) - ((S,S), R) convexity
∀	∀	∀	s(C) ⊆ r(A)	(a,a) - ((S,S), R) convexity

Table 8.1.1. Terminology for generalised convexities

There are convexity properties that cannot be obtained by means of this scheme. But they can be obtained if the order in which the elements are used is changed, namely (S, R, S). These types of properties will be called converted convexities. The same combinations as in Table 8.1.1 are possible, but some of them are identical to some of that table. This happens due to the fact that (∃ s ∈ S and ∃ r ∈ R) is identical to (∃ r ∈ R and ∃ s ∈ S) and (∀ s ∈ S and ∀ r ∈ R) is identical to (∀ r ∈ R and ∀ s ∈ S). Hence, the distinct possibilities of obtaining converted convexities are less numerous than the previous ones. In the case of the converted convexity properties, when we shall write (e, a) we shall read "for each (= <u>a</u>ll) r ∈ R there is (= <u>e</u>xists) s ∈ S ", meaning that, in the terminology of the notion we shall keep the same order (S, S, R) as in all the convexity notions. But the term "converted" will mean the change of the order between the second and the third element in the definition, but not in the name of the property. Table 8.1.2 contains all the possible combinations leading to converted convexities.

C ∈ S	r ∈ R	s ∈ S	Condition	Name
∃	∀	∃	s(C) ⊆ r(A)	converted partial (e, a) - ((S,S), R) convexity
∃	∃	∀	s(C) ⊆ r(A)	converted partial (a, e) - ((S,S), R) convexity
∀	∃	∀	s(C) ⊆ r(A)	converted (a, e) - ((S,S), R) convexity
∀	∀	∃	s(C) ⊆ r(A)	converted (e, a) - ((S,S), R) convexity

Table 8.1.2. Terminology for converted generalised convexities

With all the above-presented conventions of notation and terminology, we are able now to present the definition of generalised convexity properties.

8.2 Definitions and general remarks

In this paragraph we shall present a general description of all the convexity properties that might be obtained using the elements and the method from paragraph 8.1.

Let us consider the nonempty set X, a nonempty set Y, a nonempty subset $S \subseteq 2^X$, a family S of functions s: $S \to 2^Y$ and a family R of transformations r: $2^X \to 2^Y$.

Definition 8.2.1.

(i) *A set* A ⊆ X *is said to be partially* (e, a) - ((S, S), R) *convex if there is a subset* C ∈ S, C ⊆ A, *such that there is a function* s ∈ S *which satisfies the inclusion* s (C) ⊆ r (A) *for every transformation* r ∈ R.

(ii) *A set* A ⊆ X *is said to be partially* (e, e) - ((S, S), R) *convex if there is a subset* C ∈ S, C ⊆ A, *such that there is a function* s ∈ S *and there is a transformation* r ∈ R *satisfying the inclusion* s (C) ⊆ r (A).

(iii) *A set* A ⊆ X *is said to be partially* (a, e) - ((S, S), R) *convex if there is a subset* C ∈ S, C ⊆ A, *such that for every function* s ∈ S *there is a transformation* r ∈ R *which satisfies the inclusion* s (C) ⊆ r (A).

(iv) *A set* A ⊆ X *is said to be partially* (a, a) - ((S, S), R) *convex if there is a subset* C ∈ S, C ⊆ A, *such that for every function* s ∈ S *and for every transformation* r ∈ R *the inclusion* s (C) ⊆ r (A) *is satisfied.*

(v) *A set* A ⊆ X *is said to be* (e, a) - ((S, S), R) *convex if for every subset* C ∈ S, C ⊆ A, *there is a function* s ∈ S *which satisfies the inclusion* s (C) ⊆ r (A) *for every transformation* r ∈ R.

(vi) *A set* A ⊆ X *is said to be* (e, e) - ((S, S), R) *convex if for every subset* C ∈ S, C ⊆ A, *there is a function* s ∈ S *and there is a transformation* r ∈ R *satisfying the inclusion* s (C) ⊆ r (A).

(vii) *A set* A ⊆ X *is said to be* (a, e) - ((S, S), R) *convex if for every subset* C ∈ S, C ⊆ A, *and for every function* s ∈ S *there is a transformation* r ∈ R *which satisfies the inclusion* s (C) ⊆ r (A).

(viii) *A set* A ⊆ X *is said to be* (a, a) - ((S, S), R) *convex if for every subset* C ∈ S, C ⊆ A, *for every function* s ∈ S *and for every transformation* r ∈ R *the inclusion* s (C) ⊆ r (A) *is satisfied.*

After an attentive analysis of this definition one can notice that there are some trivial cases (as in the case of the classical convexity, see chapter 1), i.e. each subset of X may be described using a condition similar to definition 8.2.1. Also, each property may be expressed in the manner described in definition 8.2.1. The following remark describes the trivial cases and, after it, we shall refer only to non-trivial ones.

Remark 8.2.1. 1) If S = 2^X, S = {s} with s (C) = ∅ for each C ∈ S then each set A ⊆ X is (a, a) ((a, e), (e, a), (e, e)) - ((S, S), R) convex for every family R of transformations.

2) If S = 2^X, R = {r} with r (C) = Y for each C ∈ S then each set A ⊆ X is (a, a) ((a, e), (e, a), (e, e)) - ((S, S), R) convex for each family S of functions.

3) Let us suppose that (P) is a given property in 2^X. If $S = 2^X$, s: $S \to 2^Y$ and r: $2^X \to 2^Y$ are defined by

$$s(B) = \begin{cases} Y, & \text{B has not property (P)} \\ \Phi & \text{B has property (P)} \end{cases},$$

$$r(B) = \begin{cases} Y, & \text{B has property (P)} \\ \Phi & \text{B has not property (P)} \end{cases},$$

for $B \in S = 2^X$, then a set $A \subseteq X$ has the property (P) if and only if A is (a, a) ((a, e), (e, a), (e, e)) - ((S, S), R) convex for S = {s} and R = {r}.

4) If X = Y, S = {{x, y} | x ∈ X, y ∈ X} and the sets of functions are defined by

$$\mathsf{S} = \{s_x \mid s_x\colon S \to 2^X,\ s_x(\{a, b\}) = \begin{cases} \{x\}, & x \in \{a, b\} \\ \Phi, & x \notin \{a, b\} \end{cases},\ \{a, b\} \in S\},$$

$$\mathsf{R} = \{r \mid r\colon 2^X \to 2^Y,\ r(C) = \begin{cases} s(C), & C \in S \\ \Phi, & C \notin S \end{cases},\ s = s_x,\ x \in C,\ C \in 2^X\},$$

then each set $A \in 2^X$ is (a, e) (or (e, e)) - ((S, S), R) convex. It is easy to see that the empty set might be replaced by other conveniently chosen subsets of X in the definitions of the functions from the classes S and R.

The reader might find other trivial cases. The following remarks are now obvious.

Remark 8.2.2.

1) Every partially (e, a) - ((S, S), R) convex set is partially (e, e) - ((S, S), R) convex.

2) Every partially (a, e) - ((S, S), R) convex set is partially (e, e) - ((S, S), R) convex.

3) Every partially (a, a) - ((S, S), R) convex set is partially (a, e) - ((S, S), R) convex.

4) Every partially (a, a) - ((S, S), R) convex set is partially (e, a) - ((S, S), R) convex.

5) Every partially (a, a) - ((S, S), R) convex set is partially (e, e) - ((S, S), R) convex.

Remark 8.2.3.

1) Every (e, a) - ((S, S), R) convex set is (e, e) - ((S, S), R) convex.

2) Every (a, e) - ((S, S), R) convex set is (e, e) - ((S, S), R) convex.

3) Every (a, a) - ((S, S), R) convex set is (a, e) - ((S, S), R) convex.

4) Every (a, a) - ((S, S), R) convex set is (e, a) - ((S, S), R) convex.

5) Every (a, a) - ((S, S), R) convex set is (e, e) - ((S, S), R) convex.

Remark 8.2.4.

1) Every (e, a) - ((S, S), R) convex set is partially (e, a) - ((S, S), R) convex.

2) Every (e, e) - ((S, S), R) convex set is partially (e, e) - ((S, S), R) convex.

3) Every (a, e) - ((S, S), R) convex set is partially (a, e) - ((S, S), R) convex.

4) Every (a, a) - ((S, S), R) convex set is partially (a, a) - ((S, S), R) convex.

Definition 8.2.2.

(a) *A set* $A \subseteq X$ *is said to be converted partially* (e, a) - ((S, S), R) *convex if there is a subset* $C \in S$, $C \subseteq A$, *such that for every transformation* $r \in R$ *there is a function* $s \in S$ *which satisfies the inclusion* $s(C) \subseteq r(A)$.

(b) *A set* $A \subseteq X$ *is said to be converted partially* (a, e) - ((S, S), R) *convex if there is a subset* $C \in S$, $C \subseteq A$, *for which there is a transformation* $r \in R$ *such that for every function* $s \in S$ *the inclusion* $s(C) \subseteq r(A)$ *holds.*

(c) *A set* $A \subseteq X$ *is said to be converted* (e, a) - ((S, S), R) *convex if for every subset* $C \in S$, $C \subseteq A$, *and for every transformation* $r \in R$. *there is a function* $s \in S$ *which satisfies the inclusion* $s(C) \subseteq r(A)$.

(d) *A set* $A \subseteq X$ *is said to be converted* (a, e) - ((S, S), R) *convex if for every subset* $C \in S$, $C \subseteq A$, *there is a transformation* $r \in R$ *such that for every function* $s \in S$ *the inclusion* $s(C) \subseteq r(A)$ *holds.*

In the case of the converted convexities there are, also, some trivial cases, that will be presented first. We recall that a function $f: 2^X \to 2^X$ is increasing with respect to the inclusion relation if for each two subsets of X, A and B, with $A \subseteq B$ the inclusion $f(A) \subseteq f(B)$ holds. Function f is decreasing if for each two subsets of X, A and B, with $A \subseteq B$ the inclusion $f(B) \subseteq f(A)$ occurs.

Remark 8.2.5.

1) Each set $A \in 2^X$ is converted (e, a) - ((S, S), R) convex, for S, S and R defined in remark 8.2.1. 4).

2) Let us suppose that $S \subseteq \{s \mid s: S \rightarrow 2^Y\}$ and $R \subseteq \{r_A \mid r_A: 2^X \rightarrow 2^Y, \exists\, s \in S, s(C) \subseteq r_A(C), C \in S, C \subseteq A, r_A$ is increasing on $2^X \setminus \{C \in S \mid C \subseteq A\}\}$. Each set $A \subseteq X$ that contains subsets from S is (a, e) (or (e, e)) - ((S, S), R) convex. For particular choices of S and R each set $A \subseteq X$ that contains subsets from S may become (a, a) - ((S, S), R) convex or converted (a, e) (or (e, a)) - ((S, S), R) convex.

3) Let us put $S = \{\{x\} \mid x \in X\}$, $Y = \{\alpha, \beta\}$ with $\alpha \neq \beta$ and, for $A \subseteq X$, let us define the functions $s_A: S \rightarrow 2^Y$, $r_A: 2^X \rightarrow 2^Y$ defined by

$$s_A(\{x\}) = \begin{cases} \{\alpha\} & x \notin A, \\ \{\beta\} & x \in A \end{cases}$$

$$r_A(C) = \begin{cases} \{\alpha\} & C \not\subset A, \\ \{\beta\} & C \subseteq A \end{cases}$$

for $x \in X$ and $C \subseteq X$. It is obvious that each set $A \subseteq X$ satisfies the statement: for each $s \in S$ there is $r \in R$ such that $s(\{x\}) = r(\{x\}) \subseteq r(A)$, when $x \in A$. Therefore, each set $A \subseteq X$ is (a, e) - ((S, S), R) convex. Also, each set $A \subseteq X$ satisfies: for each transformation $r \in R$ there is $s \in S$ such that $s(\{x\}) = r(\{x\}) \subseteq r(A)$, when $x \in A$. Hence, each set $A \subseteq X$ is converted (e, a) - ((S, S), R) convex.

It is possible to construct other trivial situations for the converted convexities and we let this exercise to the reader. The following remarks are obvious.

Remark 8.2.6.

1) Every converted (e, a) - ((S, S), R) convex set is converted partially (e, a) - ((S, S), R) convex.

2) Every converted (a, e) - ((S, S), R) convex set is converted partially (a, e) - ((S, S), R) convex.

Remark 8.2.7.

1)Every partially (a, a) - ((S, S), R) convex set is converted partially (e, a) - ((S, S), R) convex.

2)Every partially (a, a) - ((S, S), R) convex set is converted partially (a, e) - ((S, S), R) convex.

3)Every (a, a) - ((S, S), R) convex set is converted (e, a) - ((S, S), R) convex.

4)Every (a, a) - ((S, S), R) convex set is converted (a, e) - ((S, S), R) convex.

There are some cases in which condition s (C) $\subseteq$ r (A) must be satisfied if C = A. In this situation it is necessary that A $\in$ S.

Definition 8.2.3.

A. *A set* A $\subseteq$ X *is said to be special partially* (e, a) – ((S, S), R) *convex if* A $\in$ S *and there is a function* s $\in$ S *which satisfies the inclusion* s(A) $\subseteq$ r(A) *for each transformation* r $\in$ R.

B. *A set* A $\subseteq$ X *is said to be special partially* (e, e) – ((S, S), R) *convex if* A $\in$ S *and there is a function* s $\in$ S *and there is a transformation* r $\in$ R *satisfying the inclusion* s (A) $\subseteq$ r (A).

C. *A set* A $\subseteq$ X *is said to be special partially* (a, e) – ((S, S), R) *convex if* A $\in$ S *and for each function* s $\in$ S *there is a transformation* r $\in$ R *which satisfies the inclusion* s (A) $\subseteq$ r (A).

D. *A set* A $\subseteq$ X *is said to be special partially* (a, a) – ((S, S), R) *convex if* A $\in$ S *and for each function* s $\in$ S *and for each transformation* r $\in$ R *the inclusion* s (A) $\subseteq$ r (A) *is satisfied.*

Remark 8.2.8.

1)Every special partially (e, a) - ((S, S), R) convex set is partially (e, a) - ((S, S), R) convex.

2)Every special partially (e, e) - ((S, S), R) convex set is partially (e, e) - ((S, S), R) convex.

3)Every special partially (a, e) - ((S, S), R) convex set is partially (a, e) - ((S, S), R) convex.

4)Every special partially (a, a) - ((S, S), R) convex set is partially (a, a) - ((S, S), R) convex.

The converse is not true because the condition A $\in$ S is not implied by a partial convexity, but is compulsory for a special partial convexity.

Example 8.2.1. Let be X = {1,2}, Y = X, S = 2^X, the functions

s: S $\to 2^Y$, s ($\varnothing$) = $\varnothing$, s ({1}) = {1}, s ({2}) = {2}, s ({1, 2}) = {1},

r: $2^X \to 2^Y$, r ($\varnothing$) = $\varnothing$, r ({1}) = {1}, r ({2}) = {2}, r ({1, 2}) = {2}.

Let us consider the singleton sets of functions S = {s} and R = {r}. Set A = {1, 2} is partially (a, a) - ((S, S), R) convex (for example, C = {2} $\in$ S, C $\subseteq$ A, and s (C) $\subseteq$ r (A)), but set A is not special partially (a, a) - ((S, S), R) convex (s (A) $\not\subset$ r (A)).

In fact, every special partial convexity is an (a, a) - ((S, S), R) convexity for S = {s} and R = {r}, but it is not easy to construct functions s and r, as it will be shown in the following theorem. We shall omit the pair that indicates the quantifiers. The notation ((S, s), r) will be used if the families of functions are singleton. Now let Λ and W be two nonempty sets, let Δ be a nonempty subset of 2^Λ. Let $\alpha: \Delta \to 2^W$, $\beta: 2^\Lambda \to 2^W$, be given functions.

Theorem 8.2.1 *A subset* $B \subseteq \Lambda$ *is special partially* $((\Delta, \alpha), \beta)$ *convex if and only if* $\{B\} \in S$ *and set* $\{B\}$ *is* ((S, s), r) *convex where*: $X = 2^\Lambda$, $Y = W$,

(8.2.1) $\qquad S = \{\{U\} \mid U \in \Delta\} \subseteq 2^X$,

$s: S \to 2^W$ *is the function given, for each* $V = \{U\} \in S$, *by*

(8.2.2) $\qquad s(V) = s(\{U\}) = \alpha(U)$,

and $r: 2^X \to 2^W$ *is the function given, for each* $V = \{U\} \in S$, *by*

(8.2.3) $\qquad r(V) = r(\{U\}) = \beta(U)$,

and for each $V \in 2^Y \setminus S$ *by*

(8.2.4) $\qquad r(V) = Y$ (*each value is possible*).

Proof. Necessity. Let $B \subseteq \Lambda$ be a special partial $((\Delta, \alpha), \beta)$ convex set. Then we have

(8.2.5) $\qquad \alpha(B) \subseteq \beta(B)$.

We prove that set $\{B\}$ is $((\Delta, \alpha), \beta)$ convex. Let be $C \in S$ with

(8.2.6) $\qquad C \subseteq \{B\}$.

As $C \in S$, from (8.2.1) we get that there is $U \in \Delta$ such that $C = \{U\}$. Then (8.2.6) is equivalent with $\{B\} = \{U\}$. It follows that $B = U$. Therefore

(8.2.7) $\qquad C = \{B\}$.

In view of (8.2.2) and (8.2.3) we obtain

(8.2.8) $\qquad s(C) = s(\{B\}) = \alpha(B)$

and

(8.2.9) $\qquad r(\{B\}) = \beta(B)$.

Then from (8.2.5), (8.2.8) and (8.2.9) it results

(8.2.10) $\qquad s(C) \subseteq r(\{B\})$.

Since C was arbitrary chosen in S, we get that set $\{B\}$ is ((S, s), r) convex.

Sufficiency. Let be $\{B\} \in S$ such that set $\{B\}$ a ((S, s), r) convex set. Then we have

(8.2.11) $\qquad B \in \Delta$,

and

(8.2.12) $\qquad s(\{B\}) \subseteq r(\{B\})$.

From (8.2.12), (8.2.2) and (8.2.3) we obtain

(8.2.13) $\qquad \alpha(B) = s(\{B\}) \subseteq r(\{B\}) = \beta(B)$.

Then (8.2.11) and (8.2.13) implies that set B is special partially $((\Delta, \alpha), \beta)$ convex.

Let us suppose that X is a topological space and let us consider a given convexity property for sets in X, denoted by **C**. Let $A \subseteq X$ and a point $a \in A$.

Definition 8.2.4. *Set* $A \subseteq X$ *is locally* C *- convex at point* $a \in A$ *if there is a* C *- convex neighbourhood* V *of* a *such that* $V \subseteq A$.

Definition 8.2.5. *Set* $A \subseteq X$ *is locally* C *- convex if* A *is locally* C *- convex at each point* $a \in A$.

Theorem 8.2.2. *Set* $A \subseteq X$ *is locally* C *- convex at point* $a \in A$ *if and only if* A *is* (e, a) - ((S, S), R) *convex for* $Y = X$, $S = \{a\}$,

$$\mathsf{S} = \{s \mid s: S \to 2^X, s(\{a\}) = V, V \text{ is a } \mathsf{C} \text{ - convex neighbourhood of } a\}$$

and $\mathsf{R} = \{1_{2^X}\}$.

Proof. Necessity. Let us suppose that set $A \subseteq X$ is locally C - convex at point $a \in A$. Then there is a C - convex neighbourhood U of point a such that $U \subseteq A$. We take $s \in \mathsf{S}$ defined by $s(\{a\}) = U$ and then $s(\{a\}) = U \subseteq A = r(A)$. It means that A is (e, a) - ((S, S), R) convex.

Sufficiency. Let us suppose that A is (e, a) - ((S, S), R) convex. Then for each nonempty subset of S, in our case only $\{a\}$, there is a function $s \in \mathsf{S}$ such that for each $r \in \mathsf{R}$, in our case the only element, $s(\{a\}) \subseteq r(A)$. But, according to the definition of these functions, we have $s(\{a\}) = V$, where V is a C - convex neighbourhood of a, and then $V \subseteq r(A) = A$. Hence, set A is locally C - convex at point a.

Theorem 8.2.3. *Set* $A \subseteq X$ *is locally* C *- convex if and only if* A *is* (e, a) - ((S, S), R) *convex for* $Y = X \cup \{*\}$ *where* $* \notin X$, $S = \{\{a\} \mid a \in X\}$,

$$\mathsf{S} = \{s_A: S \to 2^X \mid s_A(\{a\}) = \begin{cases} A, & x \in \text{int}(A) \\ \{*\}, & x \notin \text{int}(A) \end{cases}, A \text{ is } \mathsf{C} \text{ - convex neighbourhood of } a, \{a\} \in S\}$$

and $\mathsf{R} = \{1_{2^X}\}$.

Proof. Necessity Let A be locally C – convex. Let $C \in S$, $C \subseteq A$. There is a point $y \in A$, such that $C = \{y\}$. The property of local C – convexty of set A implies that there is a neighbourhood U of point y such that $U \subseteq A$. Taking $s = s_U$ one gets $s(C) = s_U(\{y\}) = U \subseteq A = r(A)$. Therefore A is (e, a) – ((S, S), R) convex.

Sufficiency. Let A be (e, a) – ((S, S), R) convex. Let $a \in A$. If we take $C = \{a\}$ we have $C \in S$ and $C \subseteq A$. Then there is $s \in \mathsf{S}$ with $s(\{a\}) \subseteq r(A) = A$. The definition of class S implies the existence of a set $V \in 2^X$, which is C – convex, such that $s = s_V$. As $A \neq X$ one has $s_V(\{a\}) \neq \{*\}$. It follows that $s_V(\{a\}) = V$ and $a \in \text{Int}(V)$. Then V is a C – convex neighbourhood of a, with $V \subseteq A$. Therefore A is locally C – convex at the point $a \in A$. Because point a was arbitrarily chosen, we get that the set A is locally C – convex.

As usual, it is important to study the transmissibility of a property of generalised convexity defined above from sets to subsets and from a triple (S, S, R) to another triple in some relation with the first one. Usually, the convexity is not transmissible to subsets. But there are particular cases, not trivial, in which this transmissibility holds.

Definition 8.2.6. i) *The* S *family of functions is said to be weakly dominated on* S *by the* R *family of functions if for each* $s \in S$ *there is a function* $r \in R$ *such that* $s(C) \subseteq r(C)$ *for each* $C \in S$.

ii) *The* S *family of functions is said to be dominated on* S *by the* R *family of functions if for each* $s \in S$ *and* $r \in R$ *the inclusion* $s(C) \subseteq r(C)$ *holds for each* $C \in S$.

Property 8.2.1. *If* A *is* (e, a), ((e, e), (a, e), (a, a)) - ((S, S), R) *convex and* $2^A \cap S \neq \varnothing$ *then* A *is* (a, a) - ((S, S), R) *convex.*

Proof. The result immediately occurs. Indeed, if R is a family of increasing functions and S is dominated by R then for each $C \in S$, $C \subseteq A$, the inclusion $s(C) \subseteq r(C)$ occurs for every $s \in S$ and $r \in R$. But r is increasing and $C \subseteq A$; so, $r(C) \subseteq r(A)$. From these inclusions we obtain $s(C) \subseteq r(A)$, for each $C \in S$, $C \subseteq A$. Therefore A is (a, a) – ((S, S), R) convex.

Corollary 8.2.1. *If* R *is a family of increasing functions and* S *is dominated on* S *by* R *then for* $A \subset X$, $B \subset X$, $A \subseteq B$ *and* $2^A \cap S \neq \varnothing$ *then the following implications occur:*

1. if B *is* (e, a) - ((S, S), R) *convex then* A *is* (e, a) - ((S, S), R) *convex;*

2. if B *is* (a, a) - ((S, S), R) *convex then* A *is* (a, a) - ((S, S), R) *convex.*

Proof. 1. The (e, a) - ((S, S), R) convexity of B implies that for each $C \in S$, $C \subset B$, there is $s \in S$ such that $s(C) \subseteq r(B)$ for all $r \in R$. Suppose that $C \subset A \subset B$ and $C \in S$. Then there is $s \in S$ such that for all $r \in R$ the inclusion $s(C) \subset r(C) \subseteq r(A) \subseteq r(B)$, meaning the (e, a) - ((S, S), R) convexity of A.

2) As B is (a, a) - ((S, S), R) convex, for each $C \in S$, $C \subset B$, for each functions $s \in S$ and $r \in R$ such that $s(C) \subseteq r(B)$. Let us suppose that $C \subset A \subset B$ and $C \in S$. Then for each $s \in S$ and $r \in R$, $s(C) \subset r(B)$ occurs. But the properties of the S and R families of functions imply the inclusion $s(C) \subset r(C) \subseteq r(A) \subseteq r(B)$, meaning the (a, a) - ((S, S), R) convexity of A.

Corollary 8.2.2. *If* R *is a family of increasing functions and* S *is weakly dominated on* S *by* R *then for* $A \subset X$, $B \subset X$, $A \subseteq B$ *and* $2^A \cap S \neq \varnothing$ *then the following implications occur:*

1. *if* B *is* (e, e) - ((S, S), R) *convex then* A *is* (e, e) - ((S, S), R) *convex;*

2. *if* B *is* (a, e) - ((S, S), R) *convex then* A *is* (a, e) - ((S, S), R) *convex.*

1) As B is (e, e) - ((S, S), R) convex, for each C ∈ S, C ⊂ B, there is s ∈ S and there is r ∈ R such that s (C) ⊆ r (B). Let us suppose that C ⊂ A ⊂ B and C ∈ S. There are s ∈ S and r ∈ R for which s (C) ⊂ r (B). But the properties of the S and R families of functions make true the multiple inclusion s (C) ⊂ r (C) ⊆ r (A) ⊆ r (B), meaning the (e, e) - ((S, S), R) convexity of A.

2) Since B is (a, e) - ((S, S), R) convex, for each C ∈ S, C ⊂ B and for each s ∈ S there is r ∈ R such that s (C) ⊆ r (B). Let us consider C ⊂ A ⊂ B and C ∈ S. For each s ∈ S there is r'∈ R for which s (C) ⊂ r'(B). But the hypothesis on the S and R families of functions implies the following inequality: s (C) ⊂ r'(C) ⊆ r'(A) ⊆ r'(B). Therefore, the set A is (a, e) - ((S, S), R) convex.

Corollary 8.2.3. *If* R *is a family of increasing functions and* S *is dominated by* R *then for* A ⊂ X, B ⊂ X, A ⊆ B *and* $2^A \cap S \neq \emptyset$ *then the following implications occur:*

1. if B *is converted* (e, a) - ((S, S), R) *convex then* A *is converted* (e, a) - ((S, S), R) *convex;*

2. if B *is converted* (a, e) - ((S, S), R) *convex then* A *is converted* (a, e) - ((S, S), R) *convex.*

Proof. The property of converted (e, a) - ((S, S), R) convexity of B implies that for each C ∈ S, C ⊂ B and for each r ∈ R there is s ∈ S such that s (C) ⊆ r (B). Taking C ∈ S, C ⊂ A ⊂ B, by the properties of S and R we get the inclusion s (C) ⊂ r (C) ⊆ r (A) ⊆ r (B), meaning the converted (e, a) - ((S, S), R) convexity of A.

2) Since B is converted (a, e) - ((S, S), R) convex, for each C ∈ S, C ⊂ B and there is r ∈ R such that s (C) ⊆ r (B) for each s ∈ S. If C ∈ S with C ⊂ A ⊂ B then for the function r ∈ R satisfying the above mentioned condition for B, we obtain that each s ∈ S also satisfies the same inequality for A because s (C) ⊂ r (C) ⊆ r (A) ⊆ r (B). Therefore, set A is converted (a, e) - ((S, S), R) convex.

After investigating all the 100 convexity properties of our collection we found 11 classes of convexity properties represented as in the following table:

Nr.	**Type of convexity**	**Number of members**
1	(S, s) convexity	42

2	((S, s), r) convexity	13
3	special partial ((S, s) r) convexity	7
4	(e, a) – ((S, S), r) convexity	21
5	partial (a, e) - ((S, s), R) convexity	7
6	(a, e) - ((S, s), R) convexity	1
7	special partial (a, e) - ((S, S), R) convexity	4
8	partial (a, e) - ((S, S), R) convexity	1
9	(a, e) - ((S, S), R) convexity	2
10	converted (a, e) - ((S, S), R) convexity	1
11	(a, a) - ((S, S), R) convexity	1

Table 8.2.1. Classes of convexities for sets

Notation (S, s) means that X = Y, S = {s} and R = {r} where r is the identity in 2^X.

Notation ((S, s), r) means that S = {s} and R = {r}.

Notation ((S, S), r) means that R = {r} and notation ((S, s), R) means that S = {s}.

The following paragraphs will present each class of convexity properties by means of their members, describing the manner in which the main elements of their definitions are obtained.

8.3 The class of (S, s) convexity properties

Let us consider the elements and the notations from the previous paragraph, with the condition that X = Y. set S is supposed to be singleton, the function s: $S \to 2^Y$ being its unique element. Also, the R family of transformations r: $2^X \to 2^X$ is supposed to be singleton, its unique element being the identity on 2^X, r (A) = A for $A \in 2^X$. In this case, all the partial convexity properties defined in the previous paragraph are equivalent and are called simply, *partial* (S, s) *convexity*. Also, all the convexity properties defined above are equivalent and are called (S, s) *convexity*.

Example 8.3.1. The classical convexity (Euclid (330-275 B.C.), Archimedes (287?-212 B.C.)), H. Brunn (1887), H. Minkowski (1911)

The classical framework for the convexity is a nonempty set X in which the notion of straight-line segment is defined. A subset A of X is said to be *convex* if for every two points $x \in A$ and $y \in A$, the straight-line segment joining them, denoted by <x, y>, is also contained in A. Sets having this geometric property can be found in numerous writings from the antiquity and have been studied ever since, together with their applications.

Let us take $Y = X$, $S = \{\{x, y\} \mid x \in X, y \in X\}$ and consider the function $s: S \to 2^X$ defined by $s(\{x, y\}) = <x, y>$ for $\{x, y\} \in S$.

Theorem 8.3.1. *A set* $A \subseteq X$ *is convex (classically convex) if and only if* A *is* (S, s) *convex.*

Proof. Set A is convex if and only if for each two points $x, y \in A$, the inclusion $<x, y> \subseteq A$ holds. This is equivalent to the fact that every subset $\{x, y\} \in S \cap A$ implies that $s(\{x, y\}) = <x, y> \subseteq r(A) = A$, which means that A is (S, s) convex.

Remark 8.3.1. The framework that provided the convex sets with the most properties and the most applications was the linear structure of the space X. H. Minkowski (1911) first put the linear structure of the space X in connection with the convexity. In order to record the definition of H. Minkowski, let $(X, +, \cdot)$ be a real linear space. A set $A \subseteq X$ is said to be *convex (classically convex)* if, for each two points $x \in A$ and $y \in A$, the whole straight-line segment $<x, y> = \{tx + (1-t)y \mid t \in [0, 1]\}$ lies in A.

Example 8.3.2. n-polyhedral convexity.

Let X be a set in which the straight-line segment, $<x, y>$, joining every two points $x \in X$ and $y \in X$ is defined. Then we shall define the n-polyhedron by

$< x_1, x_2, \ldots, x_n> = \cup\{<z, x_k> \mid z \in <x_1, \ldots, x_{k-1}, x_{k+1}, \ldots, x_n>, k = 1, 2, \ldots, n\}$ for $x_1, x_2, \ldots, x_n \in X$.

A set $A \subseteq X$ is said to be *n-polyhedrally convex* if for every n points $x_1, x_2, \ldots, x_n \in A$, the whole polyhedron $< x_1, x_2, \ldots, x_n>$ lies in A.

Let us consider $Y = X$, $S = \{\{x_1, x_2, \ldots, x_n\} \mid x_k \in X, k = 1, 2, \ldots, n\}$ and function $s: S \to 2^X$ is defined by $s(\{x_1, x_2, \ldots, x_n\}) = < x_1, x_2, \ldots, x_n>$.

Theorem 8.3.2. *A set* $A \subseteq X$ *is* n *– polyhedrally convex if and only if* A *is* (S, s) *convex.*

Proof. Set A is n – polyhedrally convex if and only if for every n points $x_1, x_2, \ldots, x_n \in A$, the polyhedron $<x_1, x_2, \ldots, x_n>$ lies in A. This is equivalent to $s(\{x_1, x_2, \ldots, x_n\}) = <x_1, x_2, \ldots, x_n> \subseteq r(A) = A$, meaning that A is (S, s) – convex.

Remark 8.3.2. Let $(X, +, \cdot)$ be a real linear space. A set $A \subseteq X$ is said to be *n-polyhedrally convex* if for every n points $x_1, x_2, \ldots, x_n \in A$, the whole polyhedron $< x_1, x_2, \ldots, x_n> = \{\sum_{k=1}^{n} t_k x_k \mid t_k \in [0, 1], k = 1, 2, \ldots, n,$

$\sum_{k=1}^{n} t_k = 1\}$ lies in A. In this case, a set A is n-polyhedrally convex if and only if it is classical convex. But there are more structures in which this equivalence does not occur (for example, in non-distributive lattices).

Example 8.3.3. Metric convexity (K. Menger (1928))

Let (X, d) be a metric space and for every $x \in X$ and $y \in X$ the straight-line segment $<x, y> = \{z \in X \mid d(x, z) + d(z, y) = d(x, y)\}$ is defined. A set $A \subseteq X$ is said to be *metric convex* if for every two points $x \in A$ and $y \in A$, the whole straight-line segment $<x, y>$ lies in A.

Let us take Y = X, $S = \{\{x, y\} \mid x \in X, y \in X\}$ and consider the function $s: S \to 2^X$ defined by $s(\{x, y\}) = <x, y>$ for $\{x, y\} \in S$.

Theorem 8.3.3. *A set* $A \subseteq X$ *is metric convex if and only if* A *is* (S, s) *convex.*

The proof follows the same reasoning as in the case of theorem 8.3.1.

Example 8.3.4. Order convexity (G. Birkhoff (1948))

Let $X = (X, \leq)$ be a partially ordered set and $x \in X$ and $y \in X$ be two elements such that $x \leq y$. The order segment determined by x and y is defined by $<x, y> = \{z \in X \mid x \leq z \leq y\}$. A set $A \subseteq X$ is called *order convex* if for each two points $x, y \in A$ with $x \leq y$ the condition $<x, y> \subseteq A$ is satisfied.

Let us take Y = X, $S = \{\{x, y\} \mid x \in X, y \in X, x \leq y\}$ and let us define function $s: S \to 2^X$ by $s(\{x, y\}) = <x, y>$ for $\{x, y\} \in S$.

Theorem 8.3.4. *A set* $A \subseteq X$ *is order convex if and only if* A *is* (S, s) *convex.*

The proof is similar to the case of theorem 8.3.1.

Example 8.3.5. Spherical convexity (C.V. Robinson (1942))

The space in which the spherical convexity is defined is the d-sphere in $\mathbf{R}^d$, $X = B^d$. For every non-antipodal points $x \in B^d$ and $y \in B^d$ the spherical segment determined by x and y, denoted also by $<x, y>$, is considered to be the small arc of the great circle joining them. A set $A \subseteq B^d$ is said to be *spherically convex (Robinson – convex)* if for every two non-antipodal points $x \in A$ and $y \in A$, the small arc of the great circle joining them is also contained in A. Equivalent characterisations by means of intersections were proved in L. Danzer (1982).

Let us take $Y = B^d$, $S = \{\{x, y\} \mid x \in B^d, y \in B^d$, x and y are non-antipodal points$\}$ and let us define function $s: S \to 2^{B^d}$ by $s(\{x, y\}) = \langle x, y \rangle$ for $\{x, y\} \in S$.

Theorem 8.3.5. *A set* $A \subseteq B^d$ *is spherically convex if and only if* A *is* (S, s) *convex.*

The proof is similar to the case of theorem 8.3.1.

Example 8.3.6. Strong convexity (C.V. Robinson (1942))

In this case we have also $X = B^d$, together with the previous notion of spherical convexity. A spherical convex set $A \subseteq B^d$ is said to be *strongly convex* if it does not contain antipodal points. In essence, this convexity is also a property of (S, s) convexity, due to theorem 8.3.5.

Example 8.3.7. Convexity in subsets of lattices (S.V. Ovchinnikov (1980))

Let $(L, \leq)$ be a lattice and $M \subset L$. A ternary relation on L is called a betweenness relation and is denoted by $B(x_1, x_2, x_3)$. As in the book of L.M. Blumenthal (1953), we say that $B(x, x_1, x_2)$ is true if and only if

$$x_1 \wedge x_2 \leq x \leq x_1 \vee x_2.$$

An element $x \in L$ lies between x_1 and x_2 if $B(x, x_1, x_2)$ is true. Let the points $a \in M$ and $b \in M$. An M-interval I [a, b] is the set of all the elements of M which lie between a and b,

$$I[a, b] = \{x \in M \mid a \wedge b \leq x \leq a \vee b\} = I[a \wedge b, a \vee b].$$

This ternary relation is extended to a (k+1)-ary relation B on L, called also betweenness. The relation $B(x, x_1, x_2, \ldots, x_k)$ is true if and only if

$$\bigwedge_{i=1}^{k} x_i \leq x \leq \bigvee_{i=1}^{k} x_i,$$

meaning that the point x lies between $x_1, x_2, \ldots, x_k$. A M-polyhedron $I[x_1, x_2, \ldots, x_k]$ is the set of all the elements of M which lie between $x_1, x_2, \ldots, x_k$,

$$I[x_1, x_2, \ldots, x_k] = \{x \in M \mid \bigwedge_{i=1}^{k} x_i \leq x \leq \bigvee_{i=1}^{k} x_i\} = I[\bigwedge_{i=1}^{k} x_i, \bigvee_{i=1}^{k} x_i].$$

A set $A \subseteq M$ is r-*convex with respect to* M $(r \geq 2)$ if and only if the truth of $B(x, x_1, x_2, \ldots, x_r)$ implies $x \in A$ for all $x \in M$ and all r-tuples of points $x_1, x_2, \ldots, x_r$ belonging to A.

Now, let us consider $Y = X = L$,

$$S = \{\{x_1, x_2, \ldots, x_r\} \mid x_k \in M, k = 1, 2, \ldots, r\}$$

and function s: $S \to 2^X$ is defined by s $(\{x_1, x_2, \ldots, x_r\}) = I\,[x_1, x_2, \ldots, x_r]$.

Theorem 8.3.6. *A set* $A \subseteq M$ *is* r – *convex with respect to* M *if and only if* A *is* (S, s) *convex.*

Proof. Set A is r – convex with respect to M if and only if for every r points $x_1, x_2, \ldots, x_r \in A$, the polyhedron $I\,[x_1, x_2, \ldots, x_r]$ lies in A. This is equivalent to s $(\{x_1, x_2, \ldots, x_r\}) = I\,[x_1, x_2, \ldots, x_r] \subseteq r(A) = A$, meaning that A is (S, s) – convex.

Example 8.3.8. Midpoint – convexity (J.L.W.V. Jensen (1905, 1906))

Let us suppose that $X = \mathbf{R}^n$, $n \in \mathbf{N}$. A set $A \subseteq \mathbf{R}^n$ is *midpoint convex* if for every $x, y \in A$ one has $\frac{x+y}{2} \in A$. This type of convex sets was defined in connection with the property of convexity of functions, first considered by J.L.W.V. Jensen (1905).

In order to include this convexity in the class of properties taken into account in this paragraph, let us put $X = Y = \mathbf{R}^n$, $S = \{\{x, y\} \mid x \in \mathbf{R}^n, y \in \mathbf{R}^n\}$ and let us consider function s: $S \to 2^{\mathbf{R}^n}$ defined by

$$s(\{x, y\}) = \{\frac{x+y}{2}\} \text{ for } \{x, y\} \in S.$$

Theorem 8.3.7. *A set* $A \subseteq \mathbf{R}^n$ *is midpoint convex if and only if* A *is* (S, s) *convex.*

The proof is similar to the previous ones and will be omitted.

Example 8.3.9. p – convexity (A. Aleman (1985))

The p - convexity is another non-connected convexity, obtained in the same manner as the midpoint convexity. But, in this case, the requirement that the middle point of a segment belongs to the set under consideration is replaced by a condition on another point of the segment. $(X, +, \cdot)$ is supposed to be a real or complex linear space and $p \in\,]0, 1[$. A set $A \subseteq X$ is said to be p-*convex* if for every two points $x \in A$ and $y \in A$, the element $(1\text{-}p)\,x + p\,y$ belongs to A. As in the previous example, let us put $X = Y$, $S = \{\{x, y\} \mid x \in X, y \in X\}$ and let us consider function s: $S \to 2^X$ defined by

$$s(\{x, y\}) = \{(1\text{-}p)\,x + p\,y\} \text{ for } \{x, y\} \in S.$$

Theorem 8.3.8. *A set* $A \subseteq X$ *is* p-*convex if and only if* A *is* (S, s) *convex.*

The proof is similar to the previous ones.

Example 8.3.10. Quasi-convexity (J. W. Green, W. Gustin (1950))

Let us suppose that $X = \mathbf{R}^n$, $n \in \mathbf{N}$ and $\Delta \subset [0, 1]$, $\{0, 1\} \subset \Delta$. A set $A \subseteq \mathbf{R}^n$ is said to be *quasi-convex* if for every x, y ∈ A one has

$$\{t\,x + (1\text{-}t)\,y \mid t \in \Delta\} \subset A.$$

In particular, if $\Delta = \{0, \frac{1}{2}, 1\}$ then the quasi-convexity becomes the midpoint convexity. For $\Delta = \{0, p, 1\}$, $p \in\,]0, 1[$, the quasi-convexity becomes p-convexity.

Let us put X = Y, $S = \{\{x, y\} \mid x \in \mathbf{R}^n, y \in \mathbf{R}^n\}$ and let us consider function $s: S \to 2^{\mathbf{R}^n}$ defined by

$$s(\{x, y\}) = \{t\,x + (1\text{-}t)\,y \mid t \in \Delta\} \text{ for } \{x, y\} \in S.$$

Theorem 8.3.9. *A set* $A \subseteq \mathbf{R}^n$ *is quasi-convex if and only if* A *is* (S, s) *convex.*

The proof is similar to the previous ones.

Example 8.3.11. Interval convexity (J.R. Calder (1971))

Suppose that X is a point set and let us denote by $P(X) = 2^X \setminus \varnothing$, i. e. the set of the nonempty subsets of X. A transformation $T: X^2 \to P(X)$ is said to be an interval convexity on X. A set $A \subseteq X$ is said to be *interval convex* if for every two points x ∈ A and y ∈ A, the subset T (x, y) is included in A. As in the previous example, let us put X = Y, $S = \{\{x, y\} \mid x \in X, y \in X\}$ and let us consider function $s: S \to 2^X$ defined by

$$s(\{x, y\}) = T(x, y) \text{ for } \{x, y\} \in S.$$

Theorem 8.3.10. *A set* $A \subseteq X$ *is interval convex if and only if* A *is* (S, s) *convex.*

The proof is similar to the previous ones.

Example 8.3.12. Ortho-convexity (G.J.E. Rawlins, D. Wood (1988))

The ortho-convexity has the origin in the efforts for the elaboration of algorithms to build the convex hull of a set (operation called convexification) and algorithms to detect the convexity property. Many algorithms are based on the study of the intersection between the set, which is studied, and vertical and horizontal straight-lines, together with the analysis of frontiers (see, for example, R.L. Graham (1972), C.E. Kim (1980)).

In $\mathbf{R}^2$ a set A is called *ortho-convex* if for each points x, y ∈ A, the straight –line segment (in the sense of the Euclidean geometry) <x, y> lies in A if it is either horizontal or vertical.

This is also a (S, s) convexity. To prove it, let us put $X = Y = \mathbf{R}^2$,

$$S = \{\{x, y\} \mid x \in \mathbf{R}^2, y \in \mathbf{R}^2, \langle x, y\rangle \text{ is either horizontal or vertical}\}$$

and let us consider the function s: $S \to 2^X$ defined by

$$s(\{x, y\}) = \langle x, y\rangle, \text{ for } \{x, y\} \in S.$$

The proof follows the same stages as in the previous examples.

Example 8.3.13. Convexity properties defined by generalised convex combinations (Φ-convexity, Ψ-convexity) (V. Soltan (1984))

Consider X a nonempty set and the Φ family of functions,

$$\Phi = \{\varphi_{x,y} \mid \varphi_{x,y}: [0, 1] \to X, \varphi_{x,y}(0) = y, \varphi_{x,y}(1) = x, x \in X, y \in X\}.$$

The value $\varphi_{x,y}(t)$ is called an abstract convex combination of points x and y. The straight-line segment determined by two points $x \in X$ and $y \in X$ is defined by

$$\langle x, y\rangle = \{\varphi_{x,y}(t) \mid t \in [0, 1]\}.$$

A set $A \subseteq X$ is said to be Φ - *convex* if for each two points $x \in A$ and $y \in A$, the subset $\langle x, y\rangle$ is included in A.

As in the previous examples, let us put X = Y, $S = \{\{x, y\} \mid x \in X, y \in X\}$ and let us consider function s: $S \to 2^X$ defined by

$$s(\{x, y\}) = \langle x, y\rangle \text{ for } \{x, y\} \in S.$$

With these elements we can prove as above that the property of Φ - convexity is a (S, s) convexity.

A generalisation of the abstract convex combination $\varphi_{x,y}(t)$, together with its related notions of straight-line segment and convex set is obtained by considering, for every $x \in X$ and $y \in X$, the multifunction $\psi_{x,y} : [0, 1] \to 2^X$, with the properties $y \in \psi_{x,y}(0)$ and $x \in \psi_{x,y}(1)$. Set $\psi_{x,y}(t)$ is also called an abstract convex combination. Then, for every $x \in X$ and $y \in X$, the straight-line segment determined by these points is defined by

$$\langle x, y\rangle = \bigcup_{0 \le t \le 1} \psi_{x,y}(t).$$

The related convexity property is defined as usual, is called Ψ-*convexity* and is also a (S, s) convexity.

Example 8.3.14. Baricentric approach: the superconvexity (G. Rodé (1980))

The superconvexity is another type of convexity defined using generalized convex combinations, but another type of generalization is constructed. It is a baricentric approach of the convexity.

X is supposed to be a nonempty set and by Q_X we denote the space of all the functions that are probability measures on X. Every measure $q \in Q_X$ has the form $q = \sum_{k=1}^{\infty} q_k \hat{x}_k$ where $(x_1, x_2, \ldots)$ is a sequence in X, $(q_1, q_2, \ldots)$ is a sequence of non-negative real numbers such that $\sum_{k=1}^{\infty} q_k = 1$, and $\hat{x}$ means, for each $x \in X$, the Dirac function in x,

$$\hat{x}(Y) = \begin{cases} 1, & \text{if} \quad Y \subset X \quad \text{with} \quad x \in Y \\ 0, & \text{otherwise} \end{cases}$$

Set $P_X \subset Q_X$ contains all the measure functions $q \in Q_X$ with finite support, hence $q(x) \neq 0$ only for a finite number of points $x \in X$.

A *convex space* is a set X together with an application $f: P_X \to X$ that satisfies the following axioms:

(K1) $f(\hat{x}) = x$, for every $x \in X$;

(K2) $f(\sum_{k=1}^{n} p_k m_k) = f(\sum_{k=1}^{n} p_k \widehat{f(m_k)})$ for all $n \in \mathbf{N}$, $p_1, p_2, \ldots, p_n \geq 0$ with $\sum_{k=1}^{n} p_k = 1$, and $m_1, m_2, \ldots, m_n \in P_X$.

For the points $x \in X$ and $y \in X$, the straight-line segment determined by these points is defined by

$$\langle x, y \rangle = \{f((1-t)\hat{x} + t\hat{y}) \mid 0 \leq t \leq 1\}.$$

In this framework the notion of *convex set* is defined as usual and it is easy to prove that this is a (S, s) convexity.

A *superconvex space* is a set X together with an application $f: Q_X \to X$ that satisfies the following axioms:

(SK1) $f(\hat{x}) = x$, for every $x \in X$;

(SK2) $f(\sum_{k=1}^{\infty} q_k m_k) = f(\sum_{k=1}^{\infty} q_k \widehat{f(m_k)})$ for all $q_1, q_2, \ldots \geq 0$ with $\sum_{k=1}^{\infty} q_k = 1$, and $m_1, m_2, \ldots \in Q_X$.

Every superconvex space becomes a convex space if the restriction of function f to P_X is considered.

In a superconvex space X, for the points $x \in X$ and $y \in X$, the straight-line segment determined by these points is also defined by

$$<x, y> = \{f((1-t)\hat{x} + t\hat{y}) \mid 0 \le t \le 1\}.$$

A convex set in a superconvex space is called *superconvex* set. This property is also a (S, s) convexity.

Example 8.3.15. u – convexity (A. Wieczorek (1989))

Given a nonempty set X, a *spot operator* in X is any operator

$$u: \text{Dom}(u) \to 2^X,$$

where the domain of u, denoted by Dom (u) is a family of subsets of X.

A set $A \subseteq X$ is said to be u-*convex* if $u(F) \subseteq A$ for every $F \in \text{Dom}(u)$ with $F \subseteq A$.

Let us take X = Y, S = Dom (u) and let us consider function $s: S \to 2^X$ defined by

$$s(F) = u(F), \text{ for } F \in S.$$

Theorem 8.3.11. *A set* $A \subseteq X$ *is* u-*convex if and only if* A *is* (S, s) *convex.*

Proof. The u-convexity of set A means that for every $F \in \text{Dom}(u)$ with $F \subseteq A$, the inclusion $u(F) \subseteq A$ holds. In other words, this condition is equivalent to $s(F) = u(F) \subseteq r(A) = A$ for each $F \in \text{Dom}(u) = S \cap A$, which means that A is (S, s) convex.

Example 8.3.16. The convex hull approach of the convexity (g-convexity) (V. Soltan (1984))

First, let us recall from chapter 1 that in a nonempty set X, a mapping $g: 2^X \to 2^X$ is called the convex hull operator on X if: $A \subseteq g(A)$, $g(g(A)) = g(A)$, and $g(B) \subseteq g(A)$, for any $A, B \in 2^X$, $B \subseteq A$. If *g* is a convex hull operator defined on X, and

$$G = \{A \subseteq X \mid g(A) = A\},$$

then the pair (X, G) is a convexity space on X, called the convexity space generated by g.

Conversely, if (X, G) is a convexity space on X, then a convex hull operator on X is defined by the mapping $g: 2^X \to 2^X$ given by

$$g(A) = \cap \{C \in G \mid A \subseteq C\}, \text{ for all } A \in 2^X.$$

This function g is called the convex hull operator generated by G.

If g is a convex hull operator defined on X then the straight-line segment determined by two points $x \in X$ and $y \in X$ is the set

$$<x, y> = g(\{x, y\}).$$

As usual, a set $A \subseteq X$ is said to be *g-convex* if for every two points $x \in A$ and $y \in A$, the subset $<x, y>$ is included in A.

As above, we put $X = Y$, $S = \{\{x, y\} \mid x \in X, y \in X\}$ and we consider function $s: S \to 2^X$ defined by

$$s(\{x, y\}) = <x, y> = g(\{x, y\}) \text{ for } \{x, y\} \in S.$$

With these elements we can prove as above that the property of g - convexity is a (S, s) convexity.

Example 8.3.17. Convex sets in join spaces (A. Ghika (1955), W. Prenowitz (1961), D. Voiculescu (1967 [290], [291]), V.W. Bryant and R.J. Webster (1969))

As we described in chapter 1, a join space is a nonempty set X, in which two binary operations are defined:

$$\bullet: X^2 \to 2^X \text{ and } /: X^2 \to 2^X,$$

subjected to some axioms (see chapter 1). The straight-line segment is defined by the first operation, $<x, y> = x \bullet y$, for $x \in X$ and $y \in X$. Both operations are extended to operations between subsets of X (see chapter 1) and then a set $A \subseteq X$ is said to be convex if $A \bullet A \subseteq A$. This definition is equivalent to the following statement: a set $A \subseteq X$ is said to be *convex* if for every two points $x \in A$ and $y \in A$, the subset $<x, y>$ is included in A.

As usual, if we put $X = Y$, $S = \{\{x, y\} \mid x \in X, y \in X\}$ and we consider function $s: S \to 2^X$ defined by $s(\{x, y\}) = <x, y> = x \bullet y$ for $\{x, y\} \in S$, we obtain that the convexity in a join space is a (S, s) convexity.

Example 8.3.18. Convexity in von Neumann – Morgenstern – Gudder – Schroeck convex structures (S. Gudder and F. Schroeck (1980))

Let X be a nonempty set and $F: [0, 1] \times X^2 \to X$ a function. The pair (X, F) is called a convex prestructure. A convex prestructure (X, F) is said to be a semiconvex structure if F satisfies the following conditions:

(1) $F(\alpha, x, F(\beta, y, z)) = F(\alpha + (1-\alpha)\beta, F(\alpha[\alpha + (1-\alpha)\beta]^{-1}, x, y), z)$

for $x, y, z \in X$ and $\alpha, \beta \in [0, 1]$;

(2) $F(\alpha, x, x) = x$ for each $x \in X$ and $\alpha \in [0, 1]$.

A semiconvex structure (X, F) is said to be a convex structure if F satisfies, in addition, the axioms:

(3) $F(\alpha, x, y) = F(1-\alpha, y, x)$ for each $x, y \in X$ and $\alpha \in [0, 1]$;

(4) If $\alpha \in [0, 1]$ and there is $x \in X$ such that $F(\alpha, x, y) = F(\alpha, x, z)$ then it follows that $y = x$.

Let (X, F) be a convex structure. A set $A \subseteq X$ is said to be F-*convex* if for each pair of points $x \in A$ and $y \in A$ and for each number $\alpha \in [0, 1]$, one has $F(\alpha, x, y) \in A$.

In order to prove that the F-convexity is a (S, s) convexity we put X = Y, we take $S = \{\{x, y\} \mid x \in X, y \in X\}$ and we consider function $s: S \to 2^X$ defined by $s(\{x, y\}) = \{F(\alpha, x, y) \mid \alpha \in [0, 1]\}$ for $\{x, y\} \in S$. The proof is similar to the previous ones.

Example 8.3.19. Convexity in Takahashi convex structures (W. Takahashi (1970))

A W-convex metric space is a triple (X, d, W) where:

1.(X, d) is a metric space;

2.W: $[0, 1] \times X^2 \to X$ is a function satisfying the condition

$$d(u, W(\alpha, x, y)) \leq \alpha\, d(u, x) + (1-\alpha)\, d(u, y),$$

for each $u \in X$, $x, y \in X$ and $\alpha \in [0, 1]$.

In a W-convex metric space (X, d, W), a set $A \subseteq X$ is said to be W-*convex* if for every two points $x \in A$ and $y \in A$ and for every $\alpha \in [0, 1]$, one has $W(\alpha, x, y) \in A$.

It is easy to prove that the W-convexity is a (S, s) convexity, taking X = Y, $S = \{\{x, y\} \mid x \in X, y \in X\}$ and considering function $s: S \to 2^X$ defined by $s(\{x, y\}) = \{W(\alpha, x, y) \mid \alpha \in [0, 1]\}$ for $\{x, y\} \in S$.

Remark 8.3.3. Set $s(\{x, y\}) = \{W(\alpha, x, y) \mid \alpha \in [0, 1]\}$ is called the Takahashi metric segment and it is equivalent to the Menger metric segment (see example 8.3.3). This equivalence is discussed in terms of the homeomorfism between the Menger segment, proved by K. Menger (1928), and the same property present to the Takahashi segment.

Example 8.3.20. k-strong convexity with respect to g and M (L. Blaga, L. Lupşa (1998))

Let X and M be two arbitrary nonempty sets and let g be a convex hull operator on X. Let k be a natural number, $k \geq 2$. A subset $A \subseteq X$ is called k-

strongly convex with respect to g *and* M if for any $a_1, ..., a_k$ elements of A we have $g(a_1, ..., a_k) \cap M \subseteq A$.

Let us choose Y = X, $S = \{\{x_1, x_2, \ldots, x_k\} \mid x_1, x_2, \ldots, x_k \in X\}$, $s: S \to 2^X$ defined by $s(B) = g(B) \cap M$ for all $B \in S$.

Theorem 8.3.12. *A set* $A \subseteq X$ *is* k-*strongly convex with respect to* g *and* M *if and only if* A *is* (S, s) *convex.*

The proof is similar to the previous ones.

Example 8.3.21. Strong k-convexity with respect to a given set (L. Lupşa (1981))

Let (V, +, ·) be a real linear space. In the following by M we denote a nonempty subset of V. Let n be a natural number, k > 0. A subset $A \subseteq V$ is called *strongly* k-*convex with respect to* M if $A = \emptyset$ or if for each system $x^1, ..., x^k \in A$ and all real numbers $t_1, ..., t_k$ satisfying the condition $t_1, ..., t_k \geq 0$, $t_1 + ... + t_k = 1$ and $t_1x^1 + ... + t_kx^k \in M$ we have $t_1x^1 + ... + t_kx^k \in A$.

This property of convexity is a particular case of k-strong convexity with respect to a convex hull operator and a given set M. Here, g (A) = conv (A) for each $A \subseteq X$. Therefore, if we take Y = X = V, $S = \{\{x_1, x_2, \ldots, x_k\} \mid x_1, x_2, \ldots, x_k \in X\}$, $s: S \to 2^X$ defined by $s(B) = g(B) \cap M$ for all $B \in S$, according to theorem 8.3.12, we obtain that the k-strong convexity with respect to a given set is a (S, s) convexity.

Example 8.3.22. Strong convexity with respect to g and M (L. Blaga, L. Lupşa (1998))

Let X and M be two arbitrary nonempty sets and let g be a convex hull operator on X. Let k be a natural number, $k \geq 2$. A subset $A \subseteq X$ is called *strongly convex with respect to* g *and* M if $g(A) \cap M \subseteq A$.

Let us choose Y = X, $S = 2^X$, $s: S \to 2^X$ defined by $s(B) = g(B) \cap M$ for all $B \in S$. As usual, we can prove the following result.

Theorem 8.3.13. *A set* $A \subseteq X$ *is strongly convex with respect to* g *and* M *if and only if* A *is* (S, s) *convex.*

Example 8.3.23. Slack k-convexity with respect to g and M (L. Blaga, L. Lupşa (1998))

Let X and M be two arbitrary nonempty sets and let g be a convex hull operator on X. A subset $A \subseteq X$ is called *slackly* k-*convex with respect to* g *and* M if for any $a_1,...,a_k$ elements of A we have $g(\{a_1,...,a_k\} \cap M) \cap M \subseteq A$.

Let us choose Y = X, $S = \{\{x_1, x_2, \ldots, x_k\} \mid x_1, x_2, \ldots, x_k \in X\}$, $s: S \to 2^X$ defined by $s(B) = g(B \cap M) \cap M$ for all $B \in S$.

Theorem 8.3.14. *A set* $A \subseteq X$ *is slackly* k*-convex with respect to* g *and* M *if and only if* A *is* (S, s) *convex.*

Example 8.3.24. Slack convexity with respect to g and M (L. Blaga, L. Lupşa (1998))

Let X and M be two arbitrary nonempty sets and let g be a convex hull operator on X. Let k be a natural number, $k \geq 2$. A subset $A \subseteq X$ is called *slackly convex with respect to* g *and* M if $g(A \cap M) \cap M \subseteq A$.

Let us choose $Y = X$, $S = 2^X$ and consider function $s: S \to 2^X$ defined by

$$s(B) = g(B \cap M) \cap M \text{ for all } B \in S.$$

As usual, we can prove the following result.

Theorem 8.3.15. *A set* $A \subseteq X$ *is slackly convex with respect to* g *and* M *if and only if* A *is* (S, s) *convex.*

Example 8.3.25. Slack n-convex sets with respect to M (L. Lupşa (1981))

Let $(V,+,\cdot)$ be a real linear space and M a nonempty subset of V. Let n be a natural number, $n > 0$. A subset $A \subseteq V$ is called *slackly* n*-convex with respect to* M if $A = \varnothing$ or if for all system $x^1, \ldots, x^n \in M \cap A$ and all real numbers $t_1, \ldots, t_n$ satisfying conditions $t_1 \geq 0, \ldots, t_n \geq 0, t_1 + \ldots + t_n = 1$, and $t_1x^1 + \ldots + t_nx^n \in M$ we have $t_1x^1 + \ldots + t_nx^n \in A$.

The property of slack n-convexity with respect to a given set M is a particular case of n-slack convexity with respect to g and M. According to theorem 8.3.14, it is also a (S, s) convexity.

Example 8.3.26. Strong convex sets with respect to M (L. Lupşa (1981))

Let $(V,+,\cdot)$ be a real linear space and M a nonempty subset of V. Let n be a natural number, $n > 0$. A subset $A \subseteq V$ is called *strongly convex with respect to* M if $A = \varnothing$ or if $M \cap \text{conv}(A) \subseteq A$.

The property of strong convexity with respect to a given set M is a particular case of strong convexity with respect to g and M. Therefore, according to theorem 8.3.12, it is also a (S, s) convexity.

Example 8.3.27. Integer convexity (L. Lupşa (1981))

A subset $X \subseteq \mathbf{Z}^n$ is said to be an *integer convex* set if X is strongly convex with respect to $\mathbf{Z}^n$. It is obvious, according to the previous example, that the integer convexity is also a property of (S, s) convexity.

Example 8.3.28. Slack convex sets with respect to M (L. Lupşa (1981))

Let $(V,+,\cdot)$ be a real linear space and M a nonempty subset of V. Let n be a natural number, $n > 0$. A subset $A \subseteq V$ is called *slackly convex with respect to* M if $X = \emptyset$ or if $M \cap \text{conv}(M \cap A) \subseteq A$.

The property of slack convexity with respect to a given set M is a particular case of slack convexity with respect to g and M and, according to theorem 8.3.14, a (S, s) convexity.

Example 8.3.29. Induced seg-convexity with respect to p, f and M (L. Blaga, L. Lupşa (1998))

Let X, T be two arbitrary sets, M be a nonempty subset of T and let us consider the functions $f: X \to T$ and $p: 2^T \to 2^T$. For each points $a, b \in X$ we define the set

$$<a, b>_{p,f,M} = f^{-1}(p(\{f(a), f(b)\}) \cap M) =$$
$$= \{z \in X \mid f(z) \in p(\{f(a), f(b)\}) \cap M\}.$$

A subset $A \subseteq X$ is called *induced seg-convex with respect to* p, f *and* M if $<a, b>_{p,f,M} \subseteq A$, for any $a, b \in A$.

As usual, if we put $Y = X$, $S = \{\{x, y\} \mid x \in X, y \in X\}$ and we consider the function $s: S \to 2^X$ defined by

$$s(\{x, y\}) = <x, y>_{p,f,M} = f^{-1}(p(\{f(a), f(b)\}) \cap M)$$

for $\{x, y\} \in S$, we obtain that the property of induced seg-convexity with respect to p, f and M is a (S, s) convexity.

Example 8.3.30. E – convexity (E.A. Youness (1999))

E.A. Youness (1999) defined an E-convex set as it follows: a set $A \subseteq \mathbf{R}^n$ is called E-*convex* if there is a map $E: \mathbf{R}^n \to \mathbf{R}^n$ such that $(1-t)E(x) + tE(y) \in A$ for each $x \in A$ and $y \in A$ and $t \in [0, 1]$. It is easy to see that this is a trivial case. Indeed, for each set we can define such a map E. For example, if we consider set $A \subseteq \mathbf{R}^n$ together with one of its points $a \in A$, by defining the constant function $E: \mathbf{R}^n \to \mathbf{R}^n$, $E(x) = a$ for each $x \in A$, then A is E-convex.

The situation in which function E is previously chosen and fixed is of particular interest. This case will be discussed in what follows. Let us choose function $E: \mathbf{R}^n \to \mathbf{R}^n$. A set $A \subseteq \mathbf{R}^n$ is said to be E-*convex* if for each $x \in A$ and $y \in A$ and $t \in [0, 1]$, the relation $(1-t)E(x) + tE(y) \in A$ holds.

Now, let us choose $X = Y = \mathbf{R}^n$, $S = \{\{x, y\} \mid x \in \mathbf{R}^n, y \in \mathbf{R}^n\}$ and let us consider function $s: S \to 2^{\mathbf{R}^n}$ defined by $s(\{x, y\}) = \{(1-t)E(x) + tE(y) \mid t \in [0, 1]\}$ for $\{x, y\} \in S$.

Theorem 8.3.16. *A set* $A \subseteq \mathbf{R}^n$ *is* E-*convex if and only if* A *is* (S, s) *convex.*

Proof. Set A is E-convex if and only if for each two points $x \in A$ and $y \in A$, the relation $(1-t)\, E(x) + t\, E(y) \in A$ holds. This is equivalent to the fact that each subset $\{x, y\} \in S \cap A$ implies that $s(\{x, y\}) \subseteq r(A) = A$, which means that A is (S, s) convex.

Example 8.3.31. (φ, ψ) – convexity (D. Duca, E. Duca, L. Lupşa and L. R. Blaga (2000))

Let $\varphi: \mathbf{R}^n \to \mathbf{R}^n$ and $\psi: \mathbf{R}^n \to \mathbf{R}^n$ two given functions. A subset A of $\mathbf{R}^n$ is said to be (φ, ψ) – convex if we have $(1-t)\cdot\varphi(x) + t\cdot\psi(y) \in A$, for each pair of points $x, y \in A$ and each $t \in [0, 1]$.

Now, if we choose $X = Y = \mathbf{R}^n$, $S = \{\{x, y\} \mid x \in \mathbf{R}^n, y \in \mathbf{R}^n\}$ and function $s: S \to 2^{\mathbf{R}^{2n}}$ defined by

$$s(\{x, y\}) = \{(1-t)\cdot\varphi(x) + t\cdot\psi(y) \mid t \in [0, 1]\} \text{ for } \{x, y\} \in S,$$

we can prove that a set A is (φ, ψ) – convex if and only if it is (S, s) convex.

Example 8.3.32. bi – (φ, ψ) – convexity (D. Duca, E. Duca, L. Lupşa and L. R. Blaga (2000))

Let m and n be two natural non-null numbers and let $\varphi: P \subseteq \mathbf{R}^m \to \mathbf{R}^m$, and $\psi: Q \subseteq \mathbf{R}^n \to \mathbf{R}^n$, two given non-constant functions. A set $A \subset P \times Q$ is said to be bi-(φ, ψ) - convex, if for all (x, y), (x, v), (u, y) elements of A and for each $t \in [0, 1]$ we have both

$$(\varphi(x), (1-t)\, \psi(y) + t\, \psi(v)) \in A$$

and

$$((1-t)\, \varphi(x) + t\, \varphi(u), \psi(y)) \in A.$$

Let us choose $X = Y = \mathbf{R}^{m+n}$, $S = \{\{(x, y), (x, v), (u, y)\} \mid x, u \in \mathbf{R}^m, v, y \in \mathbf{R}^n\}$ and function $s: S \to 2^{\mathbf{R}^{m+n}}$ defined by

$$s(\{x, y\}) = \{(\varphi(x), (1-t)\, \psi(y) + t\, \psi(v)) \mid t \in [0, 1]\} \cup$$
$$\cup \{((1-t)\, \varphi(x) + t\, \varphi(u), \psi(y)) \mid t \in [0, 1]\},$$

for $\{(x, y), (x, v), (u, y)\} \in S$. We can prove that a set A is bi - (φ, ψ) – convex if and only if it is (S, s) convex.

Example 8.3.33. bi - (φ, 1) – convexity (D. Duca, E. Duca, L. Lupşa and L. R. Blaga (2000))

Let $\varphi: \mathbf{R}^m \to \mathbf{R}^m$ be a given function and the nonempty sets $P \subseteq \mathbf{R}^m$ and $Q \subseteq \mathbf{R}^n$. A set $A \subset P \times Q$ is said to be bi – $(\varphi, 1)$ - convex if for each $y \in Q$, x' and x'' such that $(x', y) \in A$ and $(x'', y) \in A$ and each $t \in [0, 1]$

$$((1-t)\, \varphi(x') + t\, \varphi(x''), y) \in A.$$

Let us choose $X = Y = \mathbf{R}^{m+n}$, $S = \{\{(x, y), (u, y)\} \mid x, u \in \mathbf{R}^m, y \in \mathbf{R}^n\}$ and function $s: S \to 2^{\mathbf{R}^{m+n}}$ defined by

$$s(\{x, y\}) = \{((1-t)\,\varphi(x) + t\,\varphi(u), y) \mid t \in [0, 1]\}.$$

for $\{(x, y), (u, y)\} \in S$. As usual, set A is bi – (φ, 1) – convex if and only if A is (S, s) convex.

Example 8.3.34. bi – (1, ψ) – convexity (D. Duca, E. Duca, L. Lupşa and L. R. Blaga (2000))

Let $\psi: \mathbf{R}^n \to \mathbf{R}^n$ be a given function and two nonempty sets $P \subseteq \mathbf{R}^m$ and $Q \subseteq \mathbf{R}^n$. A set $A \subset P \times Q$ is said to be bi – (1, ψ) convex if for each $x \in P$, y' and y" such that $(x, y') \in A$ and $(x, y'') \in A$ and each $t \in [0, 1]$ we have

$$(x, (1-t)\,\psi(y') + t\,\psi(y'')) \in A.$$

Let us choose $X = Y = \mathbf{R}^{m+n}$, $S = \{\{(x, y), (x, z)\} \mid x \in \mathbf{R}^m, y, z \in \mathbf{R}^n\}$ and function $s: S \to 2^{\mathbf{R}^{m+n}}$ defined by

$$s(\{(x, y), (x, z)\}) = \{(x, (1-t)\,\psi(y) + t\,\psi(z)) \mid t \in [0, 1]\}.$$

for $\{(x, y), (x, z)\} \in S$. Set A is bi – (1, ψ) – convex if and only if A is (S, s) convex.

Example 8.3.35. Segmential multi-order convexity (J.E. Martínez-Legaz, I. Singer (1990))

A multi-ordered set is an ordered pair (X, O), where X is a set and O is a nonempty family of partial orders on X.

Let (X, O) be a multi-ordered set. Given $x, y \in X$, we define the O-segment $\langle x, y\rangle = \langle x, y\rangle_O = \cap\{\langle x, y\rangle_\rho \mid \rho \in O\}$. Here, $\langle x, y\rangle_\rho = \{z \in X \mid$ either $x\,\rho\,z\,\rho\,y$ or $y\,\rho\,z\,\rho\,x\}$ for each partial order $\rho \in O$. A set $A \subseteq X$ is said to be O – *segmentially convex* (or, briefly, O_{seg} – *convex*) if $\langle x, y\rangle \subseteq A$ for each $x, y \in A$.

As usual, one proves that the O – segmential convexity is a property of (S, s) convexity.

Example 8.3.36. ε – convexity (J. Perkal (1956))

Let (X, d) be a metric space. The ε – convex hull of $A \subseteq X$ is the set of all points $p \in X$ situated at a distance of at least ε/2 from each point situated at a distance greater than ε/2 from A. Simbol $C_\varepsilon(A)$ denotes the ε - convex hull of A. Obviously, if $A \subseteq B$ then $C_\varepsilon(A) \subseteq C_\varepsilon(B)$. On the other hand, for each $A \subseteq X$ the inclusion $A \subseteq C_\varepsilon(A)$ occurs.

A set $A \subseteq X$ is said to be ε – *convex* if $A = C_\varepsilon(A)$. Obviously, it is enough to say that a set $A \subseteq X$ is ε – convex if $C_\varepsilon(A) \subseteq A$.

If we take $X = Y$, $S = 2^X$ and $s: S \to 2^X$ is defined by $s(A) = C_\varepsilon(A)$ for each $A \subseteq X$, using the monotony of function s, it is easy to prove that a set is ε – convex if and only if it is (S, s) convex.

Example 8.3.37. B – convexity in a complete lattice (I. Singer (1997))

Let (E, $\leq$) be a complete lattice. A subset B of E is called a convexity system in E if it is closed for supremum that is if for any index set I we have the implication: $x_i \in B$ $(i \in I) \Rightarrow \sup \{x_i \mid i \in I\} \in B$.

The elements of B are called B – *convex*.

Let X = E, Y = E, S = 2^E and function s: S $\to 2^Y$, defined by

$$s(D) = \bigcup_{I \text{ index set}} \{\sup\{x_i \in D \mid i \in I\}\}, \text{ for each } D \in S,$$

Function r is the identity in 2^E.

Theorem 8.3.17. *A subset* B *of* E *is a convexity system in* E *if and only if set* B *is* (S, s) *convex.*

Proof. Necessity. Let be $B \subseteq E$ a convexity system in E. This implies that for each $D \subseteq B$, $\bigcup_{I \text{ index set}} \{\sup \{x_i \in D \mid i \in I\} \subseteq B$ occurs. In view of definitions of s and r we obtain s (B) $\subseteq$ r (B). Hence B is (S, s) convex.

Sufficiency. Let B be a (S, s) convex set. Then s (D) $\subseteq$ r (B) for each D $\subseteq$ B. According to the definitions of function s and r, one has

$$\bigcup_{I \text{ index set}} \{\sup \{x_i \in B \mid i \in I\} \subseteq B.$$

Hence, set B is closed for supremum, meaning that set B is a convexity system in E.

Example 8.3.38. Polynomial convex sets (G. Stolzenberg (1962))

Let $E \subset \mathbf{C}^n$ be a compact set. The polynomial hull of E is the set

$$\text{hull}(E) = \{z \in \mathbf{C}^n \mid |f(z)| = \max\{|f(p)| \mid p \in E\}, \forall \text{ polynomial } f \text{ on } \mathbf{C}^n\}.$$

Set E is *polynomially convex* if E = hull (E). Obviously, each set $E \subset \mathbf{C}^n$ satisfies $E \subseteq$ hull (E). Therefore, a set E is polynomially convex if and only if hull (E) $\subseteq$ E. On the other hand, it is evident that if $A \subseteq B$ then hull (A) $\subseteq$ hull (B).

Let us take X = Y = $\mathbf{C}^n$, S = 2^X, s: S $\to 2^X$ defined by s (A) = hull (A) for each $A \subseteq \mathbf{C}^n$, and $r = 1_{2^X}$. Therefore, a set $A \subset \mathbf{C}^n$ is polynomially convex if and only if A is (S, s) convex.

Example 8.3.39. Rational convex sets (G. Stolzenberg (1963))

Let $E \subset \mathbf{C}^n$ be a compact set. The rational hull of E is the set

$$\text{R-hull}(E) = \{z \in \mathbf{C}^n \mid |f(z)| = \max\{|f(p)| \mid p \in E\}, \forall \text{ rational function } f \text{ on } \mathbf{C}^n, \text{ which is analytic about } E\}.$$

Set E is *rationally convex* if E = R-hull (E). It is evident that if $A \subseteq B$ then R-hull (A) $\subseteq$ R-hull (B). On the other hand, obviously, each set $E \subset \mathbf{C}^n$ satisfies $E \subseteq$ R-hull (E). Therefore, a set E is rationally convex if and only if R-hull (E) $\subseteq$ E. Let us take $X = Y = \mathbf{C}^n$, $S = 2^X$, function s: $S \to 2^X$ defined by s (A) = R-hull (A) for each $A \subseteq \mathbf{C}^n$, and $r = 1_{2^X}$. A set $A \subset \mathbf{C}^n$ is rationally convex if and only if A is (S, s) convex.

Example 8.3.40. Discrete convexity in $\mathbf{Z}^2$ (M. Minsky, S. Papert (1969), A. Rosenfeld, A. Kak (1980))

In $\mathbf{Z}^2$ the distance $c(x, y) = \max(|x_1-y_1|, |x_2-y_2|)$, $x = (x_1, x_2) \in \mathbf{Z}^2$, $y = (y_1, y_2) \in \mathbf{Z}^2$, is used. The connectivity by means of arcs induced by this distance is considered and is called, as usual, 8-connectivity. The 8-connected component A in $\mathbf{Z}^2$ is said to be *discrete convex* if there is not a triplet (P_1, P_2, P_3) of collinear points, with P_2 situated between P_1 and P_3, such that P_1 and P_3 are situated in A and P_2 belongs to the complement of A. It means that each triplet of collinear points (P_1, P_2, P_3) with P_1 and P_3 situated in A has the property that P_2 lies in A.

Let us take $X = Y = \mathbf{Z}^2$, $S = \{\{P_1, P_2\} \mid P_1 \in \mathbf{Z}^2, P_2 \in \mathbf{Z}^2\}$, s: $S \to 2^X$ defined by $s(\{P_1, P_2\}) = \{P \mid P \in \mathbf{Z}^2, \{P_1, P, P_2\}$ collinear, P between P_1 and $P_2\}$, and $r = 1_{2^X}$.

Theorem 8.3.18. *An 8-connected component* $A \subseteq \mathbf{Z}^2$ *is discrete convex if and only if* A *is* (S, s) *convex.*

Proof. Let us suppose that A is discrete convex. Then each triplet of collinear points (P_1, P_2, P_3) with P_1 and P_3 situated in A has the property that P_2 lies in A. Therefore, for each subset $\{P_1, P_3\} \in S$, with $\{P_1, P_3\} \subseteq A$, one has $s(\{P_1, P_3\}) \subseteq A = r(A)$. It means that A is (S, s) convex.

Conversely, if A is (S, s) convex then for each $\{P_1, P_3\} \in S$, with $\{P_1, P_3\} \subseteq A$, $s(\{P_1, P_3\}) \subseteq r(A)$ occurs. It means that each point P_2 collinear with P_1 and P_3 and situated between them lies in A = r (A). Therefore, there is not a triplet (P_1, P_2, P_3) of collinear points, with P_2 situated between P_1 and P_3, such that P_1 and P_3 are situated in A and P_2 belongs to the complement of A. It means that A is discrete convex.

A. Rosenfeld, A. Kak (1980) (see also A. Rosenfeld, A. Kak (1982)) use an equivalent point of view. According to them, a set A is convex if every straight-line intersects A at most once, meaning only in a succession of 8-neighbour points. This generated investigations on the situation when the above-mentioned intersection property holds only for straight-lines with particular directions, leading to the directional convexity.

Example 8.3.41. s-convexity

In a linear space X, for a real number s, $0 < s \le 1$, a set $A \subseteq X$ is said to be s-*convex* if $x \in A$, $y \in A$ and $t \in [0,1]$ implies that

$$t^s x + (1-t)^s y \in A.$$

It is now obvious that the s-convexity is a property of (S, s) convexity if $X = Y$, $S = \{\{x, y\} \mid x \in X, y \in X\}$ and let us consider function $s: S \to 2^X$ defined by

$$s(\{x, y\}) = \{t^s x + (1-t)^s y \mid t \in [0, 1]\} \text{ for } \{x, y\} \in S.$$

Example 8.3.42. Wedge

A subset A of a linear space X is called a *wedge* if it has the following properties:

1. $x, y \in A$ implies $x + y \in A$;
2. $x \in A$ and $\alpha \in \mathbf{R}_+$, $\alpha x \in A$.

The structure of wedge of a set is a property of (S, s) convexity if $X = Y$, $S = \{\{x, y\} \mid x \in X, y \in X\}$ and function $s: S \to 2^X$ is defined by

$$s(\{x, y\}) = \{\alpha x + \beta y \mid \alpha, \beta \in \mathbf{R}_+\} \text{ for } \{x, y\} \in S.$$

8.4 The class of ((S, s), r) convexity properties

All the elements and the notations from the first paragraph are taken into account. Set S is supposed to be singleton, function $s: S \to 2^Y$ being its unique element. Also, the R family of transformations $r: 2^X \to 2^X$ is supposed to be singleton, its unique element being denoted by r. In this case, all the partial convexity properties defined in the second paragraph are equivalent and are called simply, *partial* ((S, s), r) *convexity*. Also, all the convexity properties defined in the second paragraph are equivalent and are called ((S, s), r) *convexity*. In this paragraph we shall present the elements of the class of the ((S, s), r) convexity properties.

Example 8.4.1. Directional convexity (P.L. Yu (1974[293]))

Let us consider $X = \mathbf{R}^n$ and the fixed element $u \in \mathbf{R}^n \setminus \{0\}$. A set $A \subseteq \mathbf{R}^n$ is called *directionally convex in the direction u*, if $x \in A$, $y \in A$ and $t \in [0,1]$ implies that there is the number $\mu \ge 0$, such that

$$t\,x + (1\text{-}t)\,y + \mu u \in A.$$

If we take $Y = X = \mathbf{R}^n$, $S = \{\{x, y\} \mid x \in \mathbf{R}^n, y \in \mathbf{R}^n\}$, then we define functions $s\colon S \to 2^{\mathbf{R}^n}$ and $r\colon 2^{\mathbf{R}^n} \to 2^{\mathbf{R}^n}$ by

$$s(\{x, y\}) = \{t\,x + (1\text{-}t)\,y \mid t \in [0, 1]\} \text{ for all } \{x, y\} \in S,$$

$$r(C) = \{c - \alpha u \mid c \in C, \alpha \geq 0\}, \text{ for all } C \in 2^{\mathbf{R}^n}.$$

Theorem 8.4.1. *A set* $A \subseteq \mathbf{R}^n$ *is directionally convex in direction* u, *if and only if* A *is* ((S, s), r) *convex.*

Proof. If set $A \subseteq \mathbf{R}^n$ is directionally convex in direction u then for every $x \in A$, $y \in A$ and $t \in [0, 1]$ there is the real number $\mu \geq 0$ such that the relation $t\,x + (1\text{-}t)\,y + \mu u \in A$ holds. This is equivalent to $tx + (1\text{-}t)y \in A\text{-}\mu u$ for each $t \in [0, 1]$, which means that $s(\{x, y\}) \subseteq A - \mu u \subseteq r(A)$.

Example 8.4.2. Cone convexity (P.L. Yu (1974 [293]))

Let us choose $X = \mathbf{R}^n$ and denote by K a convex cone in $\mathbf{R}^n$. A set $A \subseteq \mathbf{R}^n$ is said to be K – *cone convex*, if A + K is a convex set.

Let us put $Y = X = \mathbf{R}^n$, $S = \{\{x, y\} \mid x \in \mathbf{R}^n, y \in \mathbf{R}^n\}$ and define functions $s\colon S \to 2^{\mathbf{R}^n}$ and $r\colon 2^{\mathbf{R}^n} \to 2^{\mathbf{R}^n}$ by

$$s(\{x, y\}) = \{t\,x + (1\text{-}t)\,y + \mu \mid t \in [0, 1], \mu \in K\} \text{ for all } \{x, y\} \in S,$$

$$r(C) = C + K, \text{ for } C \in 2^{\mathbf{R}^n}.$$

Theorem 8.4.2. *A set* $A \subseteq \mathbf{R}^n$ *is* K - *cone convex if and only if* A *is* ((S, s), r) *convex.*

Proof. If set $A \subseteq \mathbf{R}^n$ is K - cone convex then for each points $x \in A + K$ and $y \in A + K$ and for every $t \in [0, 1]$ the relation $t\,x + (1\text{-}t)\,y \in A + K$ holds. But each point $z \in A + K$ can be written as $z = c + \alpha$, with $c \in A$ and $\alpha \in K$. For the points $a \in A$ and $b \in A$ one has

$$\begin{aligned}
s(\{a, b\}) &= \{t\,a + (1\text{-}t)\,b + \lambda \mid t \in [0, 1], \lambda \in K\} = \\
&= \{t\,a + (1\text{-}t)\,b + t\,\lambda + (1\text{-}t)\,\lambda \mid t \in [0, 1], \lambda \in K\} = \\
&= \{t\,(a + \lambda) + (1\text{-}t)\,(b + \lambda) \mid t \in [0, 1], \lambda \in K\} \subseteq \\
&\subseteq \{t\,(a + K) + (1\text{-}t)\,(b + K) \mid t \in [0, 1]\} \subseteq \\
&\subseteq \{t\,(A + K) + (1\text{-}t)\,(A + K) \mid t \in [0, 1]\} = A + K = r(A).
\end{aligned}$$

Therefore A is ((S, s), r) convex.

Remark 8.4.1. Both definition of the K-cone convexity and theorem 8.4.2 are valid if X is a real linear space.

Example 8.4.3. Closely convex sets (L. Blaga, I. Kolumbán (1994))

In a topological linear space X over the field **R** a set is *closely convex* if its closure is a convex set. For a set $A \subseteq X$ we denote by $\overline{A}$ its closure.

Let us take Y=X, $S = 2^X$ and define functions s: $S \to 2^X$ and r: $2^X \to 2^X$ by $s(C) = \mathrm{conv}(\overline{C})$ for $C \in S$, $r(C) = \overline{C}$ for $C \subseteq X$.

Theorem 8.4.3. *A set* $A \subseteq X$ *is closely convex if and only if* A *is* ((S, s), r) *convex.*

Proof. Let us suppose that A is closely convex. Then for every $x \in \overline{A}$ and $y \in \overline{A}$ we have $<x, y> \subseteq \overline{A}$. It means that $\mathrm{conv}(\overline{A}) = \overline{A}$ and, due to the property of monotony of function s, for each $C \subseteq A$ we have $s(C) = \mathrm{conv}(\overline{C}) \subseteq \mathrm{conv}(\overline{A}) \subseteq \overline{A} = r(A)$. Therefore A is ((S, s), r) convex.

Conversely, if A is ((S, s), r) convex then for each $C \in S$, $C \subseteq A$ one has $s(C) = \mathrm{conv}(\overline{C}) \subseteq r(A) = \overline{A}$. In particular, for C = A we obtain that $s(A) = \mathrm{conv}(\overline{A}) \subseteq r(A) = \overline{A}$. It means that $\mathrm{conv}(\overline{A}) = \overline{A}$ and $\overline{A}$ is convex. Then A is closely convex.

Example 8.4.4. Closely convex sets with respect to a cone (W.W. Breckner, G. Kassay (1997))

In a topological linear space X over **R**, we consider the convex cone K. A set $A \subseteq X$ is *closely convex with respect to cone* K if the set A + K is a closely convex set.

As in the previous example, let us take Y = X, $S = 2^X$ and define two functions s: $S \to 2^X$ and r: $2^X \to 2^X$ by

$$s(C) = \mathrm{conv}(\overline{C + K}), \text{ for } C \in S,$$

$$r(C) = \overline{C + K}, \text{ for } C \in 2^X.$$

Theorem 8.4.4. *A set* $A \subseteq X$ *is closely convex with respect to cone* K *if and only if* A *is* ((S, s), r) *convex.*

The proof is similar to the previous one.

Example 8.4.5. α-convex sets (G. Cristescu (1997[64]))

Let (X, d) be a metric space together with a notion of straight-line segment (not necessary the metric segment) joining every pair (x, y) of points of X, denoted by <x, y>. Let us choose a real number $\alpha \geq 0$.

Set $A \subset X$ is said to be α - *convex* if it is empty or if for every two points $x \in A$ and $y \in A$ every point $t \in <x, y>$ satisfies the condition

$$d(t, A) \leq \alpha.$$

Let us take $Y = X$, $S = \{\{x, y\} \mid x \in X, y \in X\}$ and define two functions $s: S \to 2^X$ and $r: 2^X \to 2^X$ by

$$s(\{x, y\}) = \langle x, y\rangle \text{ for } \{x, y\} \in S,$$

$$r(C) = \{t \in X \mid d(t, C) \leq \alpha\} \text{ for } C \subseteq X.$$

Theorem 8.4.5. *A set* $A \subseteq X$ *is* α*-convex if and only if* A *is* ((S, s), r) *convex.*

Proof. Set A is supposed to be α-convex. This means that for every $x, y \in A$ and $t \in \langle x, y\rangle$ the inequality $d(t, A) \leq \alpha$ occurs. This is equivalent to the fact that for every $\{x, y\} \in S$ with $x, y \in A$, one has

$$s(\{x, y\}) = \langle x, y\rangle \subseteq \{t \in X \mid d(t, A) \leq \alpha\} = r(A).$$

Therefore, A is also ((S, s), r) convex.

Example 8.4.6. Convexity with respect to a behaviour (G. Cristescu (1997 [64])

In the nonempty set X a notion of straight-line segment, $\langle x, y\rangle$, for $x, y \in X$, is defined. For the nonempty set B and its nonempty subset $D \subset B$, the behaviour (strict, weak or strong) $C = (D, T)$ is defined in $X \times 2^X$, with $T \in \mathsf{T} \subset \{T \mid T: X \times 2^X \to B\}$. According to definition 3.4.1, set $A \subset X$ is called *convex with respect to behaviour* C if it is empty or if for every two points $x, y \in A$ and for every $t \in \langle x, y\rangle$, the condition $(t, A) \in C(X \times 2^X)$ holds. The convexity presented in the previous example is a particular case of convexity with respect to a behaviour.

As above, let us take $Y = X$, $S = \{\{x, y\} \mid x \in X, y \in X\}$ and let us define functions $s: S \to 2^X$ and $r: 2^X \to 2^X$ by

$$s(\{x, y\}) = \langle x, y\rangle \text{ for } \{x, y\} \in S,$$

$$r(Z) = \{t \in X \mid (t, Z) \in C(X \times 2^X)\} \text{ for } Z \subseteq X.$$

Theorem 8.4.6. *A set* $A \subseteq X$ *is convex with respect to the behaviour* C *if and only if* A *is* ((S, s), r) *convex.*

Proof. Set A is supposed to be convex with respect to behaviour C. Then, for every $x, y \in A$ and $t \in \langle x, y\rangle$ the relation $(t, A) \in C(X \times 2^X)$ is satisfied. This is equivalent to the fact that for every $\{x, y\} \in S$ with $x, y \in A$, one has

$$s(\{x, y\}) = \langle x, y\rangle \subseteq \{t \in X \mid (t, A) \in C(X \times 2^X)\} = r(A).$$

Therefore, A is also ((S, s), r) convex.

Example 8.4.7. Slack (α, δ) - convexity with respect to a network (G. Cristescu (1996[57])

Let (X, d) be a totally bounded metric space and for $\varepsilon > 0$ let E' be a finite ε-network over X, and by E a finite 2ε - network over E' is denoted. For a set $A \subset X$ and a point $x \in X$, the distance between x and A is defined as usual by number $d(x, A) = \inf\{d(x, A) \mid a \in A\}$. In X it is supposed that, for each two points $x \in X$ and $y \in Y$ a notion of straight-line segment $<x, y>$ is defined (not necessarily the classical one in metric spaces). Let the real numbers $\alpha \geq 0$ and $\delta \geq 0$, and let E be a network chosen as above. Set $A \subset X$ is said to be *slackly* (α, δ) - *convex with respect to E* if it is empty or if for each pair of points $x, y \in A$ and for each $t \in <x, y>$ if the element $a \in E$ has the property that $d(t, a) \leq \alpha$ then the inequality $d(a, A) \leq \delta$ also holds.

Let us take $Y = X$, $S = \{\{x, y\} \mid x \in X, y \in X\}$ and define two functions $s: S \to 2^X$ and $r: 2^X \to 2^X$ by

$$s(\{x, y\}) = \{a \in E \mid \exists\, t \in <x, y>, d(a, t) \leq \alpha\} \text{ for } \{x, y\} \in S,$$

$$r(C) = \{t \in X \mid d(t, C) \leq \delta\} \text{ for } C \subseteq X.$$

Theorem 8.4.7. *A nonempty set* $A \subseteq X$ *is slackly* (α, δ) – *convex with respect to E if and only if* A *is* ((S, s), r) *convex.*

Proof. Set A is supposed to be slackly (α, δ)-convex. This means that for each $x, y \in A$ and $t \in <x, y>$ the inequality $d(t, a) \leq \alpha$, for $a \in E$, implies $d(a, A) \leq \delta$. This is equivalent to the fact that for each $\{x, y\} \in S$ with $x, y \in A$, one has

$$s(\{x, y\}) = \{a \in E \mid \exists\, t \in <x, y>, d(a, t) \leq \alpha\} \subseteq$$

$$\subseteq \{a \in E \mid d(a, A) \leq \delta\} \subseteq \{t \in X \mid d(t, A) \leq \delta\} = r(A).$$

Therefore the hypothesis that A is slackly (α, δ)-convex is equivalent to the property of ((S, s), r) convexity of A.

Example 8.4.8. Slack convexity with respect to a set and two behaviours (G. Cristescu (1996 [56]))

X is assumed to be a nonempty set so that for each two points x, y of X the closed straight-line segment determined by these points, denoted by $<x, y>$, is defined. Let us suppose that M is a nonempty subset of X, B' and B" two nonempty sets, $D' \subset B'$, $D'' \subset B''$ nonempty subsets, and T ' is a set of transformations, $T': X \times M \to B'$. The behaviour C' = (D', T') (or a strong one s - (D', T '), or a weak one w - (D', T ')) is defined in $X \times M$ by means of B'. For the set T " of transformations, $T'': M \times 2^X \to B''$, we define a behaviour C" = (D", T") (or s - (D', T "), or w - (D', T ")) in $M \times 2^X$ by means of B". The set $Y \subset X$ is said to be *slackly convex with respect to set*

M *and behaviours* C' *and* C" if it is empty or if for each two points $x, y \in Y$ and for each $t \in <x, y>$, if the element $(t, a) \in C'(X \times M)$, with $a \in M$, then the element $(a, Y) \in C''(M \times 2^X)$.

As above, let us take $Y = X$, $S = \{\{x, y\} \mid x \in X, y \in X\}$ and let us define functions $s: S \to 2^X$ and $r: 2^X \to 2^X$ by

$$s(\{x, y\}) = \{a \in M \mid \exists\, t \in <x, y>, (t, a) \in C'(X \times M)\} \text{ for } \{x, y\} \in S,$$

$$r(Z) = \{t \in M \mid (t, Z) \in C''(M \times 2^X)\} \text{ for } Z \subseteq X.$$

Theorem 8.4.8. *A nonempty set* $A \subseteq X$ *is convex with respect to set* M *and behaviours* C' *and* C" *if and only if* A *is* ((S, s), r) *convex.*

Proof. Set A is supposed to be convex with respect to set M and behaviours C' and C". Then, for each $x, y \in A$ and $t \in <x, y>$ the relation $(t,a) \in C'(X \times M)$, with $a \in M$, implies $(a, A) \in C''(M \times 2^X)$. This is equivalent to the fact that for each $\{x, y\} \in S$ with $x, y \in A$, one has

$$s(\{x, y\}) = \{a \in M \mid \exists\, t \in <x, y>, (t, a) \in C'(X \times M)\} \subseteq$$

$$\subseteq \{t \in M \mid (t, A) \in C''(M \times 2^X)\} = r(A).$$

meaning that A is also ((S, s), r) convex.

Example 8.4.9. Quasi-convex sets (I. Singer (1984))

Let X be a locally convex Hausdorff topological vector space and $Z \subset X$ a closed set such that $0 \in Z$. By $Q_0(Z)$ we denote the set of all lower semi-continuous quasi-convex functions $q: Z \to \mathbf{R} \cup \{+\infty\}$ with the property $q(0) = \inf\{q(x) \mid x \in Z\}$. We consider subset $J(X) \subset Q_0(X)$, which consists of all two step functions j of the following form and denoted by an ordered triple

$$j(x) = (v, c, c') = \begin{cases} c, & v(x) > 1 \\ c', & v(x) \leq 1 \end{cases},$$

where $v \in X'$, $c, c' \in \mathbf{R}$, $c' \leq c$. Here X' is the set of all continuous linear functions defined on X. Now, we consider set $\bar{J}(X)$ of two step functions $j = (v, c, c')$ with $c' = -\infty$, denoted by the pair (v, c). Operations $\oplus$ (called the sum) and $\otimes$ are defined in $\bar{J}(X)$:

$$(v_1, c_1) \oplus (v_2, c_2) = (v_1 + v_2, \min(c_1, c_2)),$$

$$\lambda \otimes (v, c) = (\lambda v, c) \text{ for } \lambda \in \mathbf{R}_+ \setminus \{0\} \text{ and } 0 \otimes (v, c) = 0.$$

Let $\lambda_i \geq 0$ and $\sum_i \lambda_i = 1$. The expression

$$\lambda_1 \otimes (v_1, c_1) \oplus \lambda_2 \otimes (v_2, c_2) \oplus \ldots \lambda_n \otimes (v_n, c_n)$$

is called a quasi-convex combination of functions $(v_1, c_1), \ldots, (v_n, c_n)$, denoted by $\sum_i' \lambda_i \otimes (v_i, c_i)$. Let $U \subseteq \bar{J}(X)$. The set of all quasi-convex combinations of elements of U is called the quasi-convex hull of this set and is denoted by co'U.

Now, assume $U \subseteq J(X)$. Let $\tilde{U} = \{(v, c) \in \bar{J}(X) \mid \exists c', (v, c, c') \in U\}$ and $m = \sup \{j(0) \mid j \in U\}$. The set

$$co'U = \{(v, c, c') \in J(X) \mid (v, c) \in co'\tilde{U}, c' \leq m\}$$

is called the quasi-convex hull of set U. A set $U \subseteq J(X)$ is called *quasi-convex* if co'U = U.

Now, it is easy to prove that the property of quasi-convexity is a ((S, s), r) convexity defined by means of a hull operator. In order to prove it we shall choose $S = 2^{J(X)}$, s = co'U and $r(U) = 1_{2^{J(X)}}$ for $U \subseteq J(X)$.

Example 8.4.10. Induced k-strong convexity with respect to p, f and M (L. Blaga, L. Lupşa (1998))

Let X, T be two arbitrary sets, M be a nonempty subset of T and let us consider functions $f: X \to T$ and $p: 2^T \to 2^T$. Let $k \geq 2$ be a natural number. A subset $A \subseteq X$ is called *induced* k-*strongly convex with respect to* p, f *and* M if $p(\{f(a_1), \ldots, f(a_k)\}) \cap M \subseteq f(A)$, for any $a_1, \ldots, a_k \in A$.

Now, let us take Y = X, $S = \{\{a_1, \ldots, a_k\} \mid a_j \in X, j \in \{1, 2, \ldots, k\}\}$ and define functions $s: S \to 2^X$ and $r: 2^X \to 2^X$ by

$$s(B) = p(\{f(a_1), \ldots, f(a_k)\}) \cap M,$$
$$r(A) = f(A),$$

for $B = \{a_1, \ldots, a_k\} \in S$ and $A \in 2^X$. It is easy to prove that a set $A \subseteq X$ is induced k-strongly convex with respect to p, f and M if and only if it is ((S, s), r) convex.

Example 8.4.11. M-convexity in a complete lattice (I. Singer (1997))

Let $(E, \leq)$ be a complete lattice and $M \subseteq E$. An element $x \in E$ is said to be *convex with respect to set* M (M – *convex*) if $x = \sup \{m \in M \mid m \leq x\}$.

Let be X = Y = E, $S = \{\{\alpha\} \mid \alpha \in E\}$, $s: S \to 2^E$,

(8.4.1) $\quad s(\{\alpha\}) = \{\sup \{m \in M \mid m \leq \alpha\}\}$,

and $r = 1_{2^E}$.

Theorem 8.4.9. *Element* $x \in E$ *is convex with respect to set* M *if and only if the set* $\{x\}$ *is* $((S, s), r)$ *– convex.*

Proof. Necessity. Let $x \in E$ be convex with respect to M. Then $\sup \{m \in M \mid m \leq x\} = x$. It implies $s(\{x\}) = \{x\} = r(\{x\})$. Let $C \in S$, with $C \subseteq \{x\}$. Then there is $y \in E$ such that $C = \{y\}$ and $\{y\} \subseteq \{x\}$. This implies that $x = y$. Then we have $s(C) = s(\{x\}) = r(\{x\})$. As C was arbitrary chosen, we get that the set $\{x\}$ is $((S, s), r)$ – convex.

Sufficiency. Suppose that set $\{x\}$ is $((S, s), r)$ – convex. Then, if we take $C = \{x\}$, we have $s(\{x\}) \subseteq r(\{x\}) = \{x\}$. In view of (8.4.1) it results that $\{\sup \{m \in M \mid m \leq x\}\} \subseteq \{x\}$. As E is a complete lattice, we get that

$$\sup \{m \in M \mid m \leq x\} = x.$$

Hence x is convex with respect to M.

Example 8.4.12. u – convexity in a complete lattice (I. Singer (1997))

Let $(E, \leq)$ be a complete lattice. Let $u: E \to E$ be a hull operator on E, i. e. an operator satisfying the conditions

i) $x \leq y \Rightarrow u(x) \leq u(y)$;

ii) $u(x) \leq x$;

iii) $u(u(x)) = u(x)$.

An element $x \in E$ is *convex with respect to* u (u – *convex*) if x is a fixed point of u, that is if $u(x) = x$.

Let $X = Y = E$, $S = \{\{\alpha\} \mid \alpha \in E\}$, $s: S \to 2^E$,

(8.4.2) $\quad s(\{\alpha\}) = \{u(\alpha)\}$, for each $\{\alpha\} \in S$,

and $r = 1_{2^E}$.

Theorem 8.4.10. *An element* $x \in E$ *is convex with respect to* u *if and only if set* $\{x\}$ *is* $((S, s), r)$ *convex.*

The proof is analogous with the proof of theorem 8.4.9.

Example 8.4.13. Δ'Δ - convexity (according to J.J.M. Evens, H. van Maaren (1985), I. Singer (1987, 1997))

Let $(E, \leq)$ and $(F, \leq)$ be two arbitrary complete lattices. A duality between E and F is a mapping $\Delta: E \to F$ satisfying

$$\Delta(\inf \{e_i \mid i \in I\}) = \sup \{\Delta(e_i) \mid i \in I\},\ e_i \in E,\ i \in I,\ I \text{ a set of indices.}$$

The mapping $\Delta': F \to E$ defined by $\Delta'(z) = \inf \{x \mid x \in E, \Delta(x) \leq z\}$, $z \in F$, is called the dual of Δ. The operator $\Delta'\Delta: E \to E$ (the compound operator) is a hull operator. The u-convexity defined by the hull operator u = Δ'Δ is called Δ'Δ *-convexity* in E. According to theorem 8.4.10 it follows that the property of Δ'Δ *-convexity* is a ((S, s), r) convexity.

8.5 The class of special partial ((S, s), r) convexities

Example 8.5.1. W – convexlike sets (I. Singer (1992))

Let X be a nonempty set and W a nonempty set of functions w: $X \to \overline{R} = [-\infty, +\infty]$. A subset $A \subseteq X$ is W – *convexlike* if

(8.5.1) $A = \{x \in X \mid w(x) < \sup \{w(A), w \in W\}\}$.

Let Y = X, $S = \{D \in 2^X \mid w(d) < \sup \{w(D), w \in W\}$, for each $d \in D\}$, function s: $S \to 2^X$ given by

(8.5.2) $s(D) = \{x \in X \mid w(x) < \sup \{w(D), w \in W\}$, for each $D \in S$,

and

(8.5.3) $r = 1_{2^X}$.

Theorem 8.5.1. *A subset* $A \subseteq X$ *is* W – *convexlike if and only if it is special partially* ((S, s), r) *convex.*

Proof. Necessity. Let A be W – convexlike. Then (8.5.1) is true. It follows that $w(a) < \sup \{w(A), w \in W\}$, for each $a \in A$. Then $A \in S$. Also, from (8.5.1), (8.5.2) and (8.5.3) we get r (A) = s (A). Therefore set A is special partially convex with respect to ((S, s), r).

Sufficiency. Let now A be special partially ((S, s), r) convex. Then

(8.5.4) $A \in S$

and

(8.5.5) $s(A) \subseteq r(A) = A$.

From (8.5.4) it results that

(8.5.6) $A \subseteq s(A)$.

Hence A = s (A) and, in view of (8.5.2) we have

$$A = \{x \in X \mid w(a) < \sup \{w(A), w \in W\}.$$

Therefore set A is W – convexlike.

Example 8.5.2. Separational multi-order convexity (J.E. Martínez-Legaz, I. Singer (1990))

A multi-ordered set is an ordered pair (X, O), where X is a set and O is a nonempty family of partial orders on X.

Let (X, O) be a multi-ordered set. A set $A \subseteq X$ is said to be O – *separationally convex* (or, briefly, O_{sep} – *convex*) if for each $x_0 \in X \setminus A$, there is an order $\rho = \rho(x_0) \in O$ such that the relation $x_0 \, \bar{\rho} \, g$ occurs for each $g \in A$, where $\bar{\rho}$ is the relation opposite to ρ.

For $\rho \in O$ and $A \subseteq X$, we denote by $A_\rho = \{x \in X \setminus A \mid x \, \bar{\rho} \, a, \forall a \in A\}$.

Let $Y = X$, $S = 2^X$, $s: S \to 2^Y$, the function given by

(8.5.7) $\quad s(D) = X \setminus D$, for each $D \in S$,

and let $r: 2^X \to 2^Y$ the function defined by

(8.5.8) $\quad r(D) = \cup \{D_\rho \mid \rho \in O\}$, for each $D \in 2^X$.

Theorem 8.5.2. *A subset* A *of* X *is* O *separationally convex if and only if it is special partially* ((S, s), r) *convex.*

Proof. Necessity. Let A be an O separationally convex subset of X. Then, for each $x \in X \setminus A$, there is $\rho \in O$, such that $x \, \bar{\rho} \, a$, for each $a \in A$. Then $x \in A_\rho$. It results $X \setminus A \subseteq \cup \{A_\rho \mid \rho \in O\}$. But, for each $\rho \in O$ we have $A_\rho \subseteq X \setminus A$. Then $\cup \{A_\rho \mid \rho \in O\} \subseteq X \setminus A$. It follows

(8.5.9) $\quad \cup \{A_\rho \mid \rho \in O\} = X \setminus A.$

From (8.5.7), (8.5.8) and (8.5.9) we obtain

(8.5.10) $\quad s(A) = r(A).$

Hence, because $A \in S$, set A is special partial ((S, s), r) – convex.

Sufficiency. Let A be a special partial ((S, s), r) – convex set. Then $A \in S$ and

(8.5.11) $\quad s(A) \subseteq r(A).$

From (8.5.7), (8.5.8), and (8.5.11), we obtain $X \setminus A \subseteq \cup \{A_\rho \mid \rho \in O\}$. But $X \setminus A \supseteq \cup \{A_\rho \mid \rho \in O\}$. Hence $X \setminus A = \cup \{A_\rho \mid \rho \in O\}$. It follows that for each $x \in X \setminus A$ it exists $\rho \in O$ such that $x \in A_\rho$. This implies that $x \, \bar{\rho} \, a$, for each $a \in A$. Therefore A is O separationally convex subset of X.

Example 8.5.3. Induced strong convexity with respect to p, f and M (L. Blaga, L. Lupşa (1998))

Let X, T be two arbitrary sets, M be a nonempty subset of T and let us consider functions $f: X \to T$ and $p: 2^T \to 2^T$. A subset $A \subseteq X$ is said to be *induced strongly - convex with respect to* p, f *and* M *if*

$$p(f(A)) \cap M \subseteq f(A).$$

Taking Y = X, S = 2^X and defining s: S → 2^X and r: 2^X → 2^X by

$$s(B) = p(f(B)) \cap M,$$

$$r(A) = f(A),$$

for B ∈ S and A ∈ 2^X, it is easy to prove that a set A ⊆ X is induced strongly convex with respect to p, f and M if and only if it is special partial ((S, s), r) convex.

Example 8.5.4. Convexity of cellular blobs (J. Sklanski (1970))

Usually, a plane figure (or, simply, a figure) is the union of a simple closed curve and its interior. A blob is any bounded figure. A map is the union of one or more figures. A figure is convex provided that it is classically convex. A cellular mosaic is a set of bounded convex figures $\{c_i\}$, called cells, such that $c_i \cap c_j$ equals either to ∅ or to a part of the boundary of c_j, for all i ≠ j, and such that the union of all the cells covers the entire plane. It was shown that every cell must be a polygon in order to satisfy the convexity requirement. A cellular map is a nonempty subset of cells of a cellular mosaic. Suppose that J is a cellular map. The boundary of J, denoted by b(J), is the boundary of the union of the cells of J. The cellular boundary of J, denoted by B(J), is the union of cells each of which belongs to J and contains a point of b(J). Let γ(J) denote a non-repeating closed chain that contains precisely the cells of B(J). Consider the map g(J) formed by the union of the boundaries of the cells of γ(J), excluding the boundaries between successive cells of γ(J). The core boundary of J, denoted by bC(J), is the set of all points in g(J) that are not connected to b(J). The union of bC(J) and its interior is the core of J, denoted by C(J). J is cored if C(J) ≠ ∅. C(J) is simply connected and touches every cell in B(J). A cellular blob is a bounded, cored cellular map.

A cellular map J is the cellular image (or briefly the image) of a figure p if and only if either

(a) the union of the members of J contains p

and

(b) every member of J containing an exterior point of p also contains a boundary point of p,

or p is the limit of a sequence of figures each of which satisfies (a) and (b).

A cellular blob is *convex* if there is at least one convex figure r of which the given cellular blob is an image.

F denotes the set of all plane figures. CB denotes the set of all cellular blobs. A digitization process is a function I: F $\to$ CB, such that I (a) = J, for a $\in$ F and J $\in$ CB is the cellular image of figure a. We denote by

$$I^{-1}(J) = \{a \in F \mid I(a) = J\} \text{ and } CF = \{a \in F \mid a \text{ is classically convex}\}$$

Now, let us take $X = 2^{R^2}$, $Y = 2^{R^2}$, S = F, s: S $\to 2^Y$, s (a) = I (a) if a $\in$ CF $\cap$ I^{-1} (CB) and s (a) = X for a $\notin$ CF and $r = 1_{2^X}$. Then it is obvious that a cellular blob J is convex if there is an element a $\in$ CF $\cap$ I^{-1} (CB) such that

$$s(I^{-1}(J)) = I(a) = J = r(J).$$

Theorem 8.5.3. *A cellular blob* J *is convex if and only if* J *is special partial* ((S, s), r) *convex.*

The proof is similar to previous ones.

Example 8.5.5. Digital convexity in $\mathbf{Z}^2$ (C.E. Kim (1981))

C.E. Kim (1980) defined a similar notion of discrete convexity in $\mathbf{Z}^2$, using the following model. Let A be a subset of plane such that $\overline{A^0} = A$ (A is the closure of its interior). This type of set is called regular. Each lattice point p $\in \mathbf{Z}^2$ is regarded as the center of an open unit square (cell), p*. The set I (A) = {p | A $\cap$ p* $\neq \varnothing$} is called the digital image of A. A set A$\subseteq \mathbf{Z}^2$ is *digitally convex* if it is the digital image of a convex regular set R$\subseteq \mathbf{R}^2$. It is evident that the digital convexity is another special partial ((S, s), r) convex.

Example 8.5.6. Discrete convexity in $\mathbf{Z}^2$ (C.E. Kim, J. Sklanski (1982))

Another discrete convexity notion is *the cellular convexity* introduced by C. Kim and J. Sklansky (1982). Even if it is equivalent to the convexity of cellular blobs, it is a distinct convexity notion referring to sets of rectangular cells in the plane, called cellular complexes, instead of sets of general cells. In order to define the convexity of discrete images in $\mathbf{Z}^2$, a network expressed by reducing the cells to some typical figures (squares) covers the plane. So, a mosaic, isomorphic to $\mathbf{Z}^2$ covers the plane. A connected component A is said to have a *cellular convexity* if and only if there is at least one convex figure in $\mathbf{R}^2$ that might become A after a digitisation process. It is obvious that this convexity belongs to the same class of properties as the convexity of cellular blobs.

Example 8.5.7. Fuzzy digital convexity (L. Janos, A. Rosenfeld (1982))

Let E be the Euclidean plane (it is possible to be a set in which the straight-line segment is defined and each straight-line is a totally ordered set). A subset M of E is said to be regular if it coincides with the closure of its interior, $M = \overline{M^\circ}$. As in example 8.5.5, each lattice point p $\in \mathbf{Z}^2$ is regarded as the center of an open unit square (cell), p*. Set I (A) = {p | A ∩ p* ≠ ∅} is called the digital image of A. A fuzzy set A in E is represented by a characteristic function χ_A: E → [0, 1]. Given a fuzzy subset A of the plane, we define a fuzzy subset A' of $\mathbf{Z}^2$ represented by the characteristic function $\chi'_A(p) = \sup\{\chi_A(x) \mid x \in p^*\}$. If A is regular then A' is its digital image (as in example 8.5.5). A fuzzy set A is fuzzily regular if all its level sets A_t, $0 \le t < 1$, are regular. A fuzzy subset A' of $\mathbf{Z}^2$ is *fuzzily digitally convex* if it is the digital image of a fuzzily regular, fuzzily convex fuzzy set A. Obviously, the fuzzy digital convexity is a special partial ((S, s), r) convexity.

8.6 The class of (e, a) - ((S, S), r) convexities

Example 8.6.1. F-convexity (R. Blind, G. Valette, T. Zamfirescu (1980))

According to R. Blind, G. Valette, T. Zamfirescu (1980), the definition of this convexity for sets was presented in a problem formulated during the meeting held in Obervolfach in 1974.

Let F be a class of sets in $\mathbf{R}^n$. A set $A \subseteq \mathbf{R}^n$ is said to be F *-convex* if, for every x ∈ A, y ∈ A, with x ≠ y, there is a set F ∈ F such that x ∈ F, y ∈ F and F ⊆ A.

For the points $x \in \mathbf{R}^n$ and $y \in \mathbf{R}^n$ we denote

$$F_{x,y} = \{F \in F \setminus \mathbf{R}^n \mid x \in F, y \in F\}.$$

Let us take $Y = X = \mathbf{R}^n$, $S = \{\{x, y\} \mid x \in \mathbf{R}^n, y \in \mathbf{R}^n\}$ and define the set S of functions and transformation $r: 2^{\mathbf{R}^n} \to 2^{\mathbf{R}^n}$ by

$$S = \{s: S \to F \cup \{\mathbf{R}^n\} \mid s(x, y) \in F_{x,y} \text{ if } F_{x,y} \neq \emptyset$$

$$\text{and } s(x, y) = \mathbf{R}^n \text{ if } F_{x,y} = \emptyset, x \in \mathbf{R}^n, y \in \mathbf{R}^n\}.$$

$$r(C) = C \text{ for } C \subseteq \mathbf{R}^n.$$

Theorem 8.6.1. *A set* $A \subset \mathbf{R}^n$ *is* F *- convex if and only if* A *is* (e, a) - ((S, S), r) *convex.*

Proof. Let us suppose that A is F – convex and consider set C = {x, y} ∈ S, with C ⊆ A. Then there is F ∈ F such that x ∈ F, y ∈ F and F ⊆ A. Therefore, $F_{x,y} \neq \emptyset$ and there is a function s ∈ S such that s ({x, y}) = s (C) = F ⊆ A = r (A). It means that A is (e, a) - ((S, S), r) convex.

Conversely, supposing that A is (e, a) - ((S, S), r) convex, let us consider x ∈ A, y ∈ A and C = {x, y} ∈ S. Then we have C ⊆ A and therefore there is $s \in \mathsf{S}$ such that $s(C) \subseteq A$. From the definition of s, as $s(C) \neq \mathbf{R}^n$, it results that $\mathsf{F}_{x,y} \neq \varnothing$ and there is a set $F \in \mathsf{F}$, with $x \in F$ and $y \in F$, such that $s(C) = F$. So, A is F – convex.

Example 8.6.2. Rectangular convexity (R. Blind, G. Valette, T. Zamfirescu (1980))

The *rectangular convexity* is obtained by taking in the definition of F - convexity the set F as the class of all 2D – rectangles in $\mathbf{R}^n$. As a consequence, the rectangular convexity is a F – convexity and, therefore, a property of (e, a) - ((S, S), r) convexity.

Example 8.6.3. Geodesic convexity (T. Rapcsák (1991))

Let $M \subseteq \mathbf{R}^n$ be a connected C^2 k-manifold. A set $A \subseteq M$ is g - *convex* if any two points of A are joined by a geodesic belonging to A.

The geodesic convexity can be seen as a F – convexity, where F is the set of all geodesics of M. As a consequence, the geodesic convexity is a property of (e, a) - ((S, S), r) convexity.

Example 8.6.4. Geodesic convexity (N.J. Hicks (1965))

In the framework provided by the differential geometry, instead of a geodesic is taken the shortest geodesic. With this change made in the example 8.6.3, we obtain that the geodesic convexity from the differential geometry is also a property of (e, a) - ((S, S), r) convexity.

Example 8.6.5. Nearly convex sets (A. Aleman (1985))

Let X be a real topological linear space. A set $A \subseteq X$ is *nearly convex* if there is a real number $t \in]0, 1[$ such that

(8.6.1) $$(1 - t) A + t A \subseteq A.$$

Let us take $S = 2^X$,

$$\mathsf{S} = \{s_t \mid s_t: S \to 2^X,\ s_t(C) = (1 - t) C + t C,\ C \in S,\ t \in]0, 1[\},$$

$$r = 1_{2^X}.$$

Theorem 8.6.2. *A set* $A \subseteq X$ *is nearly convex if and only if* A *is* (e, a) - ((S, S), r) *convex.*

Proof. Let us suppose that A is nearly convex and consider set $C \in S$, with $C \subseteq A$. Then there is $t \in]0, 1[$ such that (8.6.1) is true, that is $s_t(C) \subseteq r(A)$. Set C is arbitrarily chosen, so we get that A is (e, a) - ((S, S), r) convex.

Conversely, let us suppose that A is (e, a) - ((S, S), r) convex. Then, because $A \in S$ and $A \subseteq A$, there is $s \in \mathsf{S}$ such that $s(A) \subseteq A$. From the definition of set S we get that there is a number $t \in]0, 1[$ with $s = s_t$. It follows that $s_t(A) = (1 - t) A + t A \subseteq A = r(A)$. It means that A is nearly convex.

Example 8.6.6. Nearly convex sets with respect to a set (W.W. Breckner, G. Kassay (1997))

Let X be a real topological linear space and $K \subseteq X$. A set $A \subseteq X$ is *nearly convex with respect to* K if set A + K is nearly convex.

Let us take $S = 2^X$,

$$\mathsf{S} = \{s_t \mid s_t: S \to 2^X,\ s_t(C) = (1-t)(C+K) + t(C+K),\ C \in S,\ t \in]0, 1[\},$$

$$r: 2^X \to 2^X,\ r(B) = B + K \text{ for } B \in 2^X.$$

Theorem 8.6.3. *A set* $A \subseteq X$ *is nearly convex with respect to K if and only if* A *is* (e, a) - ((S, S), r) *convex.*

Proof. Let us suppose that A is nearly convex with respect to K and consider set $C \in S$, with $C \subseteq A$. Then there is $t \in]0, 1[$ such that is $s_t(C) = (1-t)(C+K) + t(C+K) \subseteq (1-t)(A+K) + t(A+K) \subseteq A + K = r(A)$. Set C is arbitrarily chosen, so we get that A is (e, a) - ((S, S), r) convex.

Conversely, let us suppose that A is (e, a) - ((S, S), r) convex. Then, because $A \in S$ and $A \subseteq A$, there is $s \in \mathsf{S}$ such that $s(A) \subseteq A$. From the definition of the set S we get that there is $t \in]0, 1[$ with $s = s_t$. It follows that $s_t(A) = (1-t)(A+K) + t(A+K) \subseteq A + K = r(A)$. It means that A is nearly convex with respect to K.

Example 8.6.7. Strong convexity with respect to a set and two behaviours (G. Cristescu (1996[56]))

X is assumed to be a nonempty set so that for every two points x, y of X the closed straight-line segment determined by these points, denoted by <x, y>, is defined. Let us suppose that M is a nonempty subset of X, B' and B" two nonempty sets, $D' \subset B'$, $D'' \subset B''$ nonempty subsets, and T' is a set of transformations, $T': X \times M \to B'$. Behaviour C' = (D', T') (or a strong one s - (D', T'), or a weak one w - (D', T')) is defined in $X \times M$ by means of B'. For set T" of transformations, $T'': M \times 2^X \to B''$, we define a behaviour C" = (D", T") (or s - (D', T"), or w - (D', T")) in $M \times 2^X$ by means of B".

A set $A \subset X$ is said to be *strongly convex with respect to set* M *and behaviours* C' *and* C" if it is empty or if for each pair of points $x, y \in A$ and for each $t \in$ <x, y> there is an element $a \in M$ such that the following conditions simultaneously occur: $(t,a) \in C'(X \times M)$ and $(a,A) \in C''(M \times 2^X)$.

In what follows, for each $z \in X$ we put

$$M_z = \{m \mid m \in M, (z, m) \in C' (X \times M)\}.$$

We take $S = \{\{x, y\} \mid x \in X, y \in X\}$ and we say that function $s: S \to 2^X$ has the property (P) if

$$s(\{x, y\}) = \begin{cases} X, & \text{if } < x, y >= \Phi \text{ or there is } z \in < x, y > \text{ such that } M_z = \Phi \\ \bigcup_{z \in <x,y>} \{u_z\} & \text{in other cases} \end{cases}$$

where $u_z \in M_z$ for all $z \in <x, y>$. We put $\mathbb{S} = \{s \mid s: S \to 2^X$, s has the property (P)$\}$. Let us define function $r: 2^X \to 2^X$ by

$$r(B) = \{m \mid m \in M, (m, B) \in C'' (M \times 2^X)\}, \text{ for all } B \in 2^X.$$

Theorem 8.6.4. *Set* $A \subset X$ *is strongly convex with respect to set* M *and behaviours* C' *and* C" *if and only if it is* (e, a) - ((S, $\mathbb{S}$), r) *convex.*

Proof. Let us suppose that set $A \subset X$ is strongly convex with respect to set M and behaviours C' and C". Then for each pair of points $x, y \in A$ and for each $t \in <x, y>$ there is an element $a \in M$ such that the following conditions simultaneously hold: $(t, a) \in C' (X \times M)$ and $(a, A) \in C''(M \times 2^X)$. It means that for each pair of points $x, y \in A$ and for each $t \in <x, y>$ there is an element $a \in M_t$ and then $s(\{x, y\}) = \bigcup_{t \in <x,y>} \{u_t\}$. In the same time, each element $a \in \bigcup_{t \in <x,y>} \{u_t\}$ also belongs to r (A), meaning that $s(\{x, y\}) \subseteq r(A)$. Therefore, A is (e, a) - ((S, $\mathbb{S}$), r) convex.

Conversely, if A is (e, a) - ((S, $\mathbb{S}$), r) convex then for each $\{x, y\} \in S$ and $\{x, y\} \in A$, there is a function $s \in \mathbb{S}$ such that $s(\{x, y\}) \subseteq r(A)$. It is obvious that $s(\{x, y)\} \neq X$ because $r(A) \subseteq M \subset X$. Therefore $s(\{x, y\}) = \bigcup_{t \in <x,y>} \{u_t\}$ and then for each $t \in <x, y>$ there is an element $a \in M_t$, meaning that $(t, a) \in C' (X \times M)$. The inclusion $s(\{x, y\}) \subseteq r(A)$ implies that (a, A) $\in C''(M \times 2^X)$ also occurs. Therefore, A is strongly convex with respect to set M and behaviours C' and C".

Example 8.6.8. Strong (α, δ) convexity with respect to a network (G. Cristescu (1996[57]))

Let (X, d) be a totally bounded metric space and for $\varepsilon > 0$ let E' be a finite ε-network over X, and by E a finite 2ε - network over E' is denoted. For a set $A \subset X$ and a point $x \in X$, the distance between x and A is defined as usual by the number $d(x, A) = \inf \{d(x, A) \mid a \in A\}$. In X it is supposed that, for every two points $x \in X$ and $y \in Y$ a notion of straight-line segment $<x, y>$ is defined (it might be distinct from the metric straight-line segment).

Let us fix the real numbers $\alpha \geq 0$ and $\delta \geq 0$, and let E be a network chosen as above. Set $A \subset X$ is said to be *strongly* (α, δ)*-convex with respect to* E if it is empty or for every pair of points $x, y \in A$ and for every $t \in <x, y>$ there is an element $a \in E$ such that $d(t, a) \leq \alpha$ and $d(a, A) \leq \delta$.

In what follows, for each $z \in X$ we put

$$M_z = \{m \mid m \in E, d(z, m) \leq \alpha\}.$$

We take $S = \{\{x, y\} \mid x \in X, y \in X\}$ and we say that function $s: S \rightarrow 2^X$ has the property (P) if

$$s(\{x, y\}) =$$

$$= \begin{cases} X, & \text{if } <x, y> = \Phi \text{ or there is } z \in <x, y> \text{ such that } M_z = \Phi \\ \bigcup\limits_{z \in <x,y>} \{u_z\} & \text{in other cases} \end{cases}$$

where $u_z \in M_z$ for all $z \in <x, y>$. We denote $\mathbf{S} = \{s \mid s: S \rightarrow 2^X$, s has the property (P)$\}$. Let us define function $r: 2^X \rightarrow 2^X$ by

$$r(B) = \{m \mid m \in E, d(m, B) \leq \delta\}, \text{ for all } B \in 2^X.$$

Theorem 8.6.5. *Set* $A \subset X$ *is strongly* (α, δ)*-convex with respect to* E *if and only if it is* (e, a) - ((S, $\mathbf{S}$), r) *convex.*

Proof. Let us suppose that set $A \subset X$ is strongly (α, δ)-convex with respect to E. Then for each pair of points $x, y \in A$ and for each $t \in <x, y>$ there is an element $a \in E$ such that the following conditions simultaneously hold: $d(t, a) \leq \alpha$ and $d(a, A) \leq \delta$. It means that for each pair of points $x, y \in A$ and for each $t \in <x, y>$ there is an element $a \in M_t$ and then $s(\{x, y\}) = \bigcup\limits_{t \in <x,y>} \{u_t\}$. In the same time, each element $a \in \bigcup\limits_{t \in <x,y>} \{u_t\}$ also belongs to r (A), meaning that $s(\{x, y\}) \subseteq r(A)$. Therefore, A is (e, a) - ((S, $\mathbf{S}$), r) convex.

Conversely, if A is (e, a) - ((S, $\mathbf{S}$), r) convex then for each $\{x, y\} \in S$ and $\{x, y\} \in A$, there is a function $s \in \mathbf{S}$ such that $s(\{x, y\}) \subseteq r(A)$. It is obvious that $s(\{x, y)\} \neq X$ because $r(A) \subseteq E \subset X$. Therefore $s(\{x, y\}) = \bigcup\limits_{t \in <x,y>} \{u_t\}$ and then for each $t \in <x, y>$ there is an element $a \in M_t$, meaning that $d(t, a) \leq \alpha$. The inclusion $s(\{x, y\}) \subseteq r(A)$ implies that $d(a, A) \leq \delta$ also occurs. Therefore, A is strongly (α, δ)-convex with respect to E.

Example 8.6.9. Weak convexity (A. Aleman (1985))

(X, +, ·) is supposed to be a real or complex linear space. A set $A \subseteq X$ is said to be *weakly convex* if for each two points $x \in A$ and $y \in A$, there is a number $p \in]0, 1[$ such that the element $(1-p)\,x + p\,y$ belongs to A. As in the case of p-convexity, let us put X = Y, $\mathsf{S} = \{\{x, y\} \mid x \in X, y \in X\}$ but now we consider a set of functions

$$S = \{s_p \mid s_p: \mathsf{S} \to 2^X, s_p(\{x, y\}) = \{(1-p)\,x + p\,y\}, \{x, y\} \in \mathsf{S}, p \in]0, 1[\}.$$

Function r is the identity on 2^X.

Theorem 8.6.6. *A set* $A \subseteq X$ *is weakly convex if and only if* A *is* (e, a) – ((S, S), r) *convex.*

Proof. The proof is obvious if we notice that, for each $x \in A$ and $y \in A$, the existence of $p \in]0, 1[$ such that $(1-p)\,x + p\,y \in A$ is equivalent to the existence of a function s_p such that $s_p(\{x, y\}) = \{(1-p)\,x + p\,y\} \in A = r(A)$.

Example 8.6.10. Three point convexity (F.A. Valentine (1957))

E_n is supposed to be the n-dimensional Euclidean space and $A \subseteq E_n$. Set A is said to have the three point property P_3 if for each triple of points x, y, z in A at least one of the closed segments $<x, y>$, $<x, z>$, $<y, z>$ is in A. A set A is *three point convex* if A possess the property P_3.

Let us take $X = E_n$, $Y = E_n \cup \{h\}$, where $h \notin E_n$, and $\mathsf{S} = \{\{x, y, z\} \mid x \in X, y \in X, z \in X\}$. Function r is the identity on 2^Y and the set of functions s is defined by

$$S = \{s_{x,y} \mid s_{x,y}: \mathsf{S} \to 2^Y,$$

$$s_{x,y}(\{a, b, c\}) = \begin{cases} <x, y>, & \text{if } \{x, y\} \subset \{a, b, c\} \\ \{h\}, & \text{otherwise} \end{cases}, x, y \in X\}.$$

Theorem 8.6.7. *A set* $A \subseteq E_n$ *is three point convex if and only if* A *is* (e, a) – ((S, S), r) *convex.*

Proof. The proof is evident if we notice that, for each triple $x, y, z \in A$, the existence of two of them, say a and b, such that $<a, b> \subset A$ is equivalent to the existence of a function $s = s_{a,b}$ satisfying

$$s_{a,b}(\{x, y, z\}) = <a, b> \subset A = r(A).$$

This means that A is (e, a) – ((S, S), r) convex.

Example 8.6.11. m-point convexity (F.A. Valentine (1964))

E_n is supposed to be the n-dimensional Euclidean space, m a natural number and $A \subseteq E_n$. Set A is said to be m - *point convex* if for every m-member subset of A at least one of the $\binom{m}{2}$ line segments determined by these points lies in A.

Let us take X = E_n, Y = $E_n \cup \{h\}$, where h $\notin E_n$, and S = $\{\{x_1, x_2, \ldots, x_m\} \mid x_k \in X, k \in \{1, 2, \ldots, m\}\}$. Function r is the identity on 2^Y and the set of functions s is defined by

$$\mathsf{S} = \{s_{x,y} \mid s_{x,y}: S \to 2^Y,$$

$$s_{x,y}(\{a_1, a_2, \ldots, a_m\}) = \begin{cases} <x, y>, & \text{if } \{x, y\} \subset \{a_1, a_2, \ldots, a_m\} \\ \{h\}, & \text{otherwise} \end{cases}, x, y \in X\}.$$

Theorem 8.6.8. *A set* A $\subseteq E_n$ *is* m - *point convex if and only if* A *is* (e, a) – ((S, S), r) *convex.*

The proof is similar to the previous one.

Example 8.6.12. Collinear 3-point convexity (CP_3 – convexity) (L. Latecki, A. Rosenfeld, R. Silverman (1995))

The collinear 3-point convexity is obtained by introducing the condition of collinearity in the definition of property P_3 from example 8.6.10. The collinearity was necessary in connection with the problem of recognition of the convexity by directional tests. The notion of CP_3 – convexity was defined as a plane property, even if it might be discussed in a more general framework. We shall keep the original framework. Let X = $\mathbf{R}^2$ and A $\subset \mathbf{R}^2$. We say that set A has property CP_3 (collinear P_3) if P_3 holds for all collinear triples of points of A. A set A is CP_3 – *convex* if it has property CP_3.

Let us take X = E_n, Y = $E_n \cup \{h\}$, where h $\notin E_n$, and S = $\{\{x, y, z\} \mid x \in X, y \in X, z \in X$, x, y, z collinear$\}$. Function r is the identity on 2^Y and the set of functions s is defined by

$$\mathsf{S} = \{s_{x,y} \mid s_{x,y}: S \to 2^Y,$$

$$s_{x,y}(\{a, b, c\}) = \begin{cases} <x, y>, & \text{if } \{x, y\} \subset \{a, b, c\} \\ \{h\}, & \text{otherwise} \end{cases}, x, y \in X\}.$$

Theorem 8.6.9. *A set* A $\subseteq E_n$ *is* CP_3 - *convex if and only if* A *is* (e, a) – ((S, S), r) *convex.*

The proof is similar to theorem 8.6.7.

Example 8.6.13. Arcwise convexity (F.A. Valentine (1951))

An arc C in the Euclidean plane $\mathbf{R}^2$ is said to be convex if it is contained in the boundary of its convex hull. A set A is said to be *arcwise convex* if each pair of points in A can be joined by a convex arc in A. This is a F – convexity property if we take set F containing all the convex arcs in the plane. Therefore, according to theorem 8.6.1, the arcwise convexity is an (e, a) – ((S, S), r) convexity property.

Example 8.6.14. Three point arcwise convexity (F.A. Valentine (1955))

A set A is said to be *three point arcwise convex* if each triple of points $x \in A$, $y \in A$, $z \in A$ is contained in a convex arc belonging to A. A similar argument to that from example 8.6.13 is used to prove that the property of three point arcwise convexity is another (e, a) – ((S, S), r) convexity.

Example 8.6.15. (H_0, δ) – convexity (L. Danzer, B. Grumbaum, V. Klee (1963))

Let X be a nonempty set of points, a set of parts $H_0 \subset 2^X$ and a function $\delta: H_0 \to 2^{2^X}$. A set $A \subset X$ is said to be (H_0, δ) – *convex* provided that if $h \in H_0$ and $h \subset A$ then there is an element $K \in \delta(h)$ such that $K \subset A$.

Let us take X = Y, S = H_0. Suppose that

$$\mathsf{S} = \{s_{A,K} \mid s_{A,K}: S \to 2^X,\ s_{A,K}(h) = \begin{cases} K, & K \in \delta(h) \text{ and } h \subseteq A \\ X, & K \notin \delta(h) \text{ or } h \not\subset A \end{cases},\ A, K \in 2^X \text{ and}$$

$$\exists\, h \in H_0,\ K \subset \delta(h)\}$$

and r is the identity on 2^X.

Theorem 8.6.10. *A set* $A \subset X$ *is* (H_0, δ) – *convex if and only if* A *is* (e, a) – ((S, S), r) *convex.*

Proof. Let us suppose that $A \subset X$ is (H_0, δ) – convex. It means that for each $h \subset 2^A \cap H_0$ there is $K \in \delta(h)$ such that $K \subset A$. But for $h \subset 2^A \cap H_0$ we have $s_{A,K}(h) = K$ for $K \in \delta(h)$ and then $s_{A,K}(h) = K \subset A = r(A)$. It means that A is (e, a) – ((S, S), r) convex.

Conversely, let us suppose that A is (e, a) – ((S, S), r) convex. Then for each $h \in H_0 = S$, $h \subseteq A$, there is $s \in \mathsf{S}$ such that $s(h) \subseteq r(A) = A$. It is obvious that $s(h) = X$ is impossible, therefore there is $K \in \delta(h)$ such that we have $r = r_{A,K}$. Hence, for each $h \in H_0 = S$, $h \subseteq A$, there is $K \in \delta(h)$ such that $s_{A,K}(h) = K \subset r(A) = A$, meaning the ($H_0$, δ) convexity of set A.

Example 8.6.16. Projective convexity (J. de Groot, H. de Vries (1958), T. Bisztriczky (1987))

Let P^3 be the real projective three – space. For a plane $p \subseteq \mathsf{P}^3$ we denote the affine restriction P_p^3. A subset A of P^3 is *convex* if there is a plane $p \subseteq \mathsf{P}$

^{3} disjoint from A, such that A equals to the convex hull of A in the affine restriction P_p^3.

Let us take $X = Y = P^3$, S is the set of all parts of X and

$$\mathsf{S} = \{s_p \mid s_p: S \to 2^X,\ s_p(A) = \text{convex hull of A in the affine restriction } P_p^3,$$
$$p \subseteq P^3 \text{ plane}\}.$$

Considering that r is the identity in 2^X, and using the monotony of each function belonging to S one can prove that the projective convexity is a (e, a) – ((S, S), r) convexity.

Example 8.6.17. α – convexity at a point (J.P. Crouzeix, J. Ferland, C. Zălinescu (1997))

Let $A \subseteq \mathbf{R}^n$ be a nonempty set, point $a \in \mathbf{R}^n$ and the number $\alpha \in \mathbf{R}$. Set A is α - *convex at* a if there are $\rho > 0$ and $e \in \mathbf{R}^n$, $\|e\| = 1$, such that, for all $x \in A \cap B(a, \rho)$ we have $(x\text{-}a \mid e) \geq \alpha \cdot \|x\text{-}a\|^2$.

Let us consider $X = \mathbf{R}^n$, $S = 2^X$, $\mathsf{S} = \{s_{\rho,e} \mid \rho \in]0, +\infty[,\ e \in \mathbf{R}^n, \|e\| = 1\}$, where the function $s_{\rho,e} : S \to 2^{\mathbf{R}}$ is given by

(8.6.2) $\quad s_{\rho,e}(C) = \{(x\text{-}a \mid e) - \alpha \cdot \|x - a\|^2 \mid x \in C \cap B(a, \rho)\}$.

Let $Y = \mathbf{R}$, and let $r: 2^X \to 2^R$, the function defined by

(8.6.3) $\quad r(D) = [0, +\infty[$, for each $D \in 2^X$.

Theorem 8.6.11. *A nonempty subset* A *of* $\mathbf{R}^n$ *is* α - *convex if and only if it is* (e, a) – ((S, S), r) *convex.*

Proof. Necessity. Let be $A \subseteq \mathbf{R}^n$ a nonempty α - convex set in a. Then there are $\rho_A \in]0, +\infty[$ and $e_A \in \mathbf{R}^n$, with $\|e_A\| = 1$, such that

$$(x - a \mid e) \geq \alpha \cdot \|x - a\|^2, \text{ for each } x \in A \cap B(a, e),$$

or equivalently,

$$(x - a \mid e) - \alpha \cdot \|x - a\|^2 \geq 0, \text{ for each } x \in A \cap B[a, e],$$

In view of (8.6.2) and (8.6.3) we get

(8.6.4) $\quad s_{\rho_A,e}(A) \subseteq r(A)$.

Let $C \in S$, with $C \subseteq A$. Taking $\rho = \rho_A$ and $e = e_A$,, because $C \cap B[a, e] \subseteq A \cap B[a, e_A]$, we have

$$s_{\rho,e}(C) = \{(x\text{-}a \mid e) - \alpha \cdot \|x - a\|^2 \mid x \in C \cap B(a, \rho)\} \subseteq s_{\rho_A,e}(A),$$

and, in view of (8.6.4), we obtain $s_{\rho,e}(C) \subseteq r(A)$. Hence set A is (e, a) – ((S, S), r) convex.

Sufficiency. Let A be (e, a) – ((S, S), r) convex. Then there is $s \in S$ such that

(8.6.5) $s(A) \subseteq r(A)$.

From definition of set S it results that there are: a real number ρ, $\rho > 0$, and a vector $e \in \mathbf{R}^n$, with $||e|| = 1$, such that $s = s_{\rho,e}$. Then, from (8.6.2), (8.6.3) and (8.6.5) we obtain

(8.6.6) $\{(x - a \mid e) - \alpha \cdot ||x - a||^2 \mid x \in A \cap B(a, \rho)\} \subseteq [0, +\infty[$

Then $(x - a \mid e) \geq \alpha \cdot ||x - a||^2$, for each $x \in A \cap B(a, \rho)$. Therefore, set A is α - convex in a.

Example 8.6.18. Convexity at a point

Let us suppose that X is a topological space and let us consider a given convexity property for sets in X, denoted by C. Let $A \subseteq X$ and a point $a \in A$. Set $A \subseteq X$ is locally C - convex at the point $a \in A$ if there is a C - convex neighbourhood V of a such that $V \subseteq A$. According to theorem 8.2.2, set $A \subseteq X$ is locally C - convex at the point $a \in A$ if and only if A is (e, a) - ((S, S), R) convex for Y = X, S = {a}, S = $\{s \mid s: S \rightarrow 2^X, s(\{a\}) = V$, V is a C - convex neighbourhood of a$\}$ and R = $\{1_{2^X}\}$.

Convexity properties at a point are related to the local convexity. More authors defined and used various types of convexities at a point: H. Tietze (1929), J. van Heijenoort (1952), F.A. Valentine (1964, 1965 [285, 286]), M. Breen (1974, 1976, 1977), J.P. Crouzeix, J. Ferland, C. Zălinescu (1997), G. Cristescu (2000), L. Lupşa and G. Cristescu (2000). Also, in previous chapters there are various types of convexity properties at a point.

Example 8.6.19. Local convexity

X is supposed to be a topological space. Let us denote by C a convexity property for sets in X. A set $A \subseteq X$ is said to be locally C - convex at the point $a \in A$ if there is a C - convex neighbourhood V of a such that $V \subseteq A$. Set $A \subseteq X$ is said to be locally C - convex if A is locally C - convex at each point $a \in A$. According to theorem 8.2.3, set $A \subseteq X$ is locally C - convex if and only if A is (e, a) - ((S, S), R) convex for Y = X, S = $\{\{a\} \mid a \in A\}$, S = $\{s: S \rightarrow 2^X \mid s(\{a\}) = V$, V is C - convex neighbourhood of a, $\{a\} \in S\}$ and R = $\{1_{2^X}\}$. The local convexity was studied by more authors, started from various convexities: H. Tietze (1929), J. van Heijenoort (1952), F.A. Valentine (1964, 1965 [285, 286]), M. Breen (1974, 1976, 1977), G. Cristescu (2000).

Example 8.6.20. Discrete convexity (J.M. Chassery (1978, 1983))

In the plane $\mathbf{R}^2$ we consider the chessboard distance c (x, y) = max ($|x_1 - y_1|$, $|x_2 - y_2|$) for x = $(x_1, x_2) \in \mathbf{R}^2$, y = $(y_1, y_2) \in \mathbf{R}^2$. For the real positive number h, we denote by $\mathbf{Z}^2$ (h) = {(ih, jh) | i ∈ **Z**, j ∈ **Z**} the set of the lattice points of the grid of step h over the plane. A set A ⊂ $\mathbf{Z}^2$ (h) is ε - *convex,* for ε ∈ [h/2, h[, if for every x, y ∈ A and t ∈]0, 1[, there is a point a ∈ $\mathbf{Z}^2$ (h) such that c (t x + (1 - t) y, a) ≤ ε. A set A ⊂ $\mathbf{Z}^2$ (h) is *discrete convex* if there is a real number ε ∈ [h/2, h[, such that A is ε - convex.

In what follows, for each z ∈ $\mathbf{R}^2$ and for each α ∈ [h/2, h[, we put

$$M_{z,\alpha} = \{m \mid m \in E, d(z, m) \leq \alpha\}.$$

We take X = Y = $\mathbf{Z}^2$ (h), S = {{x, y} | x ∈ $\mathbf{Z}^2$ (h), y ∈ $\mathbf{Z}^2$ (h)} and we say that function s: S → 2^X has the property (P) if

$$s(\{x, y\}) =$$

$$= \begin{cases} X, & \text{if } <x, y> = \Phi \text{ or there is } z \in <x, y> \text{ such that } M_{z,\alpha} = \Phi \\ \bigcup_{z \in <x,y>} \{u_z\} & \text{in other cases} \end{cases}$$

where $u_z \in M_{z,\alpha}$ for all z ∈ <x, y>. We denote S = {s | s: S → 2^X, s has property (P)}. Let us define function r: $2^X \to 2^X$ by

$$r(B) = B, \text{ for all } B \in 2^X.$$

Theorem 8.6.12. *Set* A ⊂ $\mathbf{Z}^2$ (h) *is discrete convex if and only if it is* (e, a) - ((S, S), r) *convex.*

The proof is similar to theorem 8.6.5.

Example 8.6.21. Discrete convexity (C.Kim and A. Rosenfeld (1980))

C.Kim and A. Rosenfeld (1980) published a constructive geometrical approach, for sets of knots of a network of type $\mathbf{Z}^2$ (h) using the chessboard distance c (x, y) = max ($|x_1-y_1|$, $|x_2-y_2|$), x = $(x_1, x_2) \in \mathbf{Z}^2$ (h), y = $(y_1, y_2) \in \mathbf{Z}^2$ (h). Let A be an 8-connected component. A' denotes the set of all the squares of side h > 0 centred in the points of S. The subset of $\mathbf{R}^2$ containing all the points of A' is denoted by p(A). Let bp(A) be its frontier. For every two points of A, P_1 and P_2, by P (A; P_1, P_2) is denoted the set of those polygons having the frontiers made both by parts of <P_1, P_2> and bp(A) and the interiors in p(A). The convexity notion defined by the two authors is: a connected component A is said to be *convex* if for every pair of points A, P_1 and P_2 the subset P(A; P_1, P_2) does not contain points from the complement of A. Therefore, we must have P (A; P_1, P_2) ∩ $\mathbf{Z}^2$ (h) ⊆ A.

Let us take X = $\mathbf{Z}^2$ (h), Y = $\mathbf{R}^2$, S = {{x, y} | x, y ∈ $\mathbf{Z}^2$ (h)}. Family S = {s_A | s_A: S → 2^Y, s_A ({x, y}) = P (A; x, y) ∩ $\mathbf{Z}^2$ (h) if {x, y} ⊂ A, s_A ({x, y})

$= X$ for $\{x, y\} \not\subset A$, $A \subseteq \mathbf{Z}^2$ (h)}, and function r: $2^X \to 2^Y$ is defined by r (A) $= A$ for each $A \subseteq \mathbf{Z}^2$ (h).

Now, it is obvious that a set $A \subseteq \mathbf{Z}^2$ (h) is convex if and only if A is (e, a)-((S, S), r) convex.

8.7 The class of partial (a, e) - ((S, s), R) convexities

Example 8.7.1. u – convexoidal sets (A. Wieczorek (1989))

We recall that in a nonempty set X, a *spot operator* in X is any operator

$$u: \mathrm{Dom}\,(u) \to 2^X,$$

where the domain of u, denoted by Dom (u) is a family of subsets of X.

A set $A \subseteq X$ is said to be u-*convexoidal* if there is $F \in$ Dom (u) such that $u(F) = A$. Of course, if $F \subseteq u$ (F) for each $F \in$ Dom (u) then every u-convex set in Dom (u) is u-convexoidal.

Let us suppose that $Y = \{*\} \cup X$, where $* \notin X$, $S = 2^X$ and let us consider the functions s: $S \to 2^Y$ and r_W: $2^X \to 2^Y$ defined by

$$s\,(B) = \begin{cases} X \cup \{*\}, & \text{if there is not } F \in \mathrm{Dom}(u) \text{ such that } B = u(F) \\ B, & \text{if } \exists F \in \mathrm{Dom}(u) \text{ such that } B = u(F) \end{cases},$$

$$r_W(A) = \begin{cases} A, & \text{if } W = A \\ \{*\}, & W \neq A \end{cases}$$

Let us take sets $S = \{s\}$ and $R = \{r_W \mid W \in 2^X\}$.

Theorem 8.7.1. *A set* $A \subseteq X$ *is* u-*convexoidal if and only if* A *is partially* (a, e) - ((S, s), R) *convex.*

Proof. Let us suppose that A is u-convexoidal. Then if we take C = A, therefore $C \in S$ and $C \subseteq A$, then there is a set $F \in$ Dom (u) such that $u(F) = A = C$. From the definition of s, we have s (A) = A. Taking W = A we obtain r_A (A) = A, which implies s (C) $\subseteq A = r_A$ (A). It means that A is partially (a, e) - ((S, s), R) convex.

Conversely, let us suppose that A is partially (a, e) - ((S, s), R) convex. Then there is $C \in S$, $C \subseteq A$, and there is $r \in R$ such that s (C) $\subseteq$ r (A). Equality s (C) $= X \cup \{*\}$ cannot hold because in this situation r (A) $= \{*\}$ and $* \notin X$, therefore s (C) $= X \cup \{*\} \subseteq$ r (A) $= \{*\}$ is not possible, X being non-empty. On the other hand, if r (A) = A then s (C) $= X \cup \{*\} \not\subset A =$ r (A).

The possibility $r_C(A) = \{*\}$ cannot occur because $C \not\subset \{*\}$. So, $r_C(A) \neq \{*\}$ implies the unique possibility $r_C(A) = A$ for $C = A$. Therefore, the definition of s implies $u(F) = A$, meaning that A is u-convexoidal.

Example 8.7.2. Convexity with respect to a family of sets (L. Danzer, B. Grunbaum, V. Klee (1963))

Let X be a nonempty set and a family of subsets $\mathsf{M} \subset 2^X$. A set $A \subseteq X$ is said to be *convex with respect to* M (or, simply, M – *convex*) if there is a subfamily M' of M such that $A = \cap \{M \mid M \in \mathsf{M}'\}$.

Let us consider $Y = X \cup \{*\}$, where $* \notin X$, $S = 2^X$ and $s: S \to 2^Y$ defined by $s(B) = Y$ if there is not $\mathsf{M}' \subset \mathsf{M}$ such that $B = \cap \{M \mid M \in \mathsf{M}'\}$, $s(B) = B$ if there is $\mathsf{M}' \subset \mathsf{M}$ such that $B = \cap \{M \mid M \in \mathsf{M}'\}$. The family of transformations $\mathsf{R} = \{r_W \mid w \in 2^X\}$ contains functions $r_W: 2^X \to 2^Y$ defined

by $r_W(A) = \begin{cases} A, & W = A \\ \{*\}, & W \neq A \end{cases}$ for $A \in 2^X$.

Theorem 8.7.2. *A set* $A \subseteq X$ *is convex with respect to* M *if and only if* A *is partially* (a, e) - ((S, s), R) *convex.*

Proof. First, let us suppose that set A is convex with respect to M. We take $C = A$, therefore $C \in S$ and $C \subseteq A$. There is $\mathsf{M}' \subset \mathsf{M}$ such that $A = \cap \{M \mid M \in \mathsf{M}'\}$. It means that $s(A) = A$. Taking $W = A$, we obtain $r_A(A) = A$ and $s(C) \subseteq A = r_A(A)$, implying that A is partially (a, e) - ((S, s), R) convex.

Conversely, suppose that A is partially (a, e) - ((S, s), R) convex. Then there is $C \in S$ and $C \subseteq A$, such that $s(C) \subseteq r(A)$. Equality $s(C) = Y$ cannot hold because in this situation $r(A) = \{*\}$ and $* \notin X$ implies $s(C) = Y \subseteq r(A) = \{*\}$, which is not possible for $X \neq \varnothing$. If $r(A) = A$ then $s(C) = Y \not\subset A = r(A)$. Possibility $r_C(A) = \{*\}$ cannot occur because $C \not\subset \{*\}$. So, relation $r_C(A) \neq \{*\}$ implies $r_C(A) = A$, situation that occurs if $C = A$. Therefore, the definition of s implies that there is $\mathsf{M}' \subset \mathsf{M}$ such that equality $A = \cap \{M \mid M \in \mathsf{M}'\}$ holds, meaning that set A is convex with respect to M.

Example 8.7.3. Convexity with respect to a set of functions (K. Fan (1963))

Let X be a nonempty set and W a given set of functions, $w: X \to \mathbf{R}$. A set $A \subseteq X$ is said to be *convex with respect to* W (or, simply, W – *convex*) if for each $x \notin A$ there is a function $w \in W$ such that

$$\sup w(A) < w(x).$$

I. Singer (1984) extends the notion of W – convexity, allowing functions of type $w: X \to \overline{R} = [-\infty, +\infty]$. He proves that the W – convexity concept obtained in this manner is equivalent to the M – convexity. According to I. Singer (1997), denoting by $S_d(w) = \{x \in X \mid w(x) \leq d\}$, then a set $A \subseteq X$ is convex with respect to W if and only if

$$A = \cap \{S_d(w) \mid (w, d) \in W \times \mathbf{R}, \sup w(A) \leq d\}.$$

It means that, for $\mathsf{M} = \{S_d(w) \mid (w, d) \in W \times \mathbf{R}\}$, a set $A \subseteq X$ is W – convex if and only if $A = \cap \{M \mid M \in \mathsf{M}'\}$ for

$$\mathsf{M}' = \{S_d(w) \mid (w, d) \in W \times \mathbf{R}, \sup w(A) \leq d\} \subseteq \mathsf{M},$$

meaning that A is convex with respect to M. Therefore, the property of convexity with respect to a set of functions W is a partial (a, e) - ((S, s), R) convexity. Elements S, s and R are defined according to the pattern from example 8.7.2.

Example 8.7.4. Convexity with respect to a pair (W, φ) (J. Scrader (1975))

Let X and W be arbitrary nonempty sets and a function $\varphi: X \times W \to \overline{R} = [-\infty, +\infty]$. J. J. Moreau (166-67) called function φ a coupling function. A set $A \subseteq X$ is said to be *convex with respect to the pair* (W, φ) (or, briefly, (W, φ) *– convex*) if for each $x \notin A$ there is an element $w \in W$ such that

$$\sup \{\varphi(g, w) \mid g \in A\} < \varphi(x, w).$$

I. Singer (1997) proved that a set $A \subseteq X$ is convex with respect to the pair (W, φ) if and only if

$$A = \cap \{S_d(\varphi(\bullet, w)) \mid (w, d) \in W \times \mathbf{R}, \sup \{\varphi(g, w) \mid g \in G\} \leq d\},$$

with $S_d(\varphi(\bullet, w)) = \{x \in X \mid \varphi(x, w) \leq d\}$.

As in the previous example, a set $A \subseteq X$ is (W, φ) – convex if and only if

$$A = \cap \{M \mid M \in \mathsf{M}'\}$$

for $\mathsf{M} = \{S_d(\varphi(\bullet, w)) \mid (w, d) \in W \times \mathbf{R}\}$, and

$$\mathsf{M}' = \{S_d(\varphi(\bullet, w)) \mid (w, d) \in W \times \mathbf{R}, \sup \{\varphi(g, w) \mid g \in G\} \leq d\} \subseteq \mathsf{M}.$$

It means that A is convex with respect to M. Therefore, the property of convexity with respect to a pair (W, φ) is a partial (a, e) - ((S, s), R) convexity. The elements S, s and R are defined as in example 8.7.2.

Example 8.7.5. Holomorphic convexity (B.A. Fuks (1962))

The holomorphic convexity of a set $A \subseteq \mathbf{C}^n$ is a W – convexity when W is the set of all functions that are holomorphic in $A \subseteq \mathbf{C}^n$. Therefore, the holomorphic convexity is also a partial (a, e) - ((S, s), R) convexity.

Example 8.7.6. Pseudoconvexity convexity (L. Hörmander (1989))

Let $A \subseteq \mathbf{C}^n$ be an open set. We recall that a function u: $A \to [-\infty, +\infty]$ is called plurisubharmonic if u is semicontinuous from above and for arbitrary z and w from $\mathbf{C}^n$, the function $\lambda \to u(z+\lambda w)$ is subharmonic in the part of $\mathbf{C}$ where it is defined. The set of all plurisubharmonic functions on A is denoted by P(A). If K is a compact subset of the open set $A \subseteq \mathbf{C}^n$ then the P(A) – hull of K is defined by

$$\hat{K}_A^P = \{z \mid z \in A, u(z) \leq \sup_K u \text{ for all } u \in P(A)\}.$$

The open set $A \subseteq \mathbf{C}^n$ is *pseudoconvex* if there is a continuous plurisubharmonic function u in A such that the set

$$A_c = \{z \mid z \in A, u(z) < c\}$$

is relatively compact in A (i.e. is contained in a compact subset of A) for every $c \in \mathbf{R}$. This property holds if and only if K relatively compact in A implies $\hat{K}_A^P$ relatively compact in A.

The pseudoconvexity of the open set $A \subseteq \mathbf{C}^n$ is another W – convexity. Here W = P (A). Therefore, the pseudoconvexity is also a partial (a, e) - ((S, s), R) convexity.

Example 8.7.7. Evenly convex subsets of a locally convex space (W. Fenchel (1952))

A subset M of a locally convex space X is said to be an open half-space if there is a linear functional $\varphi \in X^* \setminus \{0\}$ and $d \in \mathbf{R}$ such that

$$M = \{x \in X \mid \varphi(x) < d\}.$$

A subset $A \subseteq X$ is said to be *evenly convex* if A is the intersection of a family of open half-spaces. Obviously, the even convexity is another M-convexity property, if we take

$$\mathsf{M} = \{M \subseteq X \mid M = \{x \in X \mid \varphi(x) < d\}, \varphi \in X^* \setminus \{0\}, d \in \mathbf{R}\}.$$

Therefore, the even convexity is a partial (a, e) - ((S, s), R) convexity.

8.8 The class of (a, e) - ((S, s), R) convexities

Example 8.8.1. Induced strict - strong convexity with respect to p, f and M (L. Blaga, L. Lupşa (1998))

Let X, T be two arbitrary sets, M be a nonempty subset of T and let us consider functions f: $X \to T$ and p: $2^T \to 2^T$. A subset $A \subseteq X$ is said to be *induced strictly strong convex with respect to* p, f *and* M *if*

$$p\,(f\,(B)) \cap M \subseteq f\,(B), \textit{ for any } B \subseteq A.$$

Let us take Y = X, S = 2^X. We define s: $S \to 2^X$ by s (B) = p (f (B)) $\cap$ M, B $\in$ S, and denote by

$$R = \{r_C \mid r_C: 2^X \to 2^X, r_C(B) = \begin{cases} f(C), & \text{for } B = C \\ \Phi, & \text{for } B \neq C \end{cases}, B, C \in 2^X\}.$$

It is easy to prove that a set $A \subseteq X$, $A \neq \varnothing$, is induced strongly convex with respect to p, f and M if and only if it is (a, e) - ((S, s), R) convex.

8.9 The class of special partial (a, e) - ((S, S), R) convexities

Example 8.9.1. Fuzzy convexity (R. Lowen (1980))

Let E be the Euclidean plane (it is possible to be a set in which the straight-line segment is defined and each straight-line is a totally ordered set). A fuzzy set A in E is represented by a characteristic function

$$\chi_A: E \to [0, 1].$$

A fuzzy set A in E is *fuzzily convex* if for all x $\in$ E, y $\in$ E, and all z $\in$ <x, y> the inequality $\chi_A(z) \geq \min\{\chi_A(x), \chi_A(y)\}$ occurs. From the definition it follows that if A is fuzzily convex then for each pair of points x, y $\in$ E, we have

$$\chi_A(z_0) = \min\{\chi_A(t) \mid t \in \langle x, y\rangle\} \text{ if and only if } z_0 \in \{x, y\}.$$

Let us take X = E, Y = [0,1] × E × E, S = 2^E, $\mathsf{S} = \{s_{x,y} \mid x \in E, y \in E\}$, where $s_{x,y}$: $S \to 2^Y$ is defined by $s_{x,y} = \{(\chi_A(z), x, y) \mid z \in \langle x, y\rangle\}$, R = $\{r_{u,v} \mid u \in E, v \in E\}$, $r_{u,v}$: $2^E \to 2^Y$, $r_{u,v}(D) = \{(t, u, v) \mid t \in [\min\{\chi_D(u), \chi_D(v)\}, \max\{\chi_D(z) \mid z \in \langle x, y\rangle\}]\}$.

Theorem 8.9.1. *A set* $A \subseteq E$ *is fuzzily convex in* E *if and only if* A *is special partially* (a, e) - ((S, S), R) *convex.*

Proof. Necessity. Let A be fuzzily convex in E, and let s $\in$ S. Then there is x, y $\in$ E such that s = $s_{x,y}$. As A is fuzzily convex, we get

$$(8.9.1) \qquad \min\{\chi_A(x), \chi_A(y)\} \leq \chi_A(z), \text{ for each } z \in \langle x, y\rangle.$$

Taking $r = r_{x,y}$, we obtain $s_{x,y}(A) \subseteq r_{x,y}(A)$. Therefore, A is special partially (a, e) - ((S, S), R) convex.

Sufficiency. Let A be a special partial (a, e) - ((S, S), R) convex set and let $x, y \in A$. For $s = s_{x,y}$ there is $r = r_{u,v} \in \mathsf{R}$ such that

(8.9.2) $\quad s_{x,y}(A) \subseteq r_{u,v}(A).$

From the definition of functions s and r it results that x = u, y = v and

$$\chi_A(z) \in [\min\{\chi_A(x), \chi_A(y)\}, \max\{\chi_A(t) \mid t \in \langle x, y\rangle\}].$$

Hence A is fuzzily convex in E.

Example 8.9.2. Fuzzy concavity (B. B. Chaudhuri (1992))

As in the previous example, let E be the Euclidean plane. A fuzzy set A in E is *fuzzily concave* if for all $x \in E$, $y \in E$, and each $z \in \langle x, y\rangle$ the inequality $\chi_A(z) \le \max\{\chi_A(x), \chi_A(y)\}$ occurs. From the definition it follows that if A is fuzzily convex then for each $x, y \in E$, we have

$$\chi_A(z_0) = \max\{\chi_A(t) \mid t \in \langle x, y\rangle\} \text{ if and only if } z_0 \in \{x, y\}.$$

Let us take X = E, Y = [0,1] × E × E, $S = 2^E$, $\mathsf{S} = \{s_{x,y} \mid x \in E, y \in E\}$, where $s_{x,y}\colon S \to 2^Y$ is defined by $s_{x,y} = \{(\chi_A(z), x, y) \mid z \in \langle x, y\rangle\}$, $\mathsf{R} = \{r_{u,v} \mid u \in E, v \in E\}$, $r_{u,v}\colon 2^E \to 2^Y$, $r_{u,v}(D) = \{(t, u, v) \mid t \in [\min\{\chi_D(z) \mid z \in \langle x, y\rangle\}], \max\{\chi_D(u), \chi_D(v)\}\}$.

Theorem 8.9.2. *A set* $A \subseteq E$ *is fuzzily concave in* E *if and only if* A *is special partially* (a, e) - ((S, S), R) *convex.*

The proof is similar to the previous one.

Example 8.9.3. Fuzzy ortho-convexity

Let E be the Euclidean plane. A fuzzy set A in E is *fuzzily ortho-convex* if for all $x \in E$, $y \in E$, such that $\langle x, y\rangle$ is either horizontal or vertical, and all point $z \in \langle x, y\rangle$ the inequality $\chi_A(z) \ge \min\{\chi_A(x), \chi_A(y)\}$ holds.

Let us take X = E, Y = [0,1] × E × E, $S = 2^E$, $\mathsf{S} = \{s_{x,y} \mid x \in E, y \in E, \langle x, y\rangle$ is either horizontal or vertical$\}$, where $s_{x,y}\colon S \to 2^Y$ is defined by $s_{x,y} = \{(\chi_A(z), x, y) \mid z \in \langle x, y\rangle\}$, $\mathsf{R} = \{r_{u,v} \mid u \in E, v \in E\}$, $r_{u,v}\colon 2^E \to 2^Y$, $r_{u,v}(D) = \{(t, u, v) \mid t \in [\min\{\chi_D(u), \chi_D(v)\}, \max\{\chi_D(z) \mid z \in \langle x, y\rangle\}]\}$.

Theorem 8.9.3. *A set* $A \subseteq E$ *is fuzzily ortho-convex in* E *if and only if* A *is special partially* (a, e) - ((S, S), R) *convex.*

The proof is similar to the previous ones.

Example 8.9.4. Fuzzy ortho-concavity (B. B. Chaudhuri (1992))

Let E be the Euclidean plane. A fuzzy set A in E is *fuzzily ortho-concave* if for all $x \in E$, $y \in E$, which are either on a horizontal line or on a vertical one, and each $z \in \langle x, y\rangle$ the inequality $\chi_A(z) \leq \max\{\chi_A(x), \chi_A(y)\}$ occurs.

Let us take $X = E$, $Y = [0,1] \times E \times E$, $S = 2^E$, $\mathbf{S} = \{s_{x,y} \mid x \in E, y \in E, \langle x, y\rangle$ is either horizontal or vertical$\}$, where $s_{x,y}: S \rightarrow 2^Y$ is defined by $s_{x,y} = \{(\chi_A(z), x, y) \mid z \in \langle x, y\rangle\}$, $\mathbf{R} = \{r_{u,v} \mid u \in E, v \in E\}$, $r_{u,v}: 2^E \rightarrow 2^Y$, $r_{u,v}(D) = \{(t, u, v) \mid t \in [\min\{\chi_D(z) \mid z \in \langle x, y\rangle\}], \max\{\chi_D(u), \chi_D(v)\}\}$.

Theorem 8.9.4. *A set* $A \subseteq E$ *is fuzzily ortho-concave in* E *if and only if* A *is special partially* (a, e) - ((S, **S**), **R**) *convex.*

The proof is similar to proof of theorem 8.9.1.

8.10 The class of partial (a, e) - ((S, S), R) convexities

Example 8.10.1. D-convex sets in Δ-modules (A. Ghika (1952))

Let X be a left side module relative to a ring Δ which contains a subring D ordered such that the order relation is invariant by translation. This invariance means that if $\alpha, \beta, \gamma \in D$ and $\alpha < \beta$ then $\alpha + \gamma < \beta + \gamma$. In such a ring, $\alpha > 0$ and $\beta > 0$ imply that $\alpha + \beta > 0$.

A set $A \subseteq X$ is D-*convex* if for every $\alpha, \beta \in D$ with $\alpha > 0$ and $\beta > 0$ it follows that $\alpha A + \beta A \subseteq (\alpha + \beta) A$. Translations and rotations preserve this convexity property but a D-convex set does not contain, generally, the straight-line segments joining every pair of its points because, generally, this notion does not have sense in a module. In a complex linear space, the **R**-convexity with respect to the set of the real numbers, with the natural order relation, reduces to the classical convexity.

Let us choose $Y = X$, $S = 2^X$ and let us define the families of functions

$$\mathbf{S} = \{s_{\alpha,\beta} : S \rightarrow 2^X \mid s_{\alpha,\beta}(C) = \begin{cases} \alpha C + \beta C, & \text{if } \alpha C + \beta C \subseteq (\alpha+\beta)C \\ X, & \text{if } \alpha C + \beta C \not\subseteq (\alpha+\beta)C \end{cases},$$

$$C \in 2^X, \alpha > 0, \beta > 0, \alpha, \beta \in D\},$$

$$\mathbf{R} = \{r_{\alpha,\beta,C}: 2^X \rightarrow 2^X \mid r_{\alpha,\beta,C}(A) = \begin{cases} \Phi, & \text{if } C \neq A \\ (\alpha+\beta)C & \text{if } C = A \end{cases},$$

$$C \in 2^X, \alpha > 0, \beta > 0, \alpha, \beta \in D\}.$$

Theorem 8.10.1. *A set* $A \subseteq X$, $A \neq X$, *is* D-*convex if and only if* A *is partially* (a, e) - ((S, **S**), **R**) *convex.*

Proof. The D-convexity of A means that $\alpha A + \beta A \subseteq (\alpha + \beta) A$ for each $\alpha > 0$, $\beta > 0$, $\alpha, \beta \in D$. Equivalently, this definition can be written as: for each $\alpha, \beta \in D$, $\alpha > 0$, $\beta > 0$ one has

$$s_{\alpha,\beta}(A) = \alpha A + \beta A \subseteq (\alpha + \beta) A = r_{\alpha,\beta}(A).$$

Take C = A and then for each $s \in \mathsf{S}$ there are $\alpha > 0$, $\beta > 0$, $\alpha, \beta \in D$ such that $s = s_{\alpha,\beta}$. Taking function $r \in \mathsf{R}$, $r = r_{\alpha,\beta,A}$ such that

$$s_{\alpha,\beta}(A) = \alpha A + \beta A \subseteq (\alpha + \beta) A = r_{\alpha,\beta,A}(A),$$

which means that A is partially (a, e) - ((S, S), R) convex.

Conversely, let us suppose that A is partially (a, e) - ((S, S), R) convex. It means that there is a set $C \in 2^X$ with $C \subseteq A$ having the property that for each $s \in \mathsf{S}$ there is $r \in \mathsf{R}$ such that $s(C) \subseteq r(A)$.

We remark that $C \neq \varnothing$, $C \in 2^X$, $C \subseteq A$ and $s(C) \subseteq r(A)$ imply that $r(A) \neq \varnothing$ and $s(C) \neq X$. Also, $C \neq \varnothing$ implies $s(C) \neq \varnothing$ and then $s(C) \subseteq r(A)$ implies $r(A) \neq \varnothing$.

On the other hand, $r_{\alpha,\beta,C}(A) \neq \varnothing$ if and only if C = A.

Let $\alpha, \beta \in D$, $\alpha > 0$, $\beta > 0$. Therefore, $s_{\alpha,\beta}(A) \subset r(A)$.

Indeed, $s_{\alpha,\beta}(A) = X$ if and only if the relation $\alpha A + \beta A \not\subseteq (\alpha + \beta) A$ cannot hold . We have $r(A) \neq X$, because $A \neq X$ implies $(\alpha + \beta)A \neq X$. Therefore, $\alpha A + \beta A \subseteq (\alpha + \beta) A$ occurs and then A is D – convex.

8.11 The class of (a, e) - ((S, S), R) convexities

Example 8.11.1. (k, g, h, M) –convexity

Let k and n be natural numbers, $g: \mathbf{R}^n \to \mathbf{R}^n$ a function, h an element of B^n, where $B = \{0, 1\}$, and $M \subseteq \mathbf{R}^n$ a nonempty set. A set $A \subseteq \mathbf{R}^n$ is said to be (k, g, h, M) – *convex* if at least one of the following conditions is satisfied:

i) $pr(h, A) \neq \varnothing$, and $M \cap \{\Lambda(y, z) \mid z \in conv(k, g(s(y, h, A)))\} \subseteq \{\Lambda(y,z) \mid z \in s(y, h, A)\}$, for all $y \in pr(h, A)$;

ii) $pr(h, A) = \varnothing$, and $M \cap conv(k, g(A)) \subseteq A$.

We take $X = Y = \mathbf{R}^n$, $S = 2^{\mathbf{R}^n}$, $\mathsf{S} = \{ s_y \mid y \in \mathbf{R}^{n-|h|} \}$, where $s_y: S \to 2^{\mathbf{R}^n}$ with

$$s_y(A) = \begin{cases} M \cap \{ \Lambda(y,z) \mid z \in conv(k, g(s(y,h,A))) \}, \text{if } y \in pr(h,A) \\ \varnothing, \text{if } y \notin pr(h,A) \end{cases}$$

for each A $\in$ S, and R = { r_x | x $\in$ $\mathbf{R}^{n-|h|}$ }, where $r_x : 2^{\mathbf{R}^n} \to 2^{\mathbf{R}^n}$ with

$$r_x(D) = \begin{cases} \{\Lambda(x,u) \mid u \in s(x,h,D)\}, & x \in pr(h,D) \\ \varnothing, & x \notin pr(h,D) \end{cases}, \text{ for each } D \in 2^{\mathbf{R}^n}.$$

Theorem 8.11.1. *If* h $\in$ $B^n \setminus \{0_n, e_n\}$, *then a subset* A *of* $\mathbf{R}^n$ *is* (k, g, h, M) *– convex if and only if it is* (a, e) – ((S, S), R) *convex.*

Proof. Necessity. Let C $\in$ S, C $\subseteq$ A, and let s $\in$ S. There is y $\in$ $\mathbf{R}^{n-|h|}$ such that s = s_y. We take r = r_y, and we show that s(C) $\subseteq$ r(A). In view of remark 7.2.1, we can write

(8.11.1) $\quad s(C) = s_y(C) \subseteq s_y(A).$

Two cases may arise: y $\in$ pr (h, A), or y $\notin$ pr (h, A).

If y $\in$ pr (h, A), because A is (k, g, h, M) – convex, we have

$$M \cap \{\Lambda(y, z) \mid z \in \text{conv}(k, g(s(y, h, A)))\} \subseteq \{\Lambda(y, u) \mid u \in s(y, h, A)\}.$$

This implies that $s_y(A) \subseteq r_y(A)$, and from, (8.11.1) it follows that s(C) $\subseteq$ r(A).

In second case, in view of remark 7.2.1, we get that y $\notin$ pr (h, C); then s(C) = s_y(C) = $\varnothing$; then obviously s(C) $\subseteq$ r_y(A) = r(A).

In both cases we get s(C) $\subseteq$ r(A). Because C was arbitrary chosen in S, satisfying only condition C $\subseteq$ A, and because r was arbitrary chosen in S, it follows that A is (a, e) - ((S, S), R) convex.

Sufficiency. Let A $\subseteq$ $\mathbf{R}^n$ be an (a, e) - ((S, S), R) convex. We show that A is (k, g, h, M) – convex. Let y $\in$ pr (h, A). Because A is (a, e) - ((S, S), R) – convex set, if we take C = A and s = s_y, there is r $\in$ R, such that s(A) $\subseteq$ r(A). As r $\in$ R, there is z $\in$ $R^{n-|h|}$ such that r = r_z. Then we have

(8.11.2) $\quad M \cap \{\Lambda(y, z) \mid z \in \text{conv}(k, g(s(y, h, A)))\} \subseteq r_z(A).$

Two cases may arise:

i) $M \cap \{\Lambda(y, z) \mid z \in \text{conv}(k, g(s(y, h, A)))\} = \varnothing$; then evidently,

$$M \cap \{\Lambda(y, z) \mid z \in \text{conv}(k, g(s(y, h, A)))\} \subseteq \{\Lambda(y, u) \mid u \in s(y, h, A)))\}$$

ii) $M \cap \{\Lambda(y, z) \mid z \in \text{conv}(k, g(s(y, h, A)))\} \neq \varnothing$. In this case, from (8.11.2) it follows that $r_z(A) \neq \varnothing$. From definition of r it results that z $\in$ pr (h, A); hence and $r_z(A) = \{\Lambda(z, u) \mid u \in s(z, h, A)))\}$. Then, (8.11.2) implies

(8.11.3) $\quad M \cap \{\Lambda(y, z) \mid z \in \text{conv}(k, g(s(y, h, A)))\} \subseteq$

$$\subseteq \{\Lambda(z, u) \mid u \in s(z, h, A)))\}.$$

Since Λ is a bijective function, from (8.11.3) we get y = z; then (8.11.3) can be write as

$$M \cap \{\Lambda (y, z) \mid z \in \text{conv} (k, g (s (y, h, A)))\} \subseteq \{\Lambda (y, u) \mid u \in s(y, h, A)))\}.$$

This implies that A is (k, g, h, M) – convex.

Example 8.11.2. (g, h, M) – convexity (L. Lupşa, L.R. Blaga, D. Duca, E. Duca (2000))

Let n be a non-null natural number, $h \in B^n \setminus \{0_n\}$, let $g: \mathbf{R}^{|h|} \to \mathbf{R}^{|h|}$, be a given function and let $M \subseteq \mathbf{R}^n$, $M \neq \varnothing$. A set $A \subseteq \mathbf{R}^n$ is said to be (g, h, M) – *convex* if at least one of the following conditions is satisfied:

i) pr (h,A) $\neq \varnothing$ and

$$M \cap \{\Lambda (y, z) \mid z \in \text{conv} (g (s (y, h, A)))\} \subseteq$$
$$\subseteq \{\Lambda (y, z) \mid z \in s (y, h, A)\}, \text{ for all } y \in \text{pr} (h, A);$$

or

ii) pr (h, A) = $\varnothing$, and $M \cap \text{conv} (g (A)) \subseteq A$.

By analogy, if we take: $X = Y = \mathbf{R}^n$, $S = 2^{\mathbf{R}^n}$, $\mathsf{S} = \{s_y \mid y \in \mathbf{R}^{n-|h|}\}$, where $s_y : S \to 2^{\mathbf{R}^n}$ with

$$s_y(A) = \begin{cases} M \cap \{ \Lambda(y,z) \mid z \in \text{conv} (g(s(y,h,A)))\}, \text{ if } y \in \text{pr}(h,A) \\ \varnothing, \text{ if } y \notin \text{pr}(h,A) \end{cases}$$

for each $A \in S$, and $\mathsf{R} = \{ r_x \mid x \in \mathbf{R}^{n-|h|} \}$, where $r_x : 2^{\mathbf{R}^n} \to 2^{\mathbf{R}^n}$ with

$$r_x(D) = \begin{cases} \{ \Lambda(x,u) \mid u \in s(x,h,D) \}, x \in \text{pr} (h,D) \\ \varnothing, x \notin \text{pr} (h,D) \end{cases},$$

for each $D \in 2^{\mathbf{R}^n}$, we can prove:

Theorem 8.11.2. *If* $h \in B^n \setminus \{0_n, e_n\}$, *then a subset* A *of* $\mathbf{R}^n$ *is* (g, h, M) – *convex if and only if it is* (a, e) – ((S, S), R) *convex.*

8.12 The class of converted (a, e) - ((S, S), R) convexities

Example 8.12.1. H-convexity (A.M. Rubinov, B. Glover, V. Jeyakumar (1995))

Let X be an arbitrary set and $Z \subseteq X$, $Z \neq \emptyset$. Let H be a set of functions defined on X and mapping into **R**. A set $A \subseteq H$ is called H-*convex on* Z if and only if for any $h' \in H$, $h' \notin A$, there is an element $x \in Z$ such that

$$h'(x) > \sup \{h(x) \mid h \in A\}.$$

Let us take $S = 2^H$ and two distinct points h_1 and h_2 of H. For $f \in H$ we define functions $s_f: S \to 2^H$ and $r_f: 2^H \to 2^H$ by:

$$s_f(U) = \begin{cases} \{h_1\}, & f \in U \\ H, & f \notin U \text{ and there is not } z \in Z, f(z) > \sup\limits_{h \in U} h(z), \\ \{h_2\}, & f \in H \setminus U \text{ and } \exists z \in Z, f(z) > \sup\limits_{h \in U} h(z) \end{cases}$$

$$r_f(U) = \begin{cases} \{h_1\}, & f \in U \\ \Phi, & f \notin U \text{ and there is not } z \in Z, f(z) > \sup\limits_{h \in U} h(z), \\ \{h_2\}, & f \in H \setminus U \text{ and } \exists z \in Z, f(z) > \sup\limits_{h \in U} h(z) \end{cases}$$

for $U \in 2^H$. Consider the sets of functions $\mathsf{S} = \{s_f \mid f \in H\}$ and $\mathsf{R} = \{r_f \mid f \in H\}$.

Theorem 8.12.1. *A set* $A \subseteq H$ *is* H-*convex on* Z *if and only if* A *is converted* (a, e) - ((S, S), R) *convex.*

Proof. Let us suppose that A is H-convex on Z. Then for each $g \in H \setminus A$, there is $z \in Z$ such that $g(z) > \sup \{h(z) \mid h \in A\}$. For a function $r \in \mathsf{R}$ there is $g \in H$ such that $r = r_g$. Two situations are possible:

Case 1) If $g \in A$. In this situation $r_g(A) = \{h_1\}$. Then $s_g(A) = \{h_1\}$, meaning that $s_g(A) \subseteq r_g(A)$.

Case 2) If $g \notin A$, then the H-convexity of A means that there is $z \in Z$ such that $g(z) > \sup \{h(z) \mid h \in A\}$. It means that $r_g(A) = \{h_2\}$ and $s_g(A) = \{h_2\}$, therefore $s_g(A) \subseteq r_g(A)$.

Let $C \in 2^H$ with $C \subset A$. It follows that $\sup \{h(z) \mid h \in C\} \leq \sup \{h(z) \mid h \in A\}$ and $f(z) > \sup \{h(z) \mid h \in A\} \geq \sup \{h(z) \mid h \in C\}$ implies $s_f(C) = r_f(A) = \{h_2\}$. It means that A is converted (a, e) - ((S, S), R) convex.

Conversely, let us suppose that A is converted (a, e) - ((S, S), R) convex. Let $g \in H \setminus A$. It follows that there is $r \in \mathsf{R}$ such that $r = r_g$ and $r_g(A) \neq \{h_1\}$. There is $s \in \mathsf{S}$ such that, for $C = A$, we have $s(A) \subset r_g(A)$. It means that there is f such that $s_f(A) \subset r_g(A)$. But either $r_g(A) = \emptyset$ or $r_g(A) = \{h_2\}$. From the definition of the sets of functions S and R it follows that $s_f(A) \neq \varnothing$ and $r_g(A) \neq H$. From $s_f(A) \subset r_g(A)$ it follows that $s_f(A) = \{h_2\} = r_g(A)$, situation implying that there is $z \in Z$ such that $g(z) > \sup\{h(z) \mid h \in A\}$. Therefore, A is H-convex on Z.

8.13 The class of (a, a) – ((S, S), R) convexities

Example 8.13.1. n – convexity of a sequence (D.S. Mitrinović, I.B. Lacković, M.S. Stanković (1979), G. Toader (1981))

For a real sequence $(a_m)_{m=1}^{\infty}$, the n-th order difference is defined by:

(8.13.1) $\Delta^0 a_m = a_m$, $\Delta^n a_m = \Delta^{n-1} a_{m+1} - \Delta^{n-1} a_m$.

A sequence $(a_m)_{m=1}^{\infty}$ is said to be *convex of order* n (or n- *convex*) if

$$\Delta^n a_m \geqq 0, \text{ for all m.}$$

We take $X = \{x = (x_m)_{m=1}^{\infty} \mid x \text{ real sequence}\}$, $Y = \mathbf{R}$, $S = \{\{x\} \mid x \in X\}$, $S = \{s_m \mid m \in \mathbf{N}\}$, with $s_m: S \to 2^{\mathbf{R}}$, $s_m(x) = \{\Delta^n x_m\}$, for each $x \in S$, and $R = \{r\}$, where $r: 2^X \to 2^Y$, $r(A) = [0, +\infty[$, for each $A \in 2^X$.

Theorem 8.13.1. *A sequence* $x \in X$ *is* n – *convex if and only if set* $\{x\}$ *is* (a, a) – ((S, S), R) *convex.*

Proof. Necessity. Let x be a real n –convex sequence. We show that set $A = \{x\}$ is (a, a) – ((S, S), R) convex. Let be $C \in S$, with $C \subseteq A$. Then there is a real sequence y such that $C = \{y\}$. From $C \subseteq A$, we get $\{y\} \subseteq \{x\}$. Hence $y = x$. Let now $s \in S$. Then there is a natural number m such that $s = s_m$. We have $s(C) = s_m(\{y\}) = s_m(\{x\}) = \{\Delta^n x_m\}$. Since x is a real n – convex sequence, we have $\Delta^n x_m \geq 0$, implying that $s_m(\{x\}) \subseteq [0, +\infty[= r(A)$. It follows that for each $C \in S$, $C \subseteq A$, and for each $s \in S$, we have $s(C) \subseteq r(A)$. Therefore the set $A = \{x\}$ is (a, a) – ((S, S), R) convex.

Sufficiency. Let x be a real sequence with the property that set $A = \{x\}$ is (a, a) – ((S, S), R) convex. We show that x is a n – convex sequence. Let m be a natural number. If we take $C = A$ and $s = s_m$, as (a, a) – ((S, S), R) convexity of A, we get $s_m(A) \subseteq r(A)$. This implies that $\{\Delta^n x_m\} = s_m(\{x\}) = s(A) \subseteq r(A) = [0, +\infty[$, hence $\Delta^n x_m \geqq 0$. As m is a natural number arbitrarily chosen, it follows that $\Delta^n x_m \geqq 0$, for each m, meaning that sequence x is n – convex.

8.14 The classification of convexities for sets. Table of classes

The following table contains the members of all the previously described classes of convexity properties for sets. They are included respecting the order in which they are presented in the paragraph describing their class. Therefore, they are numbered exactly as inside their class, for an easier identification.

Type of convexity	Members
(S, s) convexity	1. The classical convexity (Euclid (330-275 B.C.), Archimedes (287?-212 B.C.), H. Brunn (1887), H. Minkowski (1911)) 2. n-polyhedral convexity 3. Metric convexity (K. Menger (1928)) 4. Order convexity (G. Birkhoff (1948)) 5. Spherical convexity (C.V. Robinson (1942)) 6. Strong convexity (C.V. Robinson (1942)) 7. Convexity in subsets of lattices (S.V. Ovchinnikov (1980)) 8. Midpoint – convexity (J.L.W.V. Jensen (1905, 1906)) 9. p – convexity (A. Aleman (1985)) 10. Quasi-convexity (J. W. Green, W. Gustin (1950)) 11. Interval convexity (J.R. Calder (1971)) 12. Ortho-convexity (G.J.E. Rawlins, D. Wood (1988)) 13. Convexity properties defined by generalised convex combinations (Φ-convexity, Ψ-convexity) (V. Soltan (1984)) 14. Baricentric approach: the superconvexity (G. Rodé (1980)) 15. u – convexity (A. Wieczorek (1989)) 16. The convex hull approach of the convexity (g-convexity) (V. Soltan (1984)) 17. Convex sets in join spaces (A. Ghika (1955), W. Prenowitz (1961), D. Voiculescu (1967), V.W. Bryant, R.J. Webster (1969)) 18. Convexity in von Neumann – Morgenstern – Gudder – Schroeck convex structures (S. Gudder, F. Schroeck (1980)) 19. Convexity in Takahashi convex structures (W. Takahashi (1970)) 20. k-strong convexity with respect to g and M (L. Blaga, L. Lupşa (1998)) 21. Strong k-convexity with respect to a given set (L. Lupşa (1981)) 22. Strong convexity with respect to g and M (L. Blaga, L. Lupşa (1998)) 23. Slack k-convexity with respect to g and M (L. Blaga, L. Lupşa (1998)) 24. Slack convexity with respect to g and M (L. Blaga, L. Lupşa (1998)) 25. Slack n-convex sets with respect to M (L. Lupşa (1981)) 26. Strong convex sets with respect to M (L. Lupşa (1981)) 27. Integer convexity (L. Lupşa (1981)) 28. Slack convex sets with respect to M (L. Lupşa (1981)) 29. Induced seg-convexity with respect to p, f and M (L. Blaga, L. Lupşa (1998)) 30. E – convexity (E.A. Youness (1999)) 31. (φ, ψ) – convexity (D. Duca, E. Duca, L. Lupşa and L. R. Blaga (2000)) 32. bi – (φ, ψ) – convexity (D. Duca, E. Duca, L. Lupşa and L. R. Blaga (2000))

	33. bi - (φ -1) – convexity (D. Duca, E. Duca, L. Lupşa and L. R. Blaga (2000)) 34. bi – (1-ψ) – convexity (D. Duca, E. Duca, L. Lupşa and L. R. Blaga (2000)) 35. Segmential multi-order convexity (J.E. Martínez-Legaz, I. Singer (1990)) 36. ε – convexity (J. Perkal (1956)) 37. **B** – convexity in a complete lattice (I. Singer (1997)) 38. Polynomial convex sets (G. Stolzenberg (1962)) 39. Rational convex sets (G. Stolzenberg (1963)) 40. Discrete convexity in $\mathbf{Z}^2$ (M. Minsky, S. Papert (1969), A. Rosenfeld, A. Kak (1980)) 41. s-convexity 42. Wedge

Type of convexity	Members
((S, s), r) convexity	1. Directional convexity (P.L. Yu (1974) 2. Cone convexity (P.L. Yu (1974)) 3. Closely convex sets (L. Blaga, I. Kolumbán (1994)) 4. Closely convex sets with respect to a cone (W.W. Breckner, G. Kassay (1997)) 5. α-convex sets (G. Cristescu (1997) 6. Convexity with respect to a behaviour (G. Cristescu (1997)) 7. Slack (α, δ) – convexity with respect to a network (G. Cristescu (1996) 8. Slack convexity with respect to a set and two behaviours (G. Cristescu (1996)) 9. Quasi-convex sets (I. Singer (1984)) 10. Induced k-strong convexity with respect to p, f and M (L. Blaga, L. Lupşa (1998)) 11. **M**-convexity in a complete lattice (I. Singer (1997)) 12. u – convexity in a complete lattice (I. Singer (1997)) 13. Δ'Δ - convexity (according to J.J.M. Evens, H. van Maaren (1985), I. Singer (1987, 1997))
special partial ((S, s) r) convexity	1. W – convexlike sets (I. Singer (1992)) 2. Separational multi-order convexity (J.E. Martínez- Legaz, I. Singer (1990)) 3. Induced strong convexity with respect to p, f and M (L. Blaga, L. Lupşa (1998)) 4. Convexity of cellular blobs (J. Sklanski (1970)) 5. Digital convexity in $\mathbf{Z}^2$ (C. Kim (1981)) 6. Discrete convexity in $\mathbf{Z}^2$ (C. Kim, J. Sklansky (1982)) 7. Fuzzy digital convexity (L. Janos, A. Rosenfeld (1982))
(e, a) – ((S, S), r) convexity	1. **F**-convexity (R. Blind, G. Valette, T. Zamfirescu (1980)) 2. Rectangular convexity (R. Blind, G. Valette, T. Zamfirescu (1980)) 3. Geodesic convexity (T. Rapcsák (1991)) 4. Geodesic convexity (N.J. Hicks (1965)) 5. Nearly convex sets (A. Aleman (1985)) 6. Nearly convex sets with respect to a set (W.W. Breckner, G. Kassay (1997)) 7. Strong convexity with respect to a set and two behaviours (G. Cristescu (1996)) 8. Strong (α, δ) convexity with respect to network (G. Cristescu (1996)) 9. Weak convexity (A. Aleman (1985)) 10. Three point convexity (F.A. Valentine (1957)) 11. m-point convexity (F.A. Valentine (1964))

	12. Collinear 3-point convexity (CP_3 – convexity) (L.Latecki, A. Rosenfeld, R. Silverman (1995)) 13. Arcwise convexity (F.A. Valentine (1951)) 14. Three point arcwise convexity (F.A. Valentine (1955)) 15. (H_0, δ) – convexity (L. Danzer, B. Grumbaum, V. Klee (1963)) 16. Projective convexity (J. de Groot, H. de Vries (1958), T. Bisztriczky (1987)) 17. α – convexity at a point (J.P. Crouzeix, J. Ferland, C. Zălinescu (1997)) 18. Convexity at a point 19. Local convexity 20. Discrete convexity (J.M. Chassery (1978, 1983)) 21. Discrete convexity (C.Kim, A. Rosenfeld (1980))

Type of convexity	Members
partial (a, e) - ((S, s), R) convexity	1. u – convexoidal sets (A. Wieczorek (1989)) 2. Convexity with respect to a family of sets (L. Danzer, B. Grunbaum, V. Klee (1963)) 3. Convexity with respect to a set of functions (K. Fan (1963)) 4. Convexity with respect to a pair (W, φ) (J. Scrader (1975)) 5. Holomorphic convexity (B.A. Fuks (1962)) 6. Pseudoconvexity convexity (L. Hörmander (1989)) 7. Evenly convex subsets of a locally convex space (W.Fenchel (1952))
(a, e) - ((S, s), R) convexity	Induced strict – strong convexity with respect to p, f and M (L. Blaga, L. Lupşa (1998))
special partial (a, e) - ((S, S), R) convexity	1. Fuzzy convexity (R. Lowen (1980)) 2. Fuzzy concavity (B. B. Chaudhuri (1992)) 3. Fuzzy ortho-convexity 4. Fuzzy ortho-concavity (B. B. Chaudhuri (1992))
partial (a, e) - ((S, S), R) convexity	D-convex sets in Δ-modules (A. Ghika (1952))
(a, e) - ((S, S), R) convexity	1. (k, g, h, M) –convexity 2. (g, h, M) – convexity (L. Lupşa, L.R. Blaga, D. Duca, E. Duca (2000))
converted (a, e) - ((S, S), R) convexity	H-convexity (A.M. Rubinov, B. Glover, V. Jeyakumar (1995))
(a, a) - ((S, S), R) convexity	n – convexity of a sequence (D.S. Mitrinović, I.B. Lacković, M.S. Stanković (1979), G. Toader (1981))

Table 8.14.1. Members of the classes of convexity properties

8.15 Remarks and problems related to the classification of the convexity properties for sets

As we said at the beginning of this chapter (section 8.2), all the 100 convexities for sets from our collection are obtained using the key elements of a definition in the order (S, S, R). But this order might be replaced either by (S, R, S), or (R, S, S), or (S, S, R) or (R, S, S).

Tables 8.15.1 – 8.15.4 contain all the distinct notions of convexity logically possible if another order of the appearance of the key elements in definition is used.

s ∈ S	r ∈ R	C ∈ S	Condition	Name
∃	∃	∀	s(C) ⊆ r(A)	(e.e) – (S, R, S) convexity
∃	∀	∀	s(C) ⊆ r(A)	(e,a,) – (S, R, S) convexity
∀	∃	∀	s(C) ⊆ r(A)	(a,e) – (S, R, S) convexity
∃	∀	∃	s(C) ⊆ r(A)	partial (e,a) - (S, R, S) convexity
∀	∃	∃	s(C) ⊆ r(A)	partial (a,e) - (S, R, S) convexity
∀	∀	∃	s(C) ⊆ r(A)	partial (a,a) - (S, R, S) convexity

Table 8.15.1. Terminology for (S, R, S) generalised convexities

r ∈ R	s ∈ S	C ∈ S	Condition	Name
∃	∀	∀	s(C) ⊆ r(A)	(e.a) – (R, S, S) convexity
∀	∃	∀	s(C) ⊆ r(A)	(a,e) – (R, S, S) convexity
∃	∀	∃	s(C) ⊆ r(A)	partial (e,a) – (R, S, S) convexity
∀	∃	∃	s(C) ⊆ r(A)	partial (a,e) – (R, S, S) convexity

Table 8.15.2. Terminology for (R, S, S) generalised convexities

s ∈ S	C ∈ S	r ∈ R	Condition	Name
∃	∀	∃	s(C) ⊆ r(A)	(e,a,e) – (S, S, R) convexity
∀	∃	∀	s(C) ⊆ r(A)	(a,e,a) - (S, S, R) convexity

Table 8.15.3. Terminology for (S, S, R) generalised convexities

r ∈ R	C ∈ S	s ∈ S	Condition	Name
∃	∀	∃	s(C) ⊆ r(A)	(e,a,e) – (R, S, S) convexity
∀	∃	∀	s(C) ⊆ r(A)	(a,e,a) – (R, S, S) convexity

Table 8.15.4. Terminology for (R, S, S) generalised convexities

Similar results as in section 8.2 are possible. We include some obvious remarks.

Remark 8.15.1.

1) Every (e,e) - (S, R, S) convex set is (e,e) – ((S, S), R) convex set.

2) Every (e,a) - (S, R, S) convex set is (e,a) – ((S, S), R) convex set.

3) Every (a,e) - (S, R, S) convex set is (a,e) – ((S, S), R) convex set.

The converse is not true.

Some types of convexities described in tables 8.15.1 – 8.15.4 are too restrictive to be interesting. Other types are retrieved between the above-presented classes. But, generally, they are distinct. Indeed, if we study section 8.6 from this point of view, we find that examples 8.6.1 - 8.6.4, 8.6.7, 8.6.8, 8.6.10 - 8.6.15 are (e,a) – ((S, S), r) convexities, which are not (e, a) – (S, r, S) convexities. But examples 8.6.5, 8.6.6, 8.6.9 are (e,a) – ((S, S), r) convexities that are (e, a) – (S, r, S) convexities as well.

Another problem for further study is related to the possibility of defining convexity properties for functions in connection to a certain class of convexity properties for sets.

For example, G. Cristescu (2001 [80]) presented the following geometric construction to define a (S, s) convex function.

Let us consider a nonempty set X and a totally ordered nonempty set T, denoting its order relation by <.

Let S_X be a nonempty subset of 2^X and s_X: $S_X \to 2^X$ be a function, which allows us to consider (S_X, s_X) convex subsets in X. Let S_T be a nonempty subset of 2^T, s_T: $S_T \to 2^T$ and let us take into account the (S_T, s_T) convexity property in T. We define a (S, s) convexity in X × T as it follows: we take S = $S_X \times S_T$ and consider a function s: S $\to 2^{X\times T}$ for which we apply the definition of (S, s) convexity.

For a function f: X → T we define the epigraph by the set

$$\mathrm{Epi}\,(f) = \{(x, t) \mid f(x) < t\} \subset X \times T.$$

Definition 8.15.1. *A function* f: X → T *is said to be* (S, s) *convex if the set* Epi (f) *is* (S, s) *convex in* X × T.

This manner of defining generalised convex functions might generate convexity properties with an important role in multiple objectives programming problems. Sometimes, an approach by means of operators generalising the divided differences is possible, opening ways for implications in convergence of certain approximation methods.

Part 2. Applications

Chapter 9

Applications in pattern recognition

Many approaches of the convexity property for digital sets (sets of pixels) are known, having applications in various algorithms for the detection of the convexity of a configuration appearing in an image. All of them describe convexity for a non-connected set (set of pixels), which is finite and is embedded into $\mathbf{R}^2$ or $\mathbf{R}^3$. Another purpose of the study of the discrete description of the convexity is to find a possibility of measuring the "degree of convexity" of a set in order to obtain accuracy in digital imagery. This is very important because these images might represent parts of the inside of the human body that are often investigated by physicians for diagnosis. In this chapter we will deal with both aspects. We shall discuss the connection between the convexity properties presented in chapter 5 of this book and the possibility of measuring the sizes of concavities of a set and of detecting them. Also, we shall present the way in which the detection of the convexity property appears in cytology.

9.1 Digital convexity and its connection with various non-connected convexity properties

The convexity property appears in image analysis in more stages, for example in segmentation and parameterisation: J. M. Chassery, C. Garbay, P. Cinquin (1984), J. M. Chassery and C. Garbay (1984), A. Rosenfeld, L. Davis (1979). For the detection of the convex sets an adequate description of the convexity property for the subsets of $\mathbf{Z}^2$ is necessary, because the screen of a computer is organised as a totally bounded metric space embedded in $\mathbf{Z}^2$, therefore as non-connected sets in the plane.

The first approach of the convexity without being involved a connectivity property has appeared in 1957, when F.A.Valentine (1957) has formulated the following definition:

Definition 9.1.1 *A set* A *has the property* P_3 *if for every triplet of points* P, Q, R *from* A *at least one of the straight-line segments* <P,Q>, <P,R>, <Q,R> *lies in* A.

This notion was thoroughly studied by F. A. Valentine (1964). The book also deals with a similar definition for the case when more than three points are considered. The most of the results for the general case are due to M. Breen, (1974, 1976, 1977). It was proved that in $\mathbf{R}^n$ a set that has a property of this type, called m-convexity, is a union of at most m-1 convex sets, M. Breen, (1974, 1976), J. F. Lawrence, W. R. Hare, J. M. Kenelly (1972). An analysis of frontiers for these types of sets have been made in F. A. Valentine (1964), but it was not used in image analysis until 1995, when it became the starting point for L. Latecki, A. Rosenfeld, R. Silverman (1955).

The first approach of the discrete convexity in $\mathbf{Z}^2$ has been published in M. Minsky and S. Papert (1969). They take into account 8-connected components in $\mathbf{Z}^2$, formulating the following definition.

Definition 9.1.2 *The connected component* S *is said to be convex if there is not a triplet* (P_1, P_2, P_3) *of collinear points, with* P_2 *situated between* P_1 *and* P_3, *such that* P_1 *and* P_3 *are situated in* S *and* P_2 *belongs to the complement of* S.

It is easy to see that this definition intends to imitate the classical definition of the convexity using straight-line segments, in a manner that is easy to test by a computer program.

Three years later, J. Sklanski (1972) has described the property of convexity for 8-connected components from $\mathbf{Z}^2$ as follows:

Definition 9.1.3. *A connected component* S *is said to be convex if and only if there is at least one convex figure in* $\mathbf{R}^2$ *that might become* S *after a digitisation process.*

This definition follows to the study of the same author on the recognition of the convex figures, J. Sklanski (1970), containing also a classification of the possible non-convexities for discrete sets, in order to detect them.

During the same period, precisely G. Beer (1973) put the problem of a generalisation of the properties of convexity and star-shapeness of a set in terms of the visibility function. An "index of convexity" was checked in order to obtain a classification of the convexities of the compact sets. The definition of a convexity measure for compact sets was, at the same time, the purpose of R. Schneider (1975). It has to be followed by the recognition of the convexity by means of an indicating function that might be one of the directions of development of the theory. H.I. Stern (1989) uses the polygonal entropy for this purpose and A. Held and K. Abe (1994) have formulated the problem of finding an “approximate-convexity” as a measure of the convexity. But our aim is not to follow this direction of study. We are interested in the evolution of the convexity concept in spaces isomorphic to $\mathbf{Z}^2$ (or $\mathbf{Z}^3$) and the connection with the convexity notions discussed in chapters 3, 4 and 5 and in this chapter.

Geometrically, H. Freeman and R. Shapira (1975) have described with more accuracy the convexity of a set of points from $\mathbf{Z}^2$ by the construction of the minimum perimeter convex polygon (MPCP) of an 8-connected component, S. This polygon represents a figure in $\mathbf{R}^2$ corresponding to the convex hull of the component S.

C.Kim and A. Rosenfeld (1980) published another constructive geometrical approach, for sets of knots of a network of type $\mathbf{Z}^2$ (h) using the chessboard distance c. Let S be an 8-connected component. S' denotes the set of all the squares of side h > 0 centred in the points of S. The subset of $\mathbf{R}^2$ containing all the points of S' is denoted by s (S). Let bs(S) be its frontier. For every two points of S, P_1 and P_2, by P (S; P_1, P_2) the set of those polygons having the frontiers made both by parts of $<P_1, P_2>$ and bs(S) and the interiors in s (S) is denoted. The convexity notion defined by the two authors is:

Definition 9.1.4. *A connected component* S *is said to be convex if for every pair of points* S, P_1 *and* P_2 *the subset* P(S; P_1, P_2) *does not contain points from the complement of* S.

The equivalence of definitions 9.1.1, 9.1.2 and 9.1.4 was proved by C. E. Kim and A. Rosenfeld (1980) for the class of the connected components that have a certain regularity property. In fact, definition 9.1.4 is a reconsideration of the previous point of view of the second author, expressed by A. Rosenfeld, A. Kak (1980) pp. 389 (see also A. Rosenfeld, A. Kak (1982) second ed., vol. 2, pp.287). There, a set S is convex if every straight-line intersects S at the most once, meaning only in a succession of 8-neighbour points. This caused discussions on the situation when the above-mentioned intersection property takes place only for straight-lines with particular directions, leading to the directional convexity. G.J.E. Rawlings and D. Wood (1988) have discussed the situation in which the directions under discussion are parallel with the coordinate axes. They called it *ortho-convexity*. More general cases are implied in the convexity notions defined by Liana Lupsa (1980, 1981 [152], [153], 1982, 1986) and discussed in chapter 2 of this book.

Another discrete convexity notion is *the cellular convexity* introduced by C. Kim and J. Sklansky (1982). Even if it is equivalent to definition 9.1.4, it is a distinct convexity notion, referring to sets of rectangular cells in the plane, called cellular complexes, instead of sets of grid knots from $\mathbf{Z}^2$, isomorphic to them.

But the most accurate descriptions of the discrete convexity property were geometric-analytical ones. The first of them is contained in definition 4.3.2 formulated by J.M. Chassery (1978) and generalised by J.M. Chassery (1983) in the form of definitions 3.4.3. and 3.4.4 from our book. This is the best description of the discrete convex sets in $\mathbf{Z}^2$(h) and it was used in cytology investigations, see J. M. Chassery, C. Garbay, P. Cinquin (1984), J. M. Chassery and C. Garbay (1984). In what follows we shall deal with this type of convexity, but not before a discussion on another interesting way of research.

Another method used for the recognition of the concavities and for the detection of the convex configurations is geometrical, by means of diametrical measures, A. Rosenfeld (1974, 1987), using distances of F. Hausdorff (1937) – D. Pompeiu (1905) type. Starting from geometrical properties of metrical type, L. M. Blumenthal (1953), the notion of concavity is defined, A. Rosenfeld (1985), trying to compute the concavity coarseness and so to study, from a geometrical point of view, the deviations from the convexity of a set, L. Boxer (1993). This direction of research will be followed in the next paragraph.

Now, let us come back to the discrete convexity. The first connection between the notion of (α,δ)-convexity with respect to $\mathbf{Z}^2$(h) and the notion of discrete convexity from definition 3.4.3 is obvious.

Corollary 9.1.1. *A set* A $\subset$ $\mathbf{Z}^2$ (h) *is discrete convex if there is a real number* $\alpha \in$ [h/2, h[*such that* A *is strongly* (α, 0) - *convex with respect to* $\mathbf{Z}^2$(h).

Proof. If A is strongly (α, 0)-convex with respect to $\mathbf{Z}^2$ (h) then it means that for every pair of points x, y $\in$ A and every t $\in$ <x, y> there is a $\in$ $\mathbf{Z}^2$ (h) such that c (t, a) $\leq \alpha$ and c (a, A) = 0. But c (a, A) = 0 means a $\in$ A and therefore there is a $\in$ A such that t $\in$ B_∞(a, α).

It is obvious that the property of strong (α, 0) - convexity of a set with respect to $\mathbf{Z}^2$ (h) is more general that the discrete convexity. This follows from the definitions of the two convexities: the property of discrete convexity applies to subsets from $\mathbf{Z}^2$ (h), and the property of strong (slack) (α, 0)-convexity with respect to $\mathbf{Z}^2$ (h) is defined in $\mathbf{R}^2$ (more generally in a totally bounded metric space). However, there is a strong connection between them. To obtain it the definition of a method of digitisation of $\mathbf{R}^2$ into $\mathbf{Z}^2(h)$ is needed.

Definition 9.1.5. *A function* $f: \mathbf{R}^2 \to \mathbf{Z}^2(h)$ *with the property that, for every* $x \in \mathbf{Z}^2(h)$, $f(x) = x$ *takes place is said to be a method of digitisation of* $\mathbf{R}^2$ *into* $\mathbf{Z}^2(h)$.

Theorem 9.1.1. *If a set* $A \subset \mathbf{R}^2$ *is both strongly and slackly* $(\varepsilon, 0)$ *- convex with respect to* $\mathbf{Z}^2(h)$, *for* $\varepsilon \in [h/2, h[$, *then there is a method of digitisation* $f: \mathbf{R}^2 \to \mathbf{Z}^2(h)$ *such that* $f(A)$ *is discrete convex.*

Proof. Consider the method of digitisation defined by $f(a) = (x, y)$ if $a \in]x-h/2, x+h/2] \times]y-h/2, y+h/2]$ for every $(x, y) \in \mathbf{Z}^2(h)$ and $a \in \mathbf{R}^2$. Suppose that A is both strongly and slackly $(\varepsilon, 0)$ - convex with respect to $\mathbf{Z}^2(h)$. Then for every $x, y \in A$ and $t \in]0, 1[$ there is an $a \in \mathbf{Z}^2(h)$ such that $c(tx + (1-t)y, a) \leq \varepsilon$ involves that $a \in A$. To prove that $f(A)$ is discrete convex let us suppose that x and y are two points of $f(A)$. Two situations are possible:

Case 1) $x \in A \cap f(A)$ and $y \in A \cap f(A)$;

Case 2) x or y belongs to $f(A) - A$.

Case 1) If $x, y \in A \cap f(A)$ then from the strongly $(\varepsilon, 0)$ - convexity of A with respect to $\mathbf{Z}^2(h)$ it follows that for every $t \in]0, 1[$ there is an $a \in \mathbf{Z}^2(h)$ such that $c(tx + (1-t)y, a) \leq \varepsilon$ involves that $a \in A$. But $a \in A$ and $a \in \mathbf{Z}^2(h)$ means that $a \in f(A)$ and, therefore, the definition of the discrete convexity is fulfilled.

Case 2) Now $x \in f(A) \setminus A$ and $y \in A$ are supposed. Then two situations are possible in the neighbourhood of x:

a) there is $z \in A$ such that $c(x, z) < h/2$;

b) the closest point $z \in A$ satisfies $c(x, z) = h/2$.

a) If $z \in A$ is a point such that $c(x, z) < h/2$ then there is a point $u \in \langle x, y\rangle$ satisfying $\langle z, u\rangle \subset \text{conv}(B_\infty(z, h/2))$ and $\langle u, y\rangle \not\subset \text{conv}(B_\infty(x, h/2))$. If $t \in]0, 1[$ is chosen such that $tz + (1-t)y \in \langle z, u\rangle$ then $c(tz + (1-t)y, x) < h/2 \leq \varepsilon$. Therefore, from the slackly $(\varepsilon, 0)$ - convexity of A with respect to $\mathbf{Z}^2(h)$ it follows that $x \in A$ and the hypothesis of case 2) cannot occur.

b) The slack $(\varepsilon, 0)$ - convexity of A with respect to $\mathbf{Z}^2(h)$ has as a consequence that $x \in A$, by taking $t = 0$ in the reasoning above. Therefore the hypothesis of case 2) cannot hold.

The cases $x \in f(A)$, $y \in f(A) \setminus A$ and $x \in f(A) \setminus A$, $y \in f(A) \setminus A$ are similar.

Two important consequences of the proof of theorem 9.1.1 must be registered, using also the chessboard distance.

Corollary 9.1.2. *If the set* $A \subset \mathbf{R}^2$ *is both strongly and slackly* $(\varepsilon, 0)$ *- convex with respect to* $\mathbf{Z}^2$ (h), *with* $\varepsilon \in [h/2, h[$, *then there is a method of digitisation* f: $\mathbf{R}^2 \to \mathbf{Z}^2$ (h) *such that* f (A) $\subseteq$ A.

Proof. The method of digitisation used in the proof of theorem 9.1.1 satisfies the desired condition. Indeed, if A is both strongly and slackly $(\alpha, 0)$ - convex with respect to $\mathbf{Z}^2$ (h) then it follows that c (conv (A), $\mathbf{Z}^2$ (h)) $\leq \alpha$ implies $\mathbf{Z}^2$ (h) $\cap$ conv (A) $\subseteq$ A. But f (A) $\subseteq$ $\mathbf{Z}^2$ (h) $\cap$ conv (A), which implies f (A) $\subseteq$ A. □

Corollary 9.1.3. *If the set* $A \subset \mathbf{R}^2$ *is slackly* $(\varepsilon, 0)$ *- convex with respect to* $\mathbf{Z}^2$ (h), *with* $\varepsilon \in [h/2, h[$, *then it is strongly* $(\varepsilon, 0)$ *- convex with respect to* $\mathbf{Z}^2$ (h).

Proof. The proof is immediate if the decomposition $\mathbf{R}^2 = \cup \{]x, x+h/2] \times]y-h/2, y+h/2] \mid (x, y) \in \mathbf{Z}^2 (h)\}$ is taken into account. Hence, for every m $\in$ A and n $\in$ A and every t $\in$]0, 1[it follows that there is a point a $\in$ $\mathbf{Z}^2$ (h) such that c (tm + (1-t)n, a) $\leq$ h/2 $\leq \alpha$. But, according to the slack $(\alpha, 0)$ - convexity of the set A it follows that a $\in$ A. Therefore, for every m $\in$ A and n $\in$ A and every t $\in$]0, 1[there is a point a $\in$ $\mathbf{Z}^2$ (h) such that c (tm + (1-t)n, a) $\leq \alpha$ and a $\in$ A following that A is strongly $(\alpha, 0)$ - convex with respect to $\mathbf{Z}^2$ (h). □

But if ε = h/2 and $\delta > \varepsilon$ then the situation from corollary 9.1.3 does not occur meaning that the slack (α, δ) - convexity with respect to $\mathbf{Z}^2$ (h) does not involve the strong (α, δ) - convexity with respect to $\mathbf{Z}^2$ (h). This may be seen from example 5.3.1.

9.2 Measuring the concavity

Theorems 5.5.1, 5.5.2 and corollary 5.5.1 will be used to discuss the importance of the number δ from the definition of the properties of (α, δ) – convexity in connection with the study of the possibilities of measuring the sizes of the concavities of a set. Our starting point is in the results of A. Rosenfeld (1985) which we shall quote in this paragraph because they contain a far reaching analysis of the complex problem of the elaborating a characterisation of the deviation of a set from the property of convexity.

Let us suppose that A is a simply connected region in $\mathbf{R}^2$, consisting from a rectifiable simply connected curve as frontier, fr (A), together with its interior. Then the frontier of the set conv (A) is also a simply connected curve, denoted by fr (conv (A)). Suppose that the frontier fr (conv (A)) consists from a finite sequence of arcs A_1, A_2, ... , A_n, such that A_i is either an arc of fr (A) or a straight-line segment with the extremities in fr (A) and the interior inside the exterior of A.

Definition 9.2.1. *A connected component of the set* conv (A) \ A *is called a concavity of* A.

Let S be a concavity of set A, where the frontier of S consists of an arc B and a chord K (fig. 9.2.1).

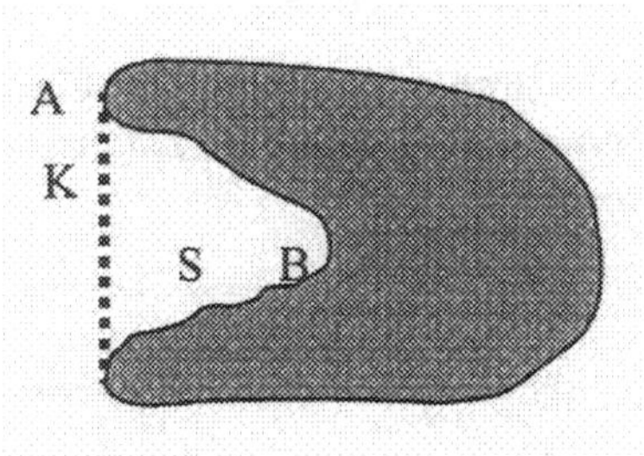

Figure 9.2.1. A concavity of the set A

Three ways of computing the size of the concavity S are considered:

(a) the aria a (S) of the set S;

(b) the maximum distance from B to K, i.e. the number

$$\Delta (B, K) = \max_{P \in B} d (P, K),$$

where $d (P, K) = \min_{Q \in K} d (P, Q)$;

(c) the maximum distance from K to B, i.e. the number

$$\Delta (K, B) = \max_{Q \in K} d (Q, B),$$

where $d (Q, B) = \min_{P \in B} d (Q, P)$.

In (b) and (c), the Euclidean distance between the points x and y is denoted by d (x, y).

Generally $\Delta (B, K) \neq \Delta (K, B)$. The relations between these measures, quoted from A. Rosenfeld (1985), are:

Proposition 9.2.1. i) Δ (K, B) $\leq \frac{l(K)}{2}$, *where* l(K) *is the length of the chord* K.

ii) Δ (K, B) $\leq \frac{l(B)}{2}$.

Obviously, l (K) < l (B), because a chord is always shorter than its arc, since its interior lies strictly outside A.

Proposition 9.2.2. Δ (K, B) $\leq$ Δ (B, K).

Proof. Let Δ (B, K) = d, and consider the set of points at distance less or equal to d from K, represented in Figure 9.2.2. Let u and v be the endpoints of K and let p be any point of K.

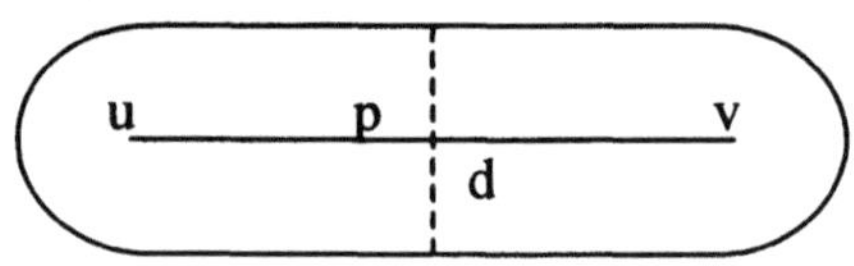

Figure 9.2.2. The set of points at distance $\leq$ d from K.

The dotted line in Figure 9.2.2 is perpendicular to K in p. Since B must stay inside this set, and must get from u to v, it must cross the dotted line. But its crossing point p of K is at a smaller or equal distance to d from B, which proves that Δ (K, B) $\leq$ d = Δ (B, K). □

Proposition 9.2.3. a (S) $\leq$ l (K) Δ (B, K) + $\pi \frac{\Delta^2(B,K)}{2}$.

Proof. Let Δ (B, K) = d. Then B lies inside the region shown in Figure 9.2.2. In fact, B must lie either in the upper or lower half of the region, since S must lie on one side of K. Thus S is a subset of half of the region and the area of the region equals to 2 l (K) Δ (B, K) + π Δ^2 (B, K). □

In A. Rosenfeld (1985) there are remarks on the information obtained by the use of one or another of these measures. It follows, from there, that everyone completes the information offered by the other two on the concavity S, and sometimes it is necessary to use all the three methods. For this reason A. Rosenfeld chose to seek for a method of describing the concavities of a region consisting of the squares of a grid, therefore isomorphic to a subset of $\mathbf{Z}^2$. His aim was to establish a connection between the concavity of such a region and a concavity of a plane region that has that union of squares as its discrete image.

Let A be a set from the plane $\mathbf{R}^2$ over which a grid having the step h is applied. Its knots are denoted by $\mathbf{Z}^2$ (h). Denote by A* ($\mathbf{Z}^2$ (h)) *the* $\mathbf{Z}^2$ (h) - *discrete image* of set A, defined as the union of all the grid squares which intersect A giving surfaces having the area at most $\frac{h^2}{2}$.

A union of grid squares, T, is said to be $\mathbf{Z}^2$ (h) - *convex*, according to A. Rosenfeld (1985), if there is a convex set A in the plane such that A* ($\mathbf{Z}^2$(h)) = T. If there is not such a set A, then T is said to be $\mathbf{Z}^2$ (h) - *concave*.

According to A. Rosenfeld (1985), a set A is said to be h - *concave* if there is a grid of step h such that A* ($\mathbf{Z}^2$ (h)) is $\mathbf{Z}^2$ (h) - concave.

Obviously, if A is h - concave then A is h' - concave, for every h' < h.

According to A. Rosenfeld (1985), the greatest number h such that set A is h - concave is called *the concavity coarseness* of set A. It shows the size of the grid step that starts to produce a discrete image h – *concave* for a set that is not convex. Hence, by digitisation in a grid of step less than the concavity coarseness of set A, the non – convexity of set A is detected. A grid of steps greater than the concavity coarseness of set A will give a convex image, therefore cannot lead to the detection of the concavity of A.

A better description of a concavity of a set needs, according to A. Rosenfeld (1985), the contribution of two descriptors: one to show how "deep" is the concavity and another one to show how "wide" is the concavity. In order to obtain the second descriptor, a set A is said to be n – *locally* h – *concave* if it is h – concave and there is a segment <x, y> of length at the most ng that detects the concaveness of A* in the sense previously presented. It follows that in order to detect a shallow concavity we need a high value of n, no matter what value of h is used. On the other hand, let S be a concavity in A that is d "deep" and w "wide", where d > w. Since S is only w wide, there is a segment <x, y> of length w whose end-points are in A on opposite sides of S. But since S is d deep, there is a point of <x, y> situated at distance d > w from A. If h ≅ w then A is h – concave and its concavity S is detectable by a segment of length on the order of h. We may remark, according to the previous discussion and to the geometrical significance of these numbers, that the two numbers Δ (B, K) and Δ (K, B) may provide us with information concerning both the depth and the width of the concavity S.

9.3 The concavity coarseness

In what follows we shall extend definition 9.2.1 for the case of the bounded plane regions that are not connected or that are multiple connected. The main results that are presented here have been published in G. Cristescu (1996 [57], [58], 1997 [65], 2000 [71]). Our aim is to find a possibility of describing the concavity coarseness in connection with the properties of (α, δ) - convexity that such a set might eventually have.

So, for a set $A \subset \mathbf{R}^2$, we shall renounce to the connectivity condition from the definition of a concavity, used by A. Rosenfeld in definition 9.2.1. Therefore, *by the concavity of set* A *is meant the set* conv (A) \ A.

Let us suppose that P is a concavity of set A. This means that there are two points $x \in A$ and $y \in A$ and a point $t \in \langle x, y \rangle$ such that $t \notin A$. Let us suppose that A is not both slackly and strongly (α, δ) - convex with respect to $\mathbf{Z}^2(h)$. From corollary 5.5.1 it follows that there are two points $x \in A$ and $y \in A$ such that, if $a \in \mathrm{Seg}(\alpha; x, y)$, then $B(a, \delta) \cap A = \emptyset$. Let us suppose, in addition, that $\alpha \leq \delta$. Then condition (1) from theorem 5.5.2 implies the existence of a point $t \in \langle x, y \rangle \cap B(a, \delta)$. It follows that there is a number $\gamma > 0$ such that $B(t, \gamma) \cap A = \emptyset$ and, applying a translation to set A such that t becomes a grid knot in $\mathbf{Z}^2(h)$, the conclusion that A is not both strongly and slackly (α, δ) - convex with respect to $\mathbf{Z}^2(h)$ is evident.

Definition 9.3.1. *The number*

$$\Delta(A) = \sup\{\delta \mid \exists\, x, y \in A, \exists\, t \in \langle x, y \rangle, d(t, A) > \delta\}$$

is called the concavity coarseness of set A.

Remark 9.3.1. Obviously, $\Delta(A) = \sup\{\delta \mid A \text{ is not } \delta\text{ - convex}\}$. Indeed, it is easy to see that the concavity coarseness of set A is the length of the radius of the greatest opened ball centred in a point t situated on a straight-line segment $\langle x, y \rangle$, with $x \in A$ and $y \in A$, which does not intersect the interior of set A.

Theorem 9.3.1. *If* $\Delta(A)$ *is the concavity coarseness of set* $A \subset \mathbf{R}^2$, *then there are three numbers* $h > 0$, $\alpha > 0$ *and* $\delta > \Delta(A)$ *with the property that* A *is strongly* (α, δ) *- convex with respect to* $\mathbf{Z}^2(h)$.

Proof. Consider $h \in [\frac{\Delta(A)}{2}, \Delta(A)]$, $\alpha = \min\{r \mid [0, h]^2 \subset B((0, 0), r) \cup B((0, h), r) \cup B((h, 0), r) \cup B((h, h), r)\}$ and a number $\Delta > \Delta(A)$. Then, for every $x \in A$ and $y \in A$ and every $t \in \langle x, y \rangle$, we have $d(t, A) \leq \Delta$. Let us fix an arbitrary point $t \in \langle x, y \rangle$. Then there is the pair of numbers $(i, j) \in \mathbf{Z}^2$ such that $t \in [ih, (i+1)h] \times [jh, (j+1)h]$. Hence, there is a point $a \in \{(ih, jh), ((i+1)h, jh), (ih, (j+1)h), ((i+1)h, (j+1)h)\} \subset \mathbf{Z}^2(h)$ such that $d(t, a) \leq \alpha$. Then

$d(a, A) \le d(a, t) + d(t, A) \le \alpha + \Delta$ and putting $\delta = \alpha + \Delta$ it follows that the set A is strongly (α, δ) - convex with respect to $\mathbf{Z}^2(h)$.

The following stronger result is obtained similarly.

Theorem 9.3.2. *If* $\Delta(A)$ *is the concavity coarseness of set* $A \subset \mathbf{R}^2$, *if*

$$h \in [\frac{\Delta(A)}{2}, \Delta(A)],$$

$$\alpha = \min\{r \mid [0, h]^2 \subset B((0, 0), r) \cup B((0, h), r) \cup$$

$$\cup B((h, 0), r) \cup B((h, h), r)\}$$

and $\Delta > \Delta(A)$

then A *is both strongly and slackly* $(\alpha, \alpha+\Delta)$ - *convex with respect to* $\mathbf{Z}^2(h)$.

Proof. Because $\Delta > \Delta(A)$ it follows that for every $x \in A$ and $y \in A$ and every $t \in \langle x, y\rangle$, $d(t, A) \le \Delta$ occurs. Let $t \in \langle x, y\rangle$ be an arbitrary point, fixed. Then there is the pair of numbers $(i, j) \in \mathbf{Z}^2$ such that $t \in [ih, (i+1)h] \times [jh, (j+1)h]$. Since the balls centred in the edge of the square $[ih, (i+1)h] \times [jh, (j+1)h]$ and having the radius r cover the square, it follows that there is at least a vertex of this square, denoted by a, such that $d(t, a) \le \alpha$. But, for every such point a, $d(a, A) \le d(a, t) + d(t, A) \le \alpha + \Delta$ takes place. In this case, both the existence of a point a having the property from definition 5.1.1.ii) and the condition that every point a, such that $d(t, a) \le \alpha$, satisfies definition 5.1.1.i) are proved. Therefore A is both strongly and slackly $(\alpha, \alpha+\Delta)$ - convex with respect to $\mathbf{Z}^2(h)$.

Theorems 9.3.1 and 9.3.2 establish the connection between the concavity coarseness of a set, defined as in definition 9.3.1, and its property of strong (or both strong and slack) (α, δ)-convexity with respect to $\mathbf{Z}^2(h)$. In order to obtain a stronger property than these theorems we need another direction of generalisation of the classical convexity, provided by the notion of α - *convexity* described by definition 5.1.1. This type of convexity defined by means of a distance function offers us the framework to prove the result that strengthens theorems 9.3.1 and 9.3.2.

Theorem 9.3.3. *If* $\Delta(A)$ *is the concavity coarseness of set* $A \subset \mathbf{R}^2$, *then* A *is* α - *convex, for every* $\alpha > \Delta(A)$.

Proof. Because $\Delta(A)$ is the concavity coarseness of set A, it follows, from definition 9.3.1, that if $\alpha > \Delta(A)$, then for every $x \in A$ and $y \in A$ and for every $t \in \langle x, y\rangle$ the inequality $d(t, A) \le \alpha$ takes place. So, set A is α - convex.

Theorem 9.3.4. *If* Δ (A) *is the concavity coarseness of set* $A \subset \mathbf{R}^2$, *then there is* $h \geq 0$ *such that* A *is strongly* $(\frac{h}{2}, \alpha)$ - *convex with respect to* $\mathbf{Z}^2(h)$, *for every* $\alpha > \Delta(A)$.

Proof. Let us consider two points x and y in A and the point $t \in \langle x, y \rangle$, such that $d(t, A) = \sup \{d(a, A) \mid a \in \langle x, y \rangle, x \in A, y \in A\}$ occurs. Then, taking the number $h \leq 2\, d(t, A)$, and the network $\mathbf{Z}^2(h)$ constructed such that $t \in \mathbf{Z}^2(h)$, it follows the existence of a point $a \in \mathbf{Z}^2(h)$ with the property that $d(t, A) \leq h/2$ and $d(a, A) \leq \alpha \leq \Delta(A)$. Therefore A is strongly $(\frac{h}{2}, \alpha)$ - convex with respect to $\mathbf{Z}^2(h)$, for every $\alpha > \Delta(A)$.

Corollary 9.3.1. i) $\inf \{\delta \mid A \text{ is } \delta \text{ - convex}\} \geq \Delta(A) \geq$

$$\geq \sup \{\delta \mid A \text{ is not } \delta - \text{convex}\}.$$

$$\text{ii) } \inf \{\delta \mid A \text{ is strongly } (\frac{h}{2}, \delta) \text{ - convex with respect to } \mathbf{Z}^2(h)\} \geq$$

$$\geq \Delta(A) \geq$$

$$\geq \sup \{\delta \mid A \text{ is not strongly } (\frac{h}{2}, \delta) - \text{convex with respect to } \mathbf{Z}^2(h)\}.$$

Proof. i) As Δ (A) is the concavity coarseness of A it follows that for every $\delta < \Delta(A)$ there are the points x and y in A for which there is a point t of the straight-line segment $\langle x, y \rangle$ such that $d(t, A) > \delta$. Hence for every $\delta < \Delta(A)$ the set A is not δ - convex. It follows that $\Delta(A) \geq \sup \{\delta \mid A \text{ is not } \delta - \text{convex}\}$.

As Δ (A) is the concavity coarseness of A it follows that if $\delta > \Delta(A)$ then for every two points x and y in A and for every point t of the straight-line segment $\langle x, y \rangle$ the condition $d(t, A) < \delta$ fulfils. Hence for every $\delta > \Delta(A)$ the set A is δ - convex. So, $\inf \{\delta \mid A \text{ is } \delta \text{ - convex}\} \geq \Delta(A)$.

Using a similar reasoning to the proof of theorem 4.1.1, by a convenient choice of the behaviours we obtain now without difficulty ii).

Theorem 9.3.5. *Suppose that* Δ (A) *is the coarseness of the concavity of set* $A \subset \mathbf{R}^2$ *and* Δ (B) *is the coarseness of the concavity of set* $B \subset \mathbf{R}^2$. *If* $\alpha > \beta$, *and:*

1. A *is* α - *convex;*
2. A *is not* β - *convex;*
3. B *is* β - *convex.*

Then $\Delta(A) \geq \Delta(B)$.

Proof. According to corollary 9.3.1. i), we have

$$\inf\{\alpha \mid A \text{ is } \alpha\text{ - convex}\} \geq \Delta(A) \geq \sup\{\beta \mid A \text{ is not } \beta - \text{convex}\} \geq \beta.$$

Hence, $\Delta(A) \geq \beta$ and, according to remark 9.3.1,

$$\beta \geq \sup\{\beta \mid B \text{ is not } \beta - \text{convex}\} = \Delta(B),$$

meaning that $\Delta(A) \geq \beta \geq \Delta(B)$.

Corollary 9.3.2. *Suppose that* $\Delta(A)$ *is the coarseness of the concavity of set* $A \subset \mathbf{R}^2$ *and* $\Delta(B)$ *is the coarseness of the concavity of set* $B \subset \mathbf{R}^2$. *If* h>k, $\alpha > \beta$, *and:*

1. A *is strongly* $(\frac{h}{2}, \alpha)$ *- convex with respect to* $\mathbf{Z}^2(h)$;
2. A *is not strongly* $(\frac{k}{2}, \beta)$ *- convex with respect to* $\mathbf{Z}^2(k)$;
3. B *is strongly* $(\frac{k}{2}, \beta)$ *- convex with respect to* $\mathbf{Z}^2(k)$.

Then $\Delta(A) \geq \Delta(B)$.

This corollary is a consequence of theorem 9.3.5 and corollary 9.3.1.

In order to present a better geometrical description of the concavity coarseness we will analyse carefully the set conv (A) \ A. Let us consider the set A of all the connected components of conv (A) \ A Let us denote by S the set the straight-line segments that belong both to the frontier of $\mathbf{R}^2 \setminus$ conv(A) ant to A, together with all the straight-line segments used to compute the diameter of every hole of A.

Theorem 9.3.6. $\Delta(A) = \sup\{d(t, A) \mid t \in \langle x, y\rangle, \langle x, y\rangle \in S\}$.

The proof consists in an immediate reasoning based on definition 9.3.1, remark 9.3.1 and theorem 9.3.5.

This means that a concavity having the coarseness $\Delta(A)$ might be detected after a digitisation of set A in a conveniently chosen network $\mathbf{Z}^2(h)$, i.e. having $h < \Delta(A)$. The strongly $(\frac{h}{2}, \alpha)$ - convex sets with respect to $\mathbf{Z}^2(h)$ are, therefore, sets with non-detectable concavities by means of network $\mathbf{Z}^2(h)$, digitising them by the method used in this chapter. Therefore, in the framework from paragraph 9.2, the concavity coarseness defined by A. Rosenfeld (1985) is retrieved between the parameters that characterise the type of convexity defined by means of a distance function, as in chapter 5 of this book. The notion of concavity coarseness of A. Rosenfeld equals to our notion from definition 9.3.1.

9.4 The concavity coarseness of some fractals

In chapter 5, as examples, were discussed some fractals having properties of strong (α,δ)-convexity with respect to a ε-network. Theorems 9.3.1, 9.3.2 and 9.3.3 put in connection these properties with the concavity coarseness of these fractals.

For example, the concavity coarseness of set A from example 5.3.1 is 0. The concavity coarseness of the Carpet of Sierpinsky and of the Sponge of Sierpinsky is 1/6 if distance c is used.

An interesting example is offered by the set called the Island of Koch. This set is geometrically generated by the following method, described in B. Mandelbrot (1975):

a) the starting point is an equilateral triangle, Figure 9.4.1;

b) the middle third of every side is replaced by two sides of the equilateral triangle having the vertex in the exterior of the initial equilateral triangle and a side on the eliminated middle third (the result is the Star of David), Figure 9.4.2;

c) the method of construction presented at b) is indefinitely repeated, applying it, every stage, to the just obtained figure, Figure 9.4.3.

The result is a polygon having the side $L = \lim_{n\to\infty} \frac{r\sqrt{3}}{3^n} = 0$, where r is the radius of the circle circumscribed to the initial equilateral triangle. Its perimeter is infinite and in every point it has no tangent. This polygon is a fractal, which, together with its interior, forms the surface named *the Island of Koch*.

If the distance c is used for measurements, then the concavity coarseness of the Island of Koch is $\frac{(\sqrt{3}-1)r}{8}$, representing the length of the side of the square drawn in Figure 9.3.4.

Figure 9.4.1 Figure 9.4.2 Figure 9.4.3

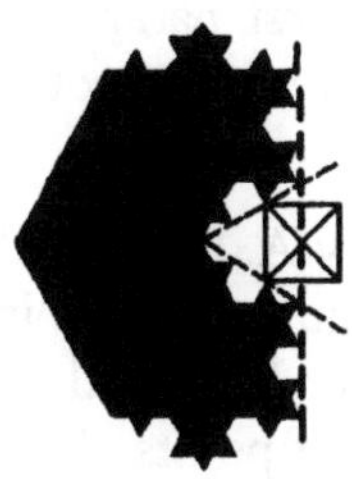

Figure 9.4.4

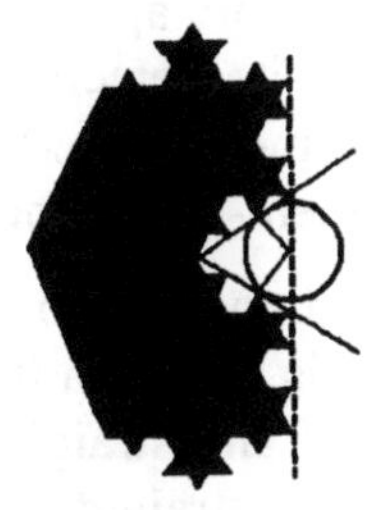

Figure 9.4.5

If the Euclidean distance is used, then the concavity coarseness of its concavity is $\frac{2r\sqrt{21}}{21}$, representing the radius of the circle drawn in Figure 9.3.5.

As it was shown in the previous examples, the concavity coarseness of a set depends on the manner of computing the distance between two points.

Theorems 9.3.1, 9.3.2 and 9.3.3 remain also valid in the $\mathbf{R}^3$ case and are easy to generalise in the framework offered by the spaces having more dimensions. It is of interest the study of the concavity coarseness of some particular fractals, for example those discussed in R. H. Bing (1961), in as much as this might lead to emphasising classes of fractals with convexity properties.

9.5 Construction of the convex hull and recognition of convex configurations

The convexity property is important basically in more stages of the process of shape analysis: segmentation, parameterisation.

Usually, the problem of shape detection deals with sets of points in $\mathbf{Z}^2(h)$ or in $\mathbf{Z}^3(h)$ that are supposed to have a sort of connectivity property (by means of digital arcs). But, we cannot forget that such a set is embedded into $\mathbf{R}^2$ or $\mathbf{R}^3$. Therefore, such a region is, in fact, non-connected and more, finite.

But it is an approximation (in the sense offered by the digitisation process) of an object of the real life.

There are more approaches to the discrete convexity in connection to the possibility of its detection, as we have seen in the first paragraph of this chapter. The most related to algorithms of detection are the directional ones and the segment-based ones.

The directional approach, didactically described by A. Rosenfeld and A. Kak (1976), is based on the property that a set A in the real plane is convex if and only if any straight-line meets A at most once. In digital context, it means that any straight-line meets A in only one run of points. This remark leads to the equivalent descriptions of the convexity in this framework.

Condition 9.5.1. *For any* $p \in A$, $q \in A$, *the straight-line segment* <p, q> *lies in* A.

Condition 9.5.2. *For any* $p = (x_p, y_p) \in A$, $q = (x_q, y_q) \in A$, *the midpoint*

$$m = \left(\frac{x_p + x_q}{2}, \frac{y_p + y_q}{2}\right) \in A.$$

For the convexity detection it is enough to choose the points p and q from the border points of A and to test one of these properties. The midpoint property is not, sometimes efficient because sometimes the middle point has not as coordinates multiples of h with integer coefficients. Then it is better to test if the points of the straight-line segment <p, q> having as coordinates multiples of h with integer coefficients belong to A.

More accuracy brings the point of view described by J.M. Chassery and C. Garbay (1984). As we have previously seen (chapter 5, chapter 8 and 9.1), Chassery's definition for discrete convexity looks for grid points around the straight line segment <p, q>, into an ε-neighborhood of this straight-line segment, with $\varepsilon \in\,]h/2, h[$. We recall that, according to this point of view, a set $A \subseteq \mathbf{Z}^2(h)$ is discrete convex if for every pair of points $p \in A$, $q \in A$ and for all t, $0 < t < 1$, there is a point $z \in A$ which satisfies the condition $c\,(tp + (1-t)q, z) < h$. As we have seen in 3.5, if A and B are two discrete convex connected components then their intersection, $A \cap B$, is discrete convex. This leads to the definition of discrete convex hull, Co (X), of a connected component X: Co (X) *is the intersection of all discrete convex sets including the component X.*

A geometric equivalent definition of the discrete convex hull was proposed by C.E. Kim and A. Rosenfeld (1981).

The construction of the minimum perimeter convex polygon (MPCP) of a 8-connected component X is, in fact, the method of determining Co (X). J.M. Chassery (1983) has proved that MPCP is discrete convex and that it coincides with the convex hull of the set X. Algorithms for the convexification of a set X can be found, for example, in the papers of J.M. Chassery (1978, 1983, 1984), R.L. Graham (1972), C.E. Kim (1980), R. Klette (1983). The convex hull is used to formulate stop criteria in algorithms for segmentation of images including convex objects.

The following algorithm for segmentation of convex coloured objects lying on a relatively homogenous background was used by J.M. Chassery and C. Garbay (1984) to detect these objects, even in the presence of poorly distinguishable frontiers or in the case of similar touching objects. They used it during their applications to bone marrow cell images. Combining the local colour similarity with global spatial cohesion controls the aggregation process. The process is stopped using a convexity predicate.

Part A. Colour Similarity and Aggregation Criterion

Colour similarity is measured by means of a colour distance. The starting point is in the acquisition of three images, R, G, B. The log-transformed images are linearly combined to yield one achromatic image, A, and two chromatic images (C_1, C_2) (according to the human colour perception). The expressions that give this combination are:

$$A = \frac{1}{3}(\log R + \log G + \log B),$$

$$C_1 = \frac{\sqrt{3}}{2}(\log R - \log G),$$

$$C_2 = \log B - \frac{1}{2}(\log R + \log G).$$

Further characterisation of a colour by its luminance L, saturation S and hue H is given by: $L = A$, $S = \left(C_1^2 + C_2^2\right)^{\frac{1}{2}}$, $H = \text{Arc}\cos\frac{C_1}{S}$. The (A, C_1, C_2) space is endowed with an Euclidean metric $D^2 = (\Delta A)^2 + (\Delta C_1)^2 + (\Delta C_2)^2$, which yields, using the polar coordinates (L, S, H)

$$D^2 = (\Delta L)^2 + (\Delta S)^2 + \left(\frac{S_1 + S_2}{2}\right)^2 (\Delta H)^2,$$

where S_1 and S_2 are the respective saturation of the points under consideration.

The aggregation criterion. The starting point is a given set of connected points. The colour similarity and the spatial cohesion is used to iteratively aggregate free pixels. Similarity between a point p of colour C_p and a given region of mean colour C_r (colour reference) is measured by computing the distance $D(C_p, C_r)$. For a given thresold α, we say that C_p is similar to C_r if the condition $D(C_p, C_r) \leq \alpha$ holds. The next step updates the color reference by considering only newly aggregated pixels in order as it remains representative as the region under interest is enlarged. It is furthermore computed if the number of newly aggregated pixels is sufficiently large. On the other hand, the threshold is incremented by a fixed amount, $\alpha = \alpha + \Delta\alpha$. The entire process is repeated until the condition fulfills, taking care to reinitialize α for further iterations. To further control the propagation and convergence, and to select candidate pixels, spatial criteria are used. They are based on convexity.

Part B. Convex Hull and Convexity Criterion

Convex hull, used to control both aggregation process propagation and convergence, is computed at each iteration from the reference region (previous iterate) and connected candidate points. Only the candidate points lying within the convex hull are finally aggregated. Reference region points together with aggregated points constitute the segmentation mask for the iteration under consideration.

The difference between segmentation mask and its convex hull area provides us with the stop criteria for aggregation process. The decrease or the increase of this difference is tested.

Part C. Algorithm

Cell segmentation is achieved by partitioning the support of the image X into four regions: X_1 corresponding to background, X_2 labeled as nucleus, X_3 representing cytoplasm and finally, X_4 corresponding to red cells. Both leukocytes and red cells are assumed to satisfy a convexity predicate, while background and nucleus satisfy a uniformity predicate. Let us suppose that the acquisition system allows us to detect only one leukocyte per image, its location being central.

The algorithm consists on the following iterative scheme.

Initial Step. Nucleus and background regions are assumed to be homogeneous and well contrasted from the rest of the image. Therefore, they are detected by simply thresholding the green image, the thresholds being extracted from the corresponding histogram. The result will consist in two regions X_1 = background, X_2 = nucleus. Before entering the iteration scheme, the centrally located connected is searched in $Y = X \setminus X_1$ and is labeled X_3. A preliminary convexity test is then performed by computing $Co(X_3)$ which allows detecting if X_3 is an isolated cell or a set of touching cells. If X_3 is convex, $Co(X_3)$ is given as segmentation mask and the processing goes on with the next image. The aggregation scheme is otherwise entered by defining $X_3^{(0)} = X_2$ as an initial set of points.

Candidate points will be searched for the first iteration in the set of free points $Y^{(1)} = X \setminus (X_1 \cup X_2)$.

Aggregation Scheme from Iteration k *to* k + 1.

The aggregation scheme deals with the construction of $X_3^{(k+1)}$ from $X_3^{(k)}$ and $Y^{(k+1)}$. The set of free points is defined by $Y^{(k+1)} = Y^{(k)} \setminus X_3^{(k)}$.

<u>If</u> [$(k \geq 10)$ or $(Y^{(k+1)} = \varnothing)$] <u>then</u>

Give Co ($X_3^{(k)}$) as segmentation mask stop

<u>Else</u>

<u>*Color similarity criterion*</u>

Compute the color reference C_r from $X_3^{(k)} \setminus X_3^{(k-1)}$ (newly aggregated pixels). Initialize α.

100 <u>do for</u> each point y in $Y^{(k+1)}$

$g(y) = 1$ if $D(C_r, C_Y) \leq \alpha$

$g(y) = 0$ otherwise

<u>enddo</u>

<u>if</u> [Num $\{y/g(y) = 1\} \leq 10$] <u>then</u>

Let $\alpha = \alpha + \Delta\alpha$

<u>goto</u> 100

<u>endif</u>

Let $E_1 = \{y \mid y \in Y^{(k+1)}, g(y) = 1\}$ (candidate pixels).

<u>*Convexity criterion*</u>

Let $H = \{y \mid y \in E_1, y \text{ connected to } X_3^{(k)}\}$

Compute Co (H)

Let $E_2 = \{y \mid y \in E_1, y \in Co(H)\}$

Let $X_3^{(k+1)} = X_3^{(k)} \cup E_2$

<u>Endif</u>

A convexity test is performed on $X_3^{(k+1)}$ before eventually repeating the entire aggregation process by letting $k = k + 1$.

Convexity Test. Computing Co $(X_3^{(k+1)})$ performs a convexity test. The difference value $D^{(k+1)} = \text{Co}\,(X_3^{(k+1)}) \setminus X_3^{(k+1)}$ is calculated and compared to previous values. If $D^{(k)}$ appears to be a local minimum or if a great increase is detected between $D^{(k)}$ and $D^{(k+1)}$, the processing stops and Co $(X_3^{(k+1)})$ is given as segmentation mask.

As can be seen, the processing stops on different considerations: $k \geq 10$, $Y^{(k+1)} = \varnothing$, or $X_3^{(k)}$ convex, giving each case Co $(X_3^{(k+1)})$ as a segmentation mask.

A method is now needed to determine whether or not such a segmentation mask is adequate. It is performed only for non-isolated cells. The segmentation of isolated cells is assumed to be always perfect. The percentage of pixels included in the convex hull and not in the region is computed and compared to a given threshold. If this percentage does not exceed the threshold then there is a reasonable agreement between $X_3^{(k)}$ and $\text{Co}(X_3^{(k)})$. An incorrect segmentation message occurs then, the corresponding cell being eliminated from further analysis.

Chapter 10

Alternative theorems and integer convex sets

The purpose of this chapter is to present some fundamental theorems for systems of linear integer inequalities. A discrete polyhedral set $SI = S \cap \mathbf{Z}^n$ corresponds to each integer polyhedral set S (i.e. a polyhedral set generated by a system of linear inequalities with integer coefficients). SI will be called the discrete polyhedral set attached to S. The set S is always classically convex. Chapter 2 proved that SI is not classically convex but it is convex with respect to $\mathbf{Z}^n$. Conversely, if $X \subseteq \mathbf{Z}^n$ is an integer convex set, then the set conv (X) is an integer polyhedral set.

We give existence theorems for linear integer systems and theorems of alternative. These theorems involve two systems, say (I) and (II), and assert that either system (I) has a solution or system (II) has a solution, but never both of them. The conclusions of these theorems have geometrical interpretations and they spotlight some relations between integer convex sets. The conditions established by alternative theorems refer to the situation when an integer convex polytope exists and another is empty, or there are two integer convex polytopes with a relationship between them. In the last paragraph of this chapter, some applications of alternative theorems in duality theory for integer programming problems are given.

10.1 Integer polyhedral sets and integer polyhedral sets with respect to $\mathbf{Z}^n$

Let m and n be natural non null numbers and a_{ij}, $i \in \{1, \ldots, m\}$, $j \in \{1, \ldots, n\}$, b_i, $i \in \{1, \ldots, n\}$, integer numbers. We consider the sets:

$$(10.1.1) \qquad S = \{x = (x_1, \ldots, x_n) \in \mathbf{R}^n_+ \mid \sum_{j=1}^{n} a_{ij}x_j \leqq b_i,\ i \in \{1, \ldots, m\}\}$$

and

$$(10.1.2) \qquad SI = S \cap \mathbf{Z}^n.$$

In the following a connection between the sets S and SI is studied. Obviously, S is an integer polyhedral set and SI is the discrete polyhedral set attached to S. We remember that if $C \subseteq \mathbf{R}^n$ is a convex set then a point $x \in S$ is an extremal point of S (in the classical theory of the convex sets) if there are no two points $u \in S \setminus \{x\}$ and $v \in S \setminus \{x\}$ such that $x \in \langle u, v \rangle$. We denote by Exr (C)the set of all extremal points of C.

Remark 10.1.1. It $S = \varnothing$, then $SI = \varnothing$. The conversely is not always true.

Example 10.1.1. Let $S = \{(x_1, x_2) \in \mathbf{R}^2 \mid 3x_1 - x_2 \leqq \frac{1}{2}, -3x_1 + x_2 \leqq -\frac{1}{2}\}$ and $SI = S \cap \mathbf{Z}^2$. Set S is an integer non-bounded polyhedral set (it is easy to see that the point $x^t = (\frac{2+t}{6}, \frac{1+t}{2}) \in S$, for all $t \in [0, +\infty[$, and $\lim_{t \to +\infty} \frac{2+t}{6} = +\infty$). But set $SI = \{(x_1, x_2) \in \mathbf{Z}^2 \mid 3x_1 - x_2 = \frac{1}{2}\}$ is empty.

The next lemmas are important for our considerations. We remember that a polyhedral set is called a pointed polyhedral set if it has at least a vertex.

Lemma 10.1.1. *If $X \subseteq \mathbf{R}^*_+$ is a nonempty polyhedral set, then it is a pointed polyhedral set.*

Proof. Suppose that X is not pointed. From theorem 35 of B. Martos (1975, chapter 1) it follows that there are two points $x = (x_1, \ldots, x_n)$ and $y = (y_1, \ldots, y_n)$ of X, $x \neq y$, such that

$$(10.1.3) \qquad (1 - t)\,x + t\,y \in X, \text{ for all } t \in \mathbf{R}.$$

Because $x \neq y$, there is $j \in \{1, \ldots, n\}$ such that $x_j \neq y_j$. For $t = - x_j \cdot (y_j - x_j)^{-1} - 1$, if $y_j - x_j > 0$, and $t = - x_j \cdot (y_j - x_j)^{-1} + 1$, if $y_j - x_j < 0$, we get, since $z_j < 0$, that $z = (1 - t) \cdot x + t \cdot y \notin X$. This contradicts (10.1.3). Therefore X is a pointed polytope.

Now, let be c_j, $j \in \{1, \ldots, n\}$, natural numbers and

$$(10.1.4) \qquad SQ = S \cap \mathbf{Q}^n,$$

where S is given by (10.1.1).

Lemma 10.1.2. *If* p *is a natural number such that*

$$(10.1.5) \qquad \{x \in \mathbf{R}_+^n \mid \sum_{j=1}^{n} c_j x_j = p\} \neq \varnothing,$$

and the polyhedral set S *is nonempty, then there exist a point* $x^p \in SQ$, *with* $\sum_{j=1}^{n} c_j x_j^p = p$, *and an edge* H_p *of* S *such that* $x^p \in H_p$.

Proof. Let us denote

$$S_{c,p} = S \cap \{x \in \mathbf{R}_+^n \mid \sum_{j=1}^{n} c_j x_j = p\},$$

$$SQ_{c,p} = \{x \in SQ \mid \sum_{j=1}^{n} c_j x_j = p\}.$$

Because a_{ij}, $i \in \{1, \ldots, m\}$, $j \in \{1, \ldots, n\}$, b_i, $i \in \{1, \ldots, m\}$, c_j, $j \in \{1, \ldots, n\}$, are integer numbers, we have

$$(10.1.6) \qquad \mathrm{Exr}\,(S_{c,p}) \subseteq SQ_{c,p}.$$

As $S \subseteq \mathbf{R}_+^n$, we get, in view of lemma 10.1.1, that the polyhedral set $S_{c,p}$ has at least a vertex. But any vertex of a polyhedral set is an extremal point of this set. Hence $\mathrm{Exr}\,(S_{c,p}) \neq \varnothing$. Let be $x^p \in \mathrm{Exr}\,(S_{c,p})$. From (10.1.6) we get that $x^p \in SQ$. As x^p is a vertex of $S_{c,p}$, there is an edge H_p of S such that $x^p \in H_p$.

Remark 10.1.2. Because a_{ij}, $i \in \{1, \ldots, m\}$, $j \in \{1, \ldots, n\}$, b_i, $i \in \{1, \ldots, n\}$, are integer numbers, if H is an edge of S, there are $x^0 \in \mathbf{Q}^n$, $d \in \mathbf{Q}^n$, and $T \subseteq]0, +\infty[$ such that

$$(10.1.7) \qquad H = \{x^0 + td \mid t \in T\}.$$

If H is an infinite edge, then

$$(10.1.8) \qquad \{x^0 + td \mid t \in [0, +\infty[\} \subseteq H.$$

Proposition 10.1.1. *If set* SQ *is non-bounded, then there is a natural number* α *such that set* SN_α *is also non-bounded, where*

$$SN_\alpha = \{x = (x_1, \ldots, x_n) \in \mathbf{N}^n \mid \sum_{j=1}^{n} a_{ij} x_j \leq \alpha b_i,\ i \in \{1, \ldots, m\}\}.$$

Proof. Because set SQ is non-bounded, it is a natural number j, $j \in \{1, \ldots, n\}$, such that set $\mathrm{pr}_j\,(SQ)$ is also non bounded. Let k be a natural number. Obviously there is a point x^k of SQ such that

(10.1.9) $x_j^k \geq k$.

Taking $p = x_j^k \in \mathbf{Q}$, and $c_i = \begin{cases} 0, \text{if } i \in \{1,...,n\} \setminus \{j\} \\ 1, \text{if } i = j \end{cases}$, and applying lemma 10.1.2, we get that there is a point $z^k \in$ SQ, with

(10.1.10) $z_j^k = x_j^k$,

and an edge H_k of S such that $z^k \in H_k$. Because S has a finite number of edges, it follows that there is a sequence $(k_h)_{h\in N}$, and an edge H of S such that $z^{k_h} \in$ H, for all natural numbers h. As $\lim_{h\to+\infty} z_j^{k_h} = +\infty$, H is an infinite edge. Then, from remark 10.1.2 we get that there is $x^0 \in \mathbf{Q}^n$, and $d \in \mathbf{Q}^n$, such that $\{x^0 + td \mid t \in [0, +\infty[\} \subseteq$ H. Since $z^{k_h} \in$ H, for all $h \in \mathbf{N}$, and $\lim_{h\to+\infty} z_j^{k_h} = +\infty$, it results that $d_j \neq 0$.

Let α be the least common divisor of the denominators of the numbers x_i^0, $i \in \{1, \ldots, n\}$ and d_i, $i \in \{1, \ldots, n\}$.

We put $u^h = \alpha\,(x^0 + hd)$, for all $h \in \mathbf{N}$. Evidently, $u^h \in \mathbf{Z}^n$, for all $h \in \mathbf{N}$.

Because $x^0 + hd \in H \subseteq S$, for all natural numbers h, we have $\sum_{j=1}^{n} a_{ij}(x_j^0 + hd) \leq b_i$, for all $i \in \{1, \ldots, m\}$. Then $\sum_{j=1}^{n} a_{ij} \cdot \alpha \cdot (x_j^0 + hd) \leq \alpha b_i$, for each $i \in \{1, \ldots, m\}$, and for each $h \in \mathbf{N}$. Hence $\sum_{j=1}^{n} a_{ij} u_j^h \leq \alpha b_i$, for all $i \in \{1, \ldots, m\}$ and for all natural numbers h. This implies that $u^h \in SN_\alpha$ for all natural numbers h.

Because sequence $(u^h)_{h\in N}$ is non-bounded (we have $\lim_{h\to+\infty} u_j^h = +\infty$), it results that the set SN_α is non-bounded.

Proposition 10.1.2. *If set* SQ, *given by* (10.1.4), *is non-bounded and set* SI, *given by* (10.1.2), *is non-empty set, then the set* SI *is also non-bounded.*

Proof. From the proof of proposition 10.1.1 we get that there is $x^0 \in \mathbf{Q}^n$, and $d \in \mathbf{Q}^n$, such that $\{x^0 + td \mid t \in]0, +\infty[\} \subseteq$ S and there is a natural number s, $s \in \{1, \ldots, n\}$ such that $\lim_{t\to+\infty}(x_s + td_s) = +\infty$. Then, it is necessary to have

(10.1.11) $a_{i1}d_1 + \ldots + a_{in}d_n \leq 0$, for each $i \in \{1, \ldots, m\}$.

Let $z \in SI$. For all natural numbers k, we take $z^k = z + k \cdot r \cdot d$, where r is the least common divisor of the denominators of the numbers $d_1, \ldots, d_n$. Evidently, $z^k \in \mathbf{N}^n$, for all $k \in \mathbf{N}$. As $z \in SI$, from (10.1.11) it results that

$$\sum_{j=1}^{n} a_{ij} z_j^k = \sum_{j=1}^{n} a_{ij} z_j + k \cdot r \cdot \sum_{j=1}^{n} a_{ij} d_j \leq \sum_{j=1}^{n} a_{ij} z_j \leq b_i, \text{ for each } i \in \{1,\ldots,m\}.$$

Hence, $z^k \in SI$, for all natural numbers k. But $\lim_{k \to +\infty} (z_s + kd_s) = +\infty$. This implies that the set SI is non-bounded.

10.2 Existence theorems for linear homogenous integer systems

We establish now some key theorems for the existence of certain types of solutions for a linear integer system and begin with an integer theorem of Tucker's lemma type.

Theorem 10.2.1. *Let* $p, n \in \mathbf{N}^*$. *For any given* $p \times n$ *matrix* $A \in M_{p \times n}(\mathbf{Z})$ *the systems*

$$\text{(I)} \qquad \begin{cases} x \cdot A^T \geqq 0_p \\ x \in Z^n \end{cases}$$

and

$$\text{(II)} \qquad \begin{cases} y \cdot A = 0_n \\ y \in N^p \end{cases}$$

possess solutions x^0 *and* y^0 *satisfying* $(x^0|A_1) + y_1^0 \geqq 1$, *where* $A_1 = (a_{11}, \ldots, a_{1n})$ *is the first row of matrix* A.

Proof. The proof is by induction on p. For p = 1, two cases are possible:

Case 1) $A_1 = 0_n$. Then $x^0 = 0_n$ is a solution of system (I), $y^0 = 1$ a solution of system (II), and $(x^0|A_1) + y_1^0 = 0 + 1$.

Case 2) $A_1 \neq 0_n$. Because $A_1 \neq 0_n$, we have $(A_1|A_1) = a_{11}^2 + \ldots + a_{1n}^2 > 0$. But $a_{11}^2 + \ldots + a_{1n}^2 \in \mathbf{N}$. Thus $a_{11}^2 + \ldots + a_{1n}^2 \geq 1$. Then we get that $x^0 = A_1$ is a solution of the system (I), $y^0 = 0$, a solution of system (II) and $(x^0|A_1) + y^0 = a_{11}^2 + \ldots + a_{1n}^2 + 0 = a_{11}^2 + \ldots + a_{1n}^2 \geqq 1$. Therefore, for p=1 the conclusion of the theorem is true.

We make the following hypothesis of induction: we assume that the theorem is true for a matrix A of p rows and proceed to prove it for a matrix

$$\overline{A} = \begin{bmatrix} A \\ A_{p+1} \end{bmatrix} = \begin{bmatrix} A_1 \\ \dots \\ A_p \\ A_{p+1} \end{bmatrix}$$

of p+1 rows, where $A_i = (a_{i1}, \dots, a_{in}) \in \mathbf{Z}^n$, for all $i \in \{1, \dots, p+1\}$. By applying the hypothesis of induction of the matrix A, we get that there are $x^0 \in \mathbf{Z}^n$, $y^0 \in \mathbf{N}^p$ satisfying:

(10.2.1) $x^0 \cdot A^T \geqq 0_p$,

(10.2.2) $y^0 \cdot A = 0_n$,

(10.2.3) $(x^0|A_1) + y_1^0 \geqq 1$.

Two cases are possible: $(x^0|A_{p+1}) \geqq 0$, or $(x^0|A_{p+1}) < 0$. If $(x^0|A_{p+1}) \geqq 0$, taking $\overline{y} = (y^0, 0)$ we have $x^0 \cdot \overline{A}^T \geqq 0_{p+1}$, $\overline{y} \in \mathbf{N}^{p+1}$, $\overline{y} \cdot \overline{A} = 0_n$, and $(x^0|A_1) + y_1^0 \geqq 1$, which states the conclusion of theorem for $\overline{A}$.

Let now $(x^0|A_{p+1}) < 0$. Then taking

(10.2.4) $c = -(x^0|A_{p+1})$, and $d_j = (x^0|A_j)$, for each $j \in \{1, \dots, p\}$,

we have

(10.2.5) $c \in \mathbf{N}$ and $d_j \in \mathbf{N}$, for each $j \in \{1, \dots, p\}$.

Let

(10.2.6) $$B = \begin{bmatrix} B_1 \\ \dots \\ B_p \end{bmatrix} = \begin{bmatrix} cA_1 + d_1 A_{p+1} \\ \dots \\ cA_p + d_p A_{p+1} \end{bmatrix}.$$

For all $j \in \{1, \dots, p\}$ we have

(10.2.7) $(x^0|B_j) = (x^0|cA_j + d_jA_{p+1}) = c(x^0|A_j) + d_j(x^0|A_{p+1}) = cd_j - d_jc = 0$.

Applying the hypothesis of induction of matrix B, we get that there are $v = (v_1, \dots, v_n) \in \mathbf{Z}^n$ and $u = (u_1, \dots, u_p) \in \mathbf{N}^p$ satisfying:

(10.2.8) $v^T \cdot B \geqq 0_p$, $u \cdot B = 0_n$, and $(v|B_1) + u_1 \geqq 1$.

Let

(10.2.9) $\overline{u} = (c \cdot u, d_1 . u_1 + \dots + d_p \cdot u_p)$.

Because $c \in \mathbf{N}$, $d_j \in \mathbf{N}$, for each $j \in \{1, ..., p\}$ and $u \in \mathbf{N}^p$, we have

(10.2.10) $\quad \bar{u} \in \mathbf{N}^{p+1}$

and

$$\bar{u} \cdot \overline{A} = c \cdot u \cdot A + (\sum_{j=1}^{n} d_j \cdot u_j) \cdot A_{p+1} \cdot$$

From (10.2.6) we have

$$u \cdot B = c \cdot u \cdot A + (\sum_{j=1}^{n} d_j u_j)\, A_{p+1}.$$

Then

(10.2.11) $\quad \bar{u} \cdot \overline{A} = u \cdot B = 0_n.$

Let $t = cv + (A_{p+1}|v)x^0$. Because $c \in \mathbf{N}$, $v \in \mathbf{Z}^n$, $A_{p+1} \in \mathbf{Z}^n$ and $x^0 \in \mathbf{Z}^n$, it follows that

(10.2.12) $\quad t \in \mathbf{Z}^n.$

We have

(10.2.13) $\quad (A_{p+1}|t) = c(A_{p+1}|v) + (A_{p+1}|v) \cdot (A_{p+1}|x^0) = (A_{p+1}|v) \cdot (c - c) = 0,$

and

$$(A_j|t) = c \cdot (A_j|v) + (A_{p+1}|v) \cdot (A_j|x^0) = c \cdot (A_j|v) + d_j (A_{p+1}|v).$$

From (10.2.6) we get

$$A_j = \frac{1}{c} B_j - \frac{d_j}{c} A_{p+1}, \text{ for all } j \in \{1, ..., n\}.$$

Hence

(10.2.14) $\quad (A_j \mid t) = c\,(\frac{1}{c} B_j \mid v) - c\frac{d_j}{c}(A_{p+1} \mid v) + d_j (A_{p+1} \mid v) = (B_j \mid v),$

for all $j \in \{1, ..., n\}$. From (10.2.14) and (10.2.13) it results that

(10.2.15) $\quad t\overline{A}^T = (t \cdot A^T, \langle A_{p+1}, t \rangle) = (v \cdot B^T, 0) \geq 0_{p+1}.$

Finally, from (10.2.8) and (10.2.14) we have

(10.2.16) $\quad (A_1|t) + u_1 = (B_1|v) + u_1 \geq 1.$

Relations (10.2.10), (10.2.11), (10.2.12), (10.2.15) and (10.2.16) state the theorem for $\overline{A}$.

Theorem 10.2.2. *Let* n, p $\in$ **N***. *For any given* p $\times$ n *matrix* A $\in M_{p\times n}(\mathbf{Z})$ *the systems*

$$\text{(I)} \quad \begin{cases} xA^T \geqq 0_p \\ x \in Z^n \end{cases}$$

and

$$\text{(II)} \quad \begin{cases} yA = 0_n \\ y \in Z^p \end{cases}$$

possess solutions x^0, *respectively* y^0, *satisfying* $x^0A^T + y^0 > 0_p$.

Proof. In theorem 10.2.1 the row A_1 played a special role. By renumbering the rows of A, any other row, say A_i, can play the same role. Hence, by theorem 10.2.1, for all i$\in$\{1, ..., p\}, there are $x^i \in \mathbf{Z}^n$ and $y^i \in \mathbf{N}^p$ such that

$$(10.2.17) \qquad x^iA^T \geqq 0_p,\ y^iA = 0_n,\ (A_i|x^i) + y^i_i > 0.$$

Consider $x^0 = x^1 + ... + x^p$ and $y^0 = y^1 + ... + y^p$. Obviously we have

$$(10.2.18) \qquad x^0 \in \mathbf{Z}^n \text{ and } y^0 \in \mathbf{N}^p.$$

From (10.2.17) we obtain

$$(10.2.19) \qquad x^0A^T = x^1A^T + ... + x^pA^T \geqq 0_p,$$

$$(10.2.20) \qquad y^0A = y^1A + ... + y^pA = 0_n,$$

and, for i $\in$\{1,...,p\},

$$(A_i|x^0) + y^0_i = (A_i|x^i) + y^i_i + \sum_{\substack{k=1 \\ k\neq i}}^{p} (A_i|x^k) + \sum_{\substack{k=1 \\ k\neq i}}^{p} y^k_i > 0.$$

Hence

$$(10.2.21) \qquad x^0A^T + y^0 > 0.$$

Relations (10.2.18) - (10.2.21) show that x^0 is a solution of system (I), y^0 a solution of system (II), and $Ax^0 + y^0 > 0$.

Theorem 10.2.3. *Let* n, p, r *be natural non-null numbers. Let* A $\in M_{p\times n}(\mathbf{Z})$, *and* B $\in M_{q\times n}(\mathbf{Z})$ *be given matrices. Then the systems*

$$\text{(I)}\qquad \begin{cases} xA^T \geqq 0_p \\ xB^T = 0_q \\ x \in Z^n \end{cases}$$

and

$$\text{(II)}\qquad \begin{cases} yA + zB = 0_n \\ y \in N^p,\ z \in Z^q \end{cases}$$

possess solutions x^0, *respectively* (y^0, z^0) *satisfying* $x^0A^T + y^0 > 0_p$.

Proof. We apply theorem 10.2.2 to the systems

$$\text{(I')}\qquad u \cdot \begin{bmatrix} A \\ B \\ -B \end{bmatrix}^T \geqq 0_{p+2q},\ u \in \mathbf{Z}^n,$$

$$\text{(II)}\qquad \begin{bmatrix} v \\ t \\ w \end{bmatrix} [A, B, -B] = 0_{3n},\ v \in \mathbf{N}^p,\ t \in \mathbf{N}^q,\ w \in \mathbf{N}^q$$

and obtain u, respectively (v, t, w) satisfying

(10.2.22) $u \cdot A^T \geqq 0_p,\ u \cdot B^T \geqq 0_q,\ u \cdot (-B)^T \geqq 0_q;$

(10.2.23) $v \cdot A + t \cdot B + (-w) \cdot B = 0_n;$

(10.2.24) $u \cdot A^T + v > 0_p,\ u \cdot B^T + t > 0_q,\ u \cdot (-B)^T + w > 0_q.$

Taking $x^0 = u$, $y^0 = v$, $z^0 = t - w$, it follows that:

(10.2.25) $x^0 \in \mathbf{Z}^n,\ y^0 \in \mathbf{N}^p,\ z^0 \in \mathbf{Z}^q$

and

(10.2.26) $x^0A^T = u \cdot A^T \geqq 0_p.$

From (10.2.22), (10.2.24) and (10.2.26) it results that

(10.2.27) $x^0 B^T = u \cdot B^T \geqq 0_q;$

(10.2.28) $y^0 \cdot A + z^0 \cdot B = v \cdot A + (t - w) \cdot B = v \cdot A + t \cdot B + w \cdot B = 0_n$

and

(10.2.29) $x^0 \cdot A^T + y^0 = u \cdot A^T + v > 0_p.$

Therefore x^0 is a solution of (I), (y^0, x^0) a solution of (II), and

$$Ax^0+y^0 > 0_p.$$

Let $A \in M_{p\times n}(\mathbf{Z})$, $B \in M_{q\times n}(\mathbf{Z})$, $C \in M_{r\times n}(\mathbf{Z})$ and $D \in M_{s\times n}(\mathbf{Z})$, where n, p, q, r, s are natural non null numbers. Taking $\overline{A} = \begin{bmatrix} A \\ B \\ C \end{bmatrix}$, $\overline{B} = D$ and applying theorem 10.2.3 for $\overline{A}, \overline{B}$ we get

Corollary 10.2.1. *If* $A \in M_{p\times n}(\mathbf{Z})$, $B \in M_{q\times n}(\mathbf{Z})$, $C \in M_{r\times n}(\mathbf{Z})$ *and* $D \in M_{s\times n}(\mathbf{Z})$, *then the systems*

$$\text{(I)} \quad \begin{cases} x \cdot A^T \geqq 0_p \\ x \cdot B^T \geqq 0_q \\ x \cdot C^T \geqq 0_r \\ x \cdot D^T = 0_s \\ x \in Z^n \end{cases}$$

$$\text{(II)} \quad \begin{cases} u \cdot A + v \cdot B + t \cdot C + w \cdot D = 0_n \\ u \in N^p, v \in N^q, t \in N^r, w \in Z^s \end{cases}$$

possess solutions x^0, *respectively* (u^0, v^0, t^0, w^0), *satisfying*:

$$x^0 \cdot A^T + u^0 > 0_p,\ x^0 \cdot B^T + v^0 > 0_q,\ x^0 \cdot C^T + t^0 > 0_r.$$

Let $G \in M_{m\times n}(\mathbf{Z})$, $C \in M_{r\times n}(\mathbf{Z})$ and $D \in M_{s\times n}(\mathbf{Z})$, where m, n, r, s are natural nun null numbers. Taking $\overline{A} = \begin{bmatrix} G \\ C \end{bmatrix}$, $\overline{B} = D$ and applying theorem 10.2.3 for $\overline{A}, \overline{B}$, we get:

Corollary 10.2.2. *If* $G \in M_{m\times n}(\mathbf{Z})$, $C \in M_{r\times n}(\mathbf{Z})$ *and* $D \in M_{s\times n}(\mathbf{Z})$, *then the systems*

$$\text{(I)} \quad \begin{cases} x \cdot G^T \geqq 0_m \\ x \cdot C^T \geqq 0_r \\ x \cdot D^T = 0_s \\ x \in Z^n \end{cases}$$

$$\text{(II)} \quad \begin{cases} y \cdot G + t \cdot C + w \cdot D = 0_n \\ y \in N^m, t \in N^r, w \in Z^s \end{cases}$$

possess solutions x^0, *respectively* (y^0, t^0, w^0), *satisfying*

$x^0 \cdot G^T + y^0 > 0_m$, $x^0 \cdot C^T + t^0 > 0_r$.

10.3 Theorems of alternative and discrete polyhedral sets

Theorem 10.3.1. *Let* n, p, r, s *be non-null natural numbers. Let* $A \in M_{p\times n}(\mathbf{Z})$, $C \in M_{r\times n}(\mathbf{Z})$ *and* $D \in M_{s\times n}(\mathbf{Z})$ *be given matrices. Then either*

$$\text{(I)} \qquad \begin{cases} x \cdot A^T > 0 \\ x \cdot C^T \geqq 0_r \\ x \cdot D^T = 0_s \\ x \in Z^n \end{cases}$$

or

$$\text{(II)} \qquad \begin{cases} u \cdot A + t \cdot C + w \cdot D = 0_n \\ u \in N^p, u \neq 0_p, t \in N^r, w \in Z^s \end{cases}$$

has a solution.

Proof. Suppose that $A \neq 0_{p\times n}$. Let the system (I) be consistent, and let x^0 be a solution of (I). Suppose that the system (II) is also consistent and let (u^0, t^0, w^0) be a solution of it. We remark that $x^0 \cdot A^T \cdot u^0 > 0$, because $x^0 \cdot A > 0_p$ and $u^0 \geq 0_p$. We have

$$(10.3.1) \qquad 0 = (x^0, u^0 \cdot A + t^0 \cdot C + w^0 \cdot D) = u^0 \cdot A \cdot (x^0)^T + t^0 \cdot A(x^0)^T + \\ + w^0 \cdot D \cdot (x^0)^T \geqq u^0 A \cdot (x^0)^T > 0,$$

which is a contradiction. Hence, if (I) is consistent, then (II) is inconsistent.

Let now system (I) be inconsistent. Applying corollary 10.2.1 for the matrices G = A, C = C, D = D, we get that there are $x^0 \in \mathbf{Z}^n$, $u^0 \in \mathbf{N}^p$, $t^0 \in \mathbf{N}^r$ and $w^0 \in \mathbf{Z}^s$ such that

$$(10.3.2) \qquad x^0 \cdot A^T \geqq 0_p,\ x^0 \cdot C^T \geqq 0_r,\ x^0 \cdot D^T = 0_s,\ x^0 \in Z^n,$$

$$(10.3.3) \qquad u^0 \cdot A + t^0 \cdot C + w^0 \cdot D = 0_n,$$

$$(10.3.4) \qquad x^0 \cdot A^T + u^0 > 0_p,\ x^0 \cdot C^T + t^0 > 0_r.$$

Two cases are possible: $u^0 \neq 0_p$ or $u^0 = 0_p$. If $u^0 \neq 0_p$, then (u^0, t^0, w^0) is a solution of (II). Let $u^0 = 0_p$. Then, from (10.3.4) it results that $x^0 \cdot A^T > 0$. This implies that x^0 is a solution of system (I). But now we have arrived at a contradiction with the assumption that (I) is inconsistent. Hence if (I) is inconsistent, then (II) is consistent.

We remark that if $A = 0_{p\times n}$, then system (I) is inconsistent. Applying theorem 10.2.3 for matrices C and D, we get that there are $x^0 \in \mathbf{Z}^n$, $t^0 \in \mathbf{N}^r$ and $w^0 \in \mathbf{Z}^s$, such that $x^0 \cdot C^T \geqq 0_r$, $x^0 \cdot D^T = 0_s$, $t^0 \cdot C + w^0 D = 0$. Then, taking $u^0 = (1, ..., 1) \in \mathbf{R}^p$, we have $u^0 A + t^0 C + w^0 D = 0_n + 0_n = 0_n$. Thus (u^0, t^0, w^0) is a solution of (II). Hence (II) is consistent.

Theorem 10.3.2. *Let* $B \in M_{q\times n}(\mathbf{Z})$, $C \in M_{r\times n}(\mathbf{Z})$ *and* $D \in M_{s\times n}(\mathbf{Z})$ *be given matrices with* $n, q, r, s \in \mathbf{N}^*$. *Then either:*

$$\text{(I)} \quad \begin{cases} x \cdot B^T \geq 0_q \\ x \cdot C^T \geqq 0_r \\ x \cdot D^T = 0_s \\ x \in Z^n \end{cases}$$

or

$$\text{(II)} \quad \begin{cases} v \cdot B + t \cdot C + w \cdot D = 0_n \\ v \in N^q \setminus \{0_q\}, t \in N^r, w \in Z^s \end{cases}$$

has a solution.

Proof. Let system (I) be consistent and let x^0 be a solution of (I). Suppose that system (II) is also consistent and let (v^0, t^0, w^0) be a solution of it. Because $x^0 \cdot B^T \geq 0$, there is $i \in \{1, ..., q\}$ such that $(B_j | x^0) \geqq 0$, for all $j \in \{1, ..., q\} \setminus \{i\}$, and $(B_i | x^0) > 0$. Since $v^0 \in \mathbf{N}^q \setminus \{0_q\}$, we get

$$v^0 B (x^0)^T = \sum_{h=1}^{q} v_h^0 \cdot (B_h \mid x^0) \geqq v_i^0 \cdot (B_i \mid x^0) > 0.$$

We have

$$0 = (x^0 | v^0 \cdot B + t^0 \cdot C + w^0 \cdot D) = v^0 \cdot B \cdot (x^0)^T + t^0 \cdot C \cdot (x^0)^T + w^0 \cdot D \cdot (x^0)^T \geqq$$
$$\geqq v^0 \cdot B \cdot (x^0)^T > 0,$$

which is a contradiction. The assumption that (II) is also consistent is false. Hence, if (I) is consistent, than (II) is inconsistent.

Let now suppose that system (I) be inconsistent. Applying corollary 10.3.2 for G = B, C = C, D = D, we get that there are $x^0 \in \mathbf{Z}^n$, $v^0 \in \mathbf{N}^q$, $t^0 \in \mathbf{N}^r$, $w^0 \in \mathbf{Z}^s$ such that

(10.3.5) $x^0 \cdot B^T \geqq 0_q$, $x^0 \cdot C^T \geqq 0_r$, $x^0 \cdot D^T = 0_s$, $v^0 \cdot B + t^0 \cdot C + w^0 \cdot D = 0_n$,

$$x^0 \cdot B^T + v^0 > 0_q,\ x^0 \cdot C^T + w^0 > 0_r.$$

If $v^0 > 0_q$, then (v^0, t^0, w^0) is a solution of system (II) and the conclusion of the theorem is true. Suppose that there is $i \in \{1,\dots,q\}$ such that $v_i^0 = 0$. Because $x^0 \cdot B^T + v^0 > 0$, we get that $(B_i|x^0) > 0$. Thus we have $x^0 \cdot B^T \geq 0_q$. Then, from (10.3.1), it follows that x^0 is a solution of system (I). This contradicts the assumption that (I) is inconsistent. Hence, if (I) is inconsistent, then (II) is consistent.

Theorem 10.3.3. *Let* n, p, q, r, s $\in$ **N***. *Let* $A \in M_{p\times n}(\mathbf{Z})$, $B \in M_{q\times n}(\mathbf{Z})$, $C \in M_{r\times n}(\mathbf{Z})$ *and* $D \in M_{s\times n}(\mathbf{Z})$, *where* $p + q \neq 0$. *Then either*

$$\text{(I)} \qquad \begin{cases} x \cdot A^T > 0_p \\ x \cdot B^T \geqq 0_q \\ x \cdot C^T \geqq 0_r \\ x \cdot D^T = 0_s \\ x \in Z^n \end{cases}$$

or

$$\text{(II)} \qquad \begin{cases} u \cdot A + v \cdot B + t \cdot C + w \cdot D = 0_n \\ u \in N^p \setminus \{0_p\},\ v \in N^q,\ t \in N^r,\ w \in Z^s \\ \text{or} \\ u \in N^p,\ v \in N^q,\ t \in N^r,\ w \in Z^s \end{cases}$$

has a solution but never both.

Proof. Suppose that A and B are not vacuous matrices. Let x^0 be a solution of system (I). Because $x^0 \cdot A^T > 0_p$, it results that $x^0 \neq 0_n$.

If we suppose that system (II) has also a solution (u^0, v^0, t^0, w^0), then we have $u^0 \cdot A + v^0 \cdot B + t^0 \cdot C + w^0 \cdot D = 0_n$. This implies that

$$(x^0 \,|\, u^0 \cdot A + v^0 \cdot B + t^0 \cdot C + w^0 \cdot D) = 0.$$

On the other hand, we have

$$\text{(10.3.6)} \qquad (x^0 \,|\, u^0 \cdot A + v^0 \cdot B + t^0 \cdot C + w^0 \cdot D) =$$
$$= u^0 A (x^0)^T + v^0 \cdot B \cdot (x^0)^T + t^0 \cdot C \cdot (x^0)^T + w^0 \cdot D \cdot (x^0)^T \geqq$$
$$\geqq \max\{u^0 \cdot A \cdot (x^0)^T,\ v^0 \cdot B \cdot (x^0)^T\}.$$

Two cases are possible:

a) $u^0 \neq 0_p$; then $u^0 \cdot A \cdot (x^0) > 0$.

b) $u^0 = 0_p$; then $v^0 > 0_q$ and, because $x^0 \cdot B^T \geq 0_q$, we get $v^0 \cdot B \cdot (x^0)^T > 0$. In both cases we have max $\{u^{0\,T} a x^0, v^{0\,T} b x^0\} > 0$. Then, from (10.3.6) it results that

(10.3.7) $(x^0 \mid Au^0 + Bv^0 + Ct^0 + Dw^0) > 0,$

which contradicts that $(x^0 \mid u^0 \cdot A + v^0 \cdot B + t^0 \cdot C + w^0 \cdot D) = 0$. Hence, if system (I) is consistent, then system (II) is inconsistent.

Let now system (I) be inconsistent. In view of corollary 10.2.1, there is $x^0 \in \mathbf{Z}^n$ and $(u^0, v^0, t^0, w^0) \in \mathbf{N}^p \times \mathbf{N}^q \times \mathbf{N}^r \times \mathbf{Z}^s$ such that:

(10.3.8) $x^0 \cdot A^T \geq 0_p, x^0 \cdot B^T \geqq 0_q, x^0 \cdot C^T \geqq 0_r, x^0 \cdot D^T = 0_s,$

(10.3.9) $u^0 \cdot A + v^0 \cdot B + t^0 \cdot C + w^0 \cdot D = 0_n,$

(10.3.10) $x^0 \cdot A^T + u^0 > 0_p, x^0 \cdot B^T + v^0 > 0_q, x^0 \cdot C^T + t^0 > 0_r.$

If $u^0 \geq 0_p$ or $v^0 > 0_q$, the conclusion of the theorem is true.

Suppose that we have not $u^0 \geq 0_p$ and we have not $v^0 > 0_q$. As $u^0 \in \mathbf{N}^p$, and u^0 is not great-less 0_p, it results that $u^0 = 0_p$. Then (10.3.10) implies

(10.3.11) $x^0 \cdot A^T > 0_p.$

Since v^0 is not great as 0_q (v^0 is not strictly greater then 0_q) and $v^0 \in \mathbf{N}_q$, it results that there is $j \in \{1,...,q\}$ with $v_j^0 = 0$. Then from (10.3.10) we get that $(B_j | x^0) > 0$. Since $x^0 B^T \geqq 0_q$, this implies

(10.3.12) $B^T \geq 0_q.$

Because $x^0 \in \mathbf{Z}^n$, from (10.3.8), (10.3.11) and (10.3.12), we get that x^0 is a solution of system (I). This contradicts the hypothesis that system (I) is inconsistent. Hence, if (I) is inconsistent, then (II) is consistent.

We remark that if A or B is a vacuous matrix, then system (I) is inconsistent. In view of corollary 10.2.1, system (II) is consistent.

Theorem 10.3.4. *Let* $A \in M_{p\times n}(\mathbf{Z})$, $B \in M_{q\times n}(\mathbf{Z})$, $C \in M_{r\times n}(\mathbf{Z})$, $D \in M_{s\times n}(\mathbf{Z})$. *Then either*

$$\text{(I)} \quad \begin{cases} x \cdot A^T \geq 0_p \\ x \cdot B^T \geqq 0_q \\ x \cdot C^T \geqq 0_r \\ x \cdot D^T = 0_s \\ x \in Z^n \end{cases} \quad \text{or} \quad \begin{cases} x \cdot A^T \geqq 0_p \\ x \cdot B^T > 0_q \\ x \cdot C^T \geqq 0_r \\ x \cdot D^T = 0_s \\ x \in Z^n \end{cases}$$

has a solution or

$$\text{(II)} \quad \begin{cases} u \cdot A + v \cdot B + t \cdot C + w \cdot D = 0_n \\ u \in (N^*)^p,\ v \in N^q \setminus \{0_q\},\ t \in N^r,\ w \in Z^s \end{cases}$$

has a solution but never both.

Proof. Suppose that A or B are not the vacuous matrix. Let system (I) be consistent and let x^0 be a solution of (I). Suppose that system (II) is also consistent and let (u^0, v^0, t^0, w^0) be a solution of (II). Because $u^0 > 0_p$, $v \geq 0_q$ and $x^0 \cdot A^T \geq 0_p$ or $x^0 \cdot B^T > 0_q$, we get $(u^0 | x^0 \cdot A^T) + (v^0 | x^0 \cdot B^T) > 0$. Since

$$u^0 \cdot Au^0 + v^0 \cdot B + t^0 \cdot C + w^0 \cdot D = 0,$$

we have

$$0 = (x^0 | u^0 \cdot A + v^0 \cdot B + t^0 \cdot C + w^0 \cdot D) =$$
$$= (u^0 | x^0 \cdot A^T) + (v^0 | x^0 \cdot B^T) + (t^0 | x^0 \cdot C^T) + (w^0 | x^0 \cdot D^T) \geqq$$
$$\geqq (u^0 | x^0 \cdot A^T) + (v^0 | x^0 \cdot B^T) > 0,$$

which is a contradiction. Thus, if (I) is consistent, then (II) is inconsistent.

Let now (I) be inconsistent. Suppose that (II) is also inconsistent. Applying corollary 10.2.1 we get that there are $x^0 \in \mathbf{Z}^n$, $u^0 \in \mathbf{N}^p$, $v^0 \in \mathbf{N}^q$, $t^0 \in \mathbf{N}^r$, $w^0 \in \mathbf{N}^s$ such that:

$$(10.3.13) \quad x^0 \cdot A^T \geqq 0_p,\ x^0 \cdot B^T \geqq 0_q,\ x^0 \cdot C^T \geqq 0_r,\ x^0 \cdot D^T = 0_s,$$

$$(10.3.14) \quad u^0 \cdot A + v^0 \cdot B + t^0 \cdot C + w^0 \cdot D = 0_n,$$

and

$$(10.3.15) \quad x^0 \cdot A^T + u^0 > 0_p,\ x^0 \cdot B^T + v^0 > 0_q,\ x^0 \cdot C^T + t^0 > 0_r.$$

Because system (II) is inconsistent, we get that or there is $i \in \{1, ..., p\}$ such that $u_j^0 = 0$ or $v^0 = 0_q$. If $u_j^0 = 0$, from (10.3.15) it results that $(A_j | x^0) > 0$. Then x^0 is a solution of (I). If $v^0 = 0$, from (10.3.15) we get $x^0 \cdot B^T > 0_q$. Then x^0 is a solution of (I) as well.

But we assumed that (I) is inconsistent. Hence if (I) is inconsistent, then (II) is consistent. If A is the vacuous matrix, then the system

$$(10.3.16) \quad \begin{cases} Ax \geq 0_p \\ Bx \geqq 0_q \\ Cx \geqq 0_r \\ Dx = 0_s \\ x \in Z^n \end{cases}$$

is anytime inconsistent. In this case theorem 10.3.4 is reverts to theorem 10.3.2.

If B is the vacuous matrix, then system (10.3.16) is anytime inconsistent. In this case theorem 10.3.4 is reverted to theorem 10.3.1.

Taking $C = 0_{r\times n}$, $D = 0_{s\times n}$, from theorem 10.3.1 we get

Theorem 10.3.5. *For each given matrix* $A \in M_{p\times n}(\mathbf{Z})$, *either*

(I) $x \cdot A^T > 0_p$

has a solution or

(II) $$\begin{cases} uA = 0_n \\ u \in Z^p, u \neq o_p \end{cases}$$

has a solution but never both.

Taking in theorem 10.3.2, $C = 0_{r\times n}$, $D = 0_{s\times n}$, we get

Theorem 10.3.6. *For each given matrix* $B \in M_{q\times n}(\mathbf{Z})$, *either*

(I) $x \cdot B^T \geq 0_q$

has a solution or

(II) $$\begin{cases} vB = 0_n \\ v \in (N^*)^q \end{cases}$$

has a solution but never both.

Chapter 11

Various types of generalised convex functions

Various types of convexity for functions were defined related to various notions of convexity for sets. The reason of this correspondence is in thc behaviour of the graph of the function or of a region of the space related to this graph. Even if the aim of this book is not to study convexities for functions, we dedicate this chapter to some classes of convex functions related to some above discussed convexities for sets, in order to prepare the results contained in the chapters dealing with applications in mathematical programming. Also, we give some interesting models of construction of some generalisations of convex functions.

11.1 Convex functions with respect to a given set

Let $(V, +, \cdot)$ be a linear space, $(U, \leq)$ an ordered space, $M \subseteq V \times U$ a nonempty set, and p a natural number.

Definition 11.1.1. *A mapping* $f\colon A \subseteq V \to B \subseteq U$ *is called strongly convex with respect to* $M \times P$ *if the set*

$$\text{epi } f = \{(x, y) \in A \times U \mid x \in A,\ f(x) \leq y\}$$

is strongly convex with respect to M.

Example 11.1.1. Let be $V = \mathbf{R}$, $U = \mathbf{R}$, $M = \mathbf{Z}^2$, and $f\colon \mathbf{R} \to \mathbf{R}$, the mapping given by $f(x) = [|x|]$, for each $x \in \mathbf{R}$. Because

$$\text{conv (epi f)} = \{(x, y) \in \mathbf{R}^2 \mid x - y < 1, x + y > 1, y \geq 0\},$$

and

$$\mathbf{Z}^2 \cap \text{conv (epi f)} = \bigcup_{n=1}^{\infty}(\bigcup_{j=n}^{\infty}\{(n, j)\}) =$$

$$= \{(x, y) \in \mathbf{N}^2 \mid x - y \leq 0, x + y \geq 0\} \subseteq \text{epi f},$$

we get that function f is strongly convex with respect to $\mathbf{Z}^2$.

Example 11.1.2. Let V = U = **R**, M = $\{(x,y) \in \mathbf{R}^2 \mid xy+1, x+y \geq -1, x\leq 1\}$, and f: **R** → **R**, the mapping given by

$$f(x) = \begin{cases} 1, \text{if } x \in]-\infty, 0[\cup]1, +\infty[\\ -1, \text{if } x \in [0, 1] \end{cases}$$

As

$$\text{conv (epi f)} = \{(x, y) \mid x \in [0, 1], y \geq -1\} \cup \{(x, y) \mid x \in \mathbf{R} \setminus [0, 1], y > -1\}$$

we get that point $(\frac{-3}{2}, \frac{1}{2}) \in M \cap$ conv (epi f), but it is not an element of epi f. Therefore f is not strongly convex with respect to M.

Example 11.1.3. Let V = U = **R**, M = $\mathbf{Z}^2$, and f: {1, 2, 3, 4, 5, 6, 7, 8} → **R**, the mapping given by f (x) = $\log_2 x$, for each x ∈ {1, 2, 3, 4, 5, 6, 7, 8}. It is easy to see that (5, 2) ∈ conv (epi f) ∩ $\mathbf{Z}^2$, but (5, 2) ∉ epi f. Then, the function f is not strongly convex with respect to M.

11.2 Properties of the convex functions with respect to a given set

Let (V, +, ·) be a real linear space and (U, +, ·, ≤) an other linear space, ≤ is a preorder compatible with the linear space (i.e. ≤ is transitive and reflexive binary relation and

- for each x, y, u, v ∈ U, with x ≤ y and u ≤ v, we have x + u ≤ y + v,

and

- for each x, y ∈ U, with x ≤ y, and each real number t, t ≥ 0, we have t·x ≤ t·y.)

Let M ⊆ V and P ⊆ U be nonempty sets.

Theorem 11.2.1. *If subset* A *of* V *is nonempty and strongly convex with respect to* M, *and mapping* f: A ⊆ V → U *is strongly convex with respect to*

$M \times P$, *then for any natural number* p, $p \geq 2$, *any elements* $x^1, ..., x^p$ *of* A, *and any* p *non negative real numbers* $t_1, ..., t_p$, *with* $t_1 + ... + t_p = 1$, *satisfying the condition that*

$$(11.2.1) \qquad (t_1x^1 + ... + t_px^p, t_1f(x^1) + ... + t_pf(x^p)) \in M \times P,$$

we have

$$(11.2.2) \qquad f(t_1x^1 + ... + t_px^p) \leq t_1f(x^1) + ... + t_pf(x^p).$$

Proof. Because $(t_1x^1 + ... + t_px^p, t_1f(x^1) + ... + t_pf(x^p)) \in M \times P$, and $(x^k, f(x^k)) \in \text{epi } f$, for each $k \in \{1, ..., p\}$, in view of strong convexity of function f, we get $(t_1x^1 + ... + t_px^p, t_1f(x^1) + ... + t_pf(x^p)) \in \text{epi } f$. That implies

$$f(t_1x^1 + ... + t_px^p) \leq t_1 \cdot f(x^1) + ... + t_p \cdot f(x^p).$$

We remark that the converse theorem is not always true (see example 11.1.3).

Theorem 11.2.2. *If subset* A *of* V *is nonempty and strongly convex with respect to* M, *then mapping* $f: A \subseteq V \to U$ *is strongly convex with respect to* $M \times U$, *if and only if for any natural number* p, $p \geq 2$, *any elements* $x^1, ..., x^p$ *of* A, *and any* p *nonnegative real numbers* $t_1, ..., t_p$, *with* $t_1 + ... + t_p = 1$, *satisfying the condition* (11.2.1) *we have* (11.2.2).

Proof. Necessity results by theorem 11.2.1.

Sufficiency. Let $(x, y) \in M \cap \text{conv}(\text{epi } f)$. Then there are: a natural number $p \geq 2$, p elements $(x^1, y^1), ..., (x^p, y^p)$ of epi f and p nonnegative real numbers $t_1, ..., t_p$ with $t_1 + ... + t_p = 1$, such that

$$(11.2.3) \qquad (x, y) = t_1(x^1, y^1) + ... + t_p(x^p, y^p) = \\ = (t_1x^1 + ... + t_px^p, t_1y^1 + ... + t_py^p).$$

Because $(x, y) \in (M \times U) \cap \text{conv}(\text{epi } f) \subseteq \text{epi } f$, from (11.2.3) it results

$$(11.2.4) \qquad t_1x^1 + ... + t_px^p \in A \cap M.$$

On the other hand, we have $(t_1x^1 + ... + t_px^p, t_1f(x^1) + ... + t_pf(x^p)) \in M \times U$. Applying (11.2.4) we get

$$(11.2.5) \qquad f(t_1x^1 + ... + t_px^p) \leq t_1f(x^1) + ... + t_pf(x^p).$$

As $(x^k, f(x^k)) \in \text{epi } f$ for each $k \in \{1, ..., p\}$, it follows

$$(11.2.6) \qquad f(x^k) \leq y^k, k \in \{1, ..., p\}.$$

Because $t_k \geq 0$, from (11.2.5) it results

$$(11.2.7) \qquad t_1f(x^1) + ... + t_pf(x^p) \leq t_1y^1 + ... + t_py^p.$$

Then (9.2.7) and (9.2.6) imply $f(t_1x^1 + ... + t_px^p) \le t_1y^1 + ... + t_py^p$. Then $(x, y) \in$ epi f. Therefore epi f is a strongly convex set with respect to M×U.

11.3 Characterisation of convex functions and strongly convex functions with respect to a given set by divided differences

It is known that if $A \subseteq \mathbf{R}$ is a convex nonempty set, then a function $f:A \to R$ is convex if and only if for all $x, y, z \in A$ we have $[x, y, z; f] \ge 0$, where $[x, y, z; f]$ is the divided difference of the function f on the points x, y, z.

In this paragraph we give a generalisation of the divided differences to characterise the metric convex functions.

Let (X, d) be a metric space. If $x \in X$ and $y \in X$, we denote by $[x, y]_d$ the metric segment, i.e. $[x, y]_d = \{z \in X \mid d(x, z) + d(z, y) = d(x, y)\}$. Let < be a total strict order on X, satisfying the conditions that if $x \in X$, $y \in X$, $x < y$, then:

i) $x < z < y$, for each $z \in [x, y]_d \setminus \{x, y\}$;

ii) for all $z \in X$ such that $y \in [x, z]_d \setminus \{x, z\}$, we have $y < z$;

iii) for all $z \in X$ such that $x \in [z, y]_d \setminus \{z, y\}$, we have $z < x$.

If (X, d) is a metric space and < is a total strict order relation defined on X, satisfying the condition i), ii) and iii), then we say that (X, d, <) is a strictly ordered metric space. We shall write $x \le y$ if $x < y$ or $x = y$.

Let (X, d, ≤) be a metric ordered set, E a nonempty subset of X and let a function $f: E \to \mathbf{R}$. We generalise the divided differences as it follows:

$$[x; f] = f(x), \text{ for all } x \in X \tag{11.3.1}$$

and, if n is a natural number n, $n \ge 2$, we give the following recurrence formula

$$[x_1, ..., x_n, x_{n+1}; f] = \frac{[x_2, ..., x_{n+1}; f] - [x_1, ..., x_n; f]}{e(x_1, x_{n+1})d(x_{n+1}, x_1)} \tag{11.3.2}$$

where

$$e(x_1, x_{n+1}) = \begin{cases} 1 & \text{if } x_1 < x_{n+1} \\ -1 & \text{if } x_{n+1} < x_1 \end{cases} \tag{11.3.3}$$

for each $x_1, ..., x_{n+1}$ of E.

Remark 11.3.1. If X = **R** and < is the usual strict order on **R**, then we obtain the usual notion of divided difference because:

- if $x_1 < x_{n+1}$, then $e(x_1, x_{n+1}) = 1$;

- if $x_{n+1} < x_1$, then $e(x_1, x_{n+1}) = -1$

and $e(x_1, x_{n+1})\, d(x_{n+1}, x_1) = x_{n+1} - x_1$.

Lemma 11.3.1. *If* x *and* z *are two distinct points of* E, *then for any* $y \in ([x, z] \setminus \{x, z\}) \cap E$ *we have*

$$(11.3.4) \qquad [x,y,z;f] = f(x)\frac{d(y,z)}{d(x,z)} + f(z)\frac{d(x,y)}{d(x,z)} - f(y).$$

Proof. Let $y \in [x, z] \cap E$. Then

$$(11.3.5) \qquad [x,y,z] = \frac{\dfrac{f(z)-f(y)}{e(y,z)d(z,y)} - \dfrac{f(y)-f(x)}{e(x,y)d(y,x)}}{e(x,z)d(z,x)}$$

Two cases may arise: $x < z$ or $z < x$.

If $x < z$, then, from i), it results that $e(x, z) = e(y, z) = e(x, z) = 1$, and in view of (11.3.5) we obtain

$$[x,y,z;f] = \frac{f(z)d(y,x) - f(y)(d(y,x)+d(z,y)) + f(x)d(z,y)}{d(z,x)} =$$

$$\frac{f(x)d(y,z) - f(y)d(x,z) + f(z)d(x,y)}{d(x,z)} = f(x)\frac{d(y,z)}{d(x,z)} + f(z)\frac{d(x,y)}{d(x,z)} - f(y).$$

If $z < x$, then $e(x, y) = e(y, z) = e(x, z) = -1$ and hence

$$[x,y,z;f] = \frac{-f(z)d(y,x) + f(y)(d(y,x)+d(z,y)) - f(x)d(z,y)}{-d(z,x)} =$$

$$= f(x)\frac{d(y,z)}{d(x,z)} + f(z)\frac{d(x,y)}{d(x,z)} - f(y).$$

In both cases we get that (11.3.4) is true.

We remember that a subset E of a metric space (X, d) is said to be metric convex if $[x, z]_d \subseteq E$, for all x, z points of E. A function f: $A \rightarrow$ **R**, where E is a metric convex subset of X, is said to be metric convex if we have

$$f(y) \le f(x)\frac{d(y,z)}{d(x,z)} + f(z)\frac{d(x,y)}{d(x,z)},$$

for all points x, z of E and for each $y \in [x, z]$.

Let (X, d, <) be a strictly ordered metric space. From lemma 11.3.1 it results immediately that:

Theorem 11.3.2. *If* E *is a metrically convex subset of* X, *then the function* f: E → **R** *is a metric convex function is and only if we have*

(11.3.6) $[x, y, z; f] \ge 0$,

for all x, z *points of* E *and for each* $y \in [x, z]$.

Theorem 11.3.3. *If* $X \subseteq V$ *is strongly* 2*-convex with respect to* M, *then the function* f: X→**R** *is strongly* 2*-convex function with respect to* $M \times \mathbf{R}$ *if and only if for any distinct points* a, b, c *of* X, *with* $b \in [a, c] \cap M$ *we have*

$$[a, b, c; f] \ge 0.$$

Proof. Necessity. Let a, b, c be distinct points of X with $b \in [a, c]$. Then it is $0 < t < 1$ such that $b = (1 - t)\, a + t\, c$. Obviously $(a, f(a)) \in \text{epi } f$ and $(c, f(c)) \in \text{epi } f$. Because $(1 - t)\, a + t\, c = b \in M$, we get that

$$(1 - t)(a, f(a)) + t(c, f(c)) = ((1 - t)\, a + tc, (1 - t)\, f(a) + r\, f(c)) \in M \times \mathbf{R}.$$

Then, since epi f is strongly 2-convex with respect to $M \times \mathbf{R}$, we have

$$((1 - t)\, a + t\, c, (1 - t)\, f(a) + t\, f(c)) \in \text{epi } f.$$

That implies that

(11.3.7) $f((1 - t)\, a + t\, c) \le (1 - t)\, f(a) + t\, f(c)$.

From (11.3.4) it follows that

$$[a, b, c; f] = [a, (1-t)\cdot a + t\cdot c, c; f] = \frac{(1-t)f(a) + tf(c) - f((1-t)a + tc)}{(1-t)t\, d^2(c,a)}.$$

Sufficiency. Let (a, x), (c, y) be two points of epi f. Then

(11.3.8) $a \in X, c \in X, x \ge f(a)$ and $y \ge f(c)$.

Let $0 < t < 1$ such that $(1 - t)(a, x) + t(c, y) \in M \times \mathbf{R}$. Then $(1-t)\, a + t\, c \in M$ and, because X is a strongly 2-convex set with respect to M, we have $(1-t)\, a + t\, c \in X$. Hence $(1-t)\, a + t\, c \in [a, c] \cap X \cap M$. That implies that

(11.3.9) $0 \le [a, (1-t)\, a + t\, c, c; f]$.

It follows that

(11.3.10) $f((1-t)\, a + t\, c) \leq (1-t)\, f(a) + t\, f(c)$.

From (11.3.8) we obtain $(1-t)\, x + t\, y \geq (1-t)\, f(a) + t\, f(c)$ and if we take into account (11.3.10) we get $(1-t)\, x + t\, y \geq f((1-t)\, a + t\, c)$. It implies that $(1-t)(a, x) + t\,(c, y) \in \text{epi } f$. Therefore, since (a, x), (c, y) and $0 < t < 1$ was arbitrarily chosen, it results that f is strongly 2-convex with respect to $M \times \mathbf{R}$.

11.4 Optimum points of real strongly convex functions with respect to a given set

In this paragraph we study the properties of minimum points of a strongly convex function with respect to a given set.

Let $M \subseteq \mathbf{R}^n$ be and $P \subseteq \mathbf{R}$ nonempty sets, $A \subseteq \mathbf{R}^n$ a nonempty strongly convex set with respect to M, and $f: A \to \mathbf{R}$ a function. We denote by

$$B = \{x \in A \mid (x, f(x)) \in M \times P\} \text{ and } C = \{(x, f(x)) \mid x \in B\}.$$

Remark 11.4.1. It is easy to see that:

i) $\text{conv}(B) = \{x \in \mathbf{R}^n \mid \exists\, y \in \mathbf{R}, (x, y) \in \text{conv}(C)\}$.

ii) If A is a compact set, then B and conv (B) are compact sets.

Theorem 11.4.1. *If* B *is a compact set and function* $f: A \to \mathbf{R}$ *satisfies the conditions*:

i) f *is continuous and*

ii) f *is strongly convex with respect to* M,

then there is a convex function $g: \text{conv}(B) \to \mathbf{R}$, *such that*

(11.4.1) $g(x) \leq f(x)$, *for all* $x \in B$;

(11.4.2) *if* x^0 *is a minimum point of* f *on* B, *then* x^0 *is also a minimum point of* g *on* conv (B);

(11.4.3) *if* z^0 *is a minimum point of* g *on* conv (B), *then there is a point* $x^0 \in B$ *such that* $g(z^0) = f(x^0)$, *and* x^0 *is a minimum point of* f *on* B.

Proof. Let $x \in \text{conv}(B)$, and $U_x = \{y \in \mathbf{R} \mid (x, y) \in \text{conv}(C)\}$. Because C is a compact set, set B is also compact. As f is continuous on B, f is lower bounded on B. Then there is y^0 such that $y^0 \leq f(x)$, for each $x \in B$.

Let $y \in U_x$. Because $(x, y) \in$ conv (C), there is a natural number m, m elements $(x^1, f(x^1))$, ..., $(x^m, f(x^m))$ of C and m nonnegative real numbers t_1, ..., t_m, with $t_1 + ... + t_m = 1$, such that

$$(x, y) = t_1 (x^1, f(x^1)) + ... + t^m (x^m, f(x^m)).$$

But $y^0 \leq f(x^k)$, for all $k \in \{1,...,m\}$. Hence

$$y = t_1 \cdot f(x^1) + ... + t_m \cdot f(x^m) \geq t_1 y^0 + ... + t_m y^0 = y^0.$$

Because y was arbitrarily chosen, it follows that the set U_x is lower bounded. Let be $y^x = \inf U_x$.

We show that $y^x \in U_x$. From the definition of greatest lower bound it follows that for each natural number n, $n \geq 0$, there is an element $y^n \in U_x$, such that

(11.4.4) $\quad y^x \leq y^n \leq y^x + 1/n$

Obviously $\lim y^n = y^x$. Because $y^n \in U_x$ for all $n \in \mathbf{N}^*$, we get that

(11.4.5) $\quad (x, y^n) \in$ conv (C), for each $n \in \mathbf{N}^*$.

From (11.4.4) and (11.4.5) it follows $(x, y^x) \in$ conv (C). Hence $y^x \in U_x$. Therefore $y^x = \min U_x$.

Now we construct function g: conv (B) $\to \mathbf{R}$, given by

(11.4.6) $\quad g(x) = \min U_x$, for each $x \in B$.

First we show that the function g is convex. Let be x', x" two elements of conv (B). From (11.4.6), in view of definition of U_x, we get $(x', g(x')) \in$ conv (C) and $(x'', g(x'')) \in$ conv (C). Then we have $t (x', g(c')) + (1-t) (x'', g(x'')) \in$ conv (C), for all $t \in [0, 1]$. But

$$(t x' + (1-t) x'', t g(x') + (1-t) g(x'')) = t (x', g(x')) + (1-t) (x'', g(x'')).$$

Hence $t g(x') + (1-t) g(x'') \in U_{tx'+(1-t)x''}$, implying that

$$t g(x') + (1-t) g(x'') \geq g(t x' + (1-t) g(x'')).$$

It follows that g is a convex function. Let $x \in B$. Because $(x, f(x)) \in C \subseteq$ conv (C), we get $f(x) \in U_x$. Then $g(x) \leq f(x)$ and (11.4.1) is true.

In what follows we prove that (11.4.2) is true. Let x^0 be a minimum point of f on B. If we suppose that x^0 is not a minimum point of g on B, then there is a point $x \in B$, such that

(11.4.7) $\quad g(x) < g(x^0)$.

Because $(x, g(x)) \in$ conv (C) then there are: a natural number m, m elements $(x^k, f(x^k))$, $k \in \{1, ..., m\}$ and m non-negative real numbers t_1, t_2,

..., t_m, with $t_1+...+t_m=1$, such that $(x, g(x)) = t_1 (x^1, f(x^1)) + ... + t_m (x_m, f(x^m))$. Then

$$(11.4.8) \qquad g(x) = t_1 f(x^1) + ... + t_m f(x^m) \geq \min \{f(x^k) \mid k \in \{1, ..., m\}\}$$

From (11.4.7) and (11.4.8) we get that $g(x^0) > \min \{f(x^k) \mid k \in \{1, ..., m\}\}$. But $g(x^0) \leq f(x^0)$ and then

$$(11.4.9) \qquad f(x^0) > \min \{f(x^k) \mid k \in \{1, ..., m\}\}.$$

As $(x^k, f(x^k)) \in C$, for each $k \in \{1, ..., m\}$, it results that $x^k \in B$, for each $k \in \{1, ..., m\}$. Then (11.4.9) contradicts the fact that x^0 is a minimum point of f on B. Hence (11.4.2) is true.

Now let z^0 be a minimum point of g on conv (B). If $z^0 \in B$, then (11.4.3) is true. Suppose that $z^0 \in (\text{conv}(B)) \setminus B$. Then we have $(z^0, g(x^0)) \in \text{conv}(C)$. Then there are: a natural number m, m elements $(x^1, f(x^1))$, ..., $(x^m, f(x^m))$ of C, and m non-negative real numbers $t_1, t_2, \ldots, t_m$, with $t_1 + ... + t_m = 1$, such that $(z^0, g(z^0)) = t_1 (x^1, f(x^1)) + ... + t_m f(x^m)$. Then $g(z^0) = t_1 f(x^1) + ... + t_m f(x^m)$, and therefore

$$\min \{f(x^k) \mid k \in \{1, ..., m\}\} \leq g(z^0) \leq \max \{f(x^k) \mid k \in \{1, ..., m\}\}.$$

Without loosing the generality, we suppose that

$$f(x^1) = \min \{f(x^k) \mid k \in \{1, ..., m\}\}.$$

Then

$$(11.4.10) \qquad f(x^1) \leq g(z^0).$$

But from (11.4.1) we get

$$(11.4.11) \qquad g(x^1) \leq f(x^1).$$

As z^0 is a minimum point of g on conv (B), from (11.4.10) and (11.4.11) it results

$$(11.4.12) \qquad g(x^1) = f(x^1) = g(z^0) \leq g(x), \text{ for each } x \in \text{conv}(B).$$

Hence x^1 is a minimum point of g on conv (B). Because $(x^1, f(x^1)) \in C$, we have $x^1 \in B$. Obviously x^1 is a minimum point of f on B (from (11.4.1) and (11.4.12) it results $f(x^1) \leq g(x) \leq f(x)$, for all $x \in B$). Then (11.4.3) is true.

By analogy we can prove the following theorem

Theorem 11.4.2. *If* B *is a compact set and function* f: A → **R** *satisfies the conditions*:

i) f *is continuous,*

ii) f *is strongly convex with respect to* M,

then there is a concave function h: conv (B) → **R**, *such that*

(11.4.13) $f(x) \leq h(x)$, *for all* $x \in B$;

(11.4.14) *if* x^0 *is a maximum point of* f *on* B, *then* x^0 *is also a maximum point of* h *on* conv (B);

(11.4.15) *if* z^0 *is a maximum point of* h *on* conv (B), *then there is a point* $x^0 \in B$ *such that* $h(z^0) = f(x^0)$, *and* x^0 *is a maximum point of* f *on* B.

Corollary 11.4.1. *If* B *is a compact set and function* f: A → **R** *satisfies the conditions*:

i) f *is continuous,*

ii) f *is strongly convex with respect to* M,

then we have $\Omega_B = B \cap \Omega_{conv\,(B)}$, *where the set* Ω_B *is the set of all minimum points of* f *on* B, *and* $\Omega_{conv\,(B)}$ *denotes the set of all minimum points of* g *on* conv (B).

Corollary 11.4.2. *If function* f: A → **R** *satisfies the conditions*:

i) f *is continuous,*

ii) f *is strongly convex with respect to* M,

then set Ω_B *is strongly convex with respect to* M.

Proof. We apply remark 11.4.1 and theorem 11.4.2.

Theorem 11.4.3. *If* B *is a compact set and function* f: A → **R** *satisfies the conditions*:

i) f *is continuous,*

ii) f *is strongly convex with respect to* M,

then the function f *has a unique minimum point on* B *if and only if the function* g *has a single minimum point of* conv (B).

Corollary 11.4.3. *If* (B) *is a compact set and function* f: A → **R** *satisfies the conditions*:

i) f *is continuous,*

ii) f *is strongly convex with respect to* M,

and if function g *is uniformly convex, then function* f *has a unique minimum point on* B.

Let (V, +, ·) be a real linear space of a finite dimension, and let $M \subseteq V$ a nonempty set.

Theorem 11.4.4. *If* $A \subseteq V$ *is a strongly convex set with respect to* M *and function* $f: A \to \mathbf{R}$ *is strongly convex with respect to* $M \times \mathbf{R}$ *then set* Ω *of all minimum points of* f *on* A *is strongly convex with respect to* M.

11.5 Duality theorems

Let X be a nonempty subset of $\mathbf{Z}^n$ and $f: X \to \mathbf{R}$ a function. We denote by gr f the graph of the function f, i.e.

(11.5.1) $\quad \text{gr } f = \{(x, f(x)) \mid x \in X\},$

and, for each $x \in X$, we put

(11.5.2) $\quad M_x = \{y \in \mathbf{R} \mid (x, y) \in \text{conv}(\text{gr } f)\}.$

Let (P) be the following mathematical programming problem:

$$\text{(P)} \quad \begin{cases} f(x) \to \sup \\ x \in X \end{cases}.$$

Lemma 11.5.1. *If* $X \subseteq \mathbf{Z}^n$ *is a finite set, then for each* $z \in X$, *sets* conv (X), conv (gr f) *and* M_z *are compact.*

Proof. Because X is finite, set gr f is also finite. Then sets conv(X) and conv (gr f) are compact sets. Now, let $z = (z_1, ..., z_n) \in X$. In order to prove that set M_z is compact, we notice that

$$M_z = pr_{n+1}(\{x = (x_1, ..., x_n, x_{n+1}) \in \mathbf{R}^{n+1} \mid (x_1, ..., x_n) = z\} \cap \text{conv}(\text{gr } f)),$$

where pr_{n+1} is the projector on (n+1) - axis. As conv (gr f) is a compact set, M_z is bounded. Since sets $\{x = (x_1, ..., x_n, x_{n+1}) \in \mathbf{R}^{n+1} \mid (x_1, ..., x_n) = z\}$ and conv (gr f) are closed, M_z is also closed. Therefore M_z is a compact set.

Let X be a nonempty finite subset of $\mathbf{Z}^n$, and let $h: X \to \mathbf{R}$, the function defined by

(11.5.3) $\quad h(x) = \max M_x$, for each $x \in X$,

and $g: X \to \mathbf{R}$, the function defined by $g(x) = \min M_x$, for each $x \in X$.

Lemma 11.5.2. *If* X *is a nonempty finite subset of* $\mathbf{Z}^n$, *then*:

a) *function* h *is concave*;

b) *we have* $f(x) \leq h(x)$, *for all* $x \in \text{conv}(X)$.

Proof. a) Let x, z be elements of conv (X), and let $t \in [0, 1]$. From (11.5.3) we get that (x, h (x)) $\in$ conv (gr f), and (z, h (z)) $\in$ conv (gr f). Then we have

$$(t\,x + (1\text{-}t)\,z,\ t\,h(x) + (1\text{-}t)\,h(z)) = t\,(x, h(x)) + (1\text{-}t)\,(z, h(z)) \in \text{conv}\,(\text{gr } f),$$

which implies $t\,h(x) + (1-t)\,h(z) \in M_{tx+(1-t)z}$. Hence

$$(11.5.4) \qquad h\,(t\,x + (1-t)\,z\,) \geq t\,h(x) + (1-t)\,h(z).$$

Therefore function h is concave.

b) Let x be an arbitrary point of X. As $(x, f(x)) \in \text{gr } f \subseteq \text{conv}\,(\text{gr } f)$, it follows that $f(x) \in M_x$. Hence $h(x) \geq f(x)$.

Our purpose is to give a duality theorem for the following mathematical programming problem:

$$(\text{P}) \qquad \begin{cases} f(x) \to \max \\ x \in X \end{cases},$$

where X is a finite convex subset of $\mathbf{Z}^n$, and $f: X \to \mathbf{R}$ is a function.

Applying theorem 2.7.16, it results that there are: a natural number m, a matrix $A \in M_{m\times n}(\mathbf{Z})$ and a vector $b \in \mathbf{Z}^m$, such that conv (X) = $\{x \in \mathbf{Z}^n \mid x\cdot A^T \leq b\}$.

By (PC) we denote the following problem

$$(\text{PC}) \qquad \begin{cases} h(x) \to \max \\ x \in \text{conv}\,(X) \end{cases}.$$

Theorem 11.5.1. *If* X *is a finite subset of* $\mathbf{Z}^n$ *then the following propositions are true*:

a) *if* x^0 *is an optimal solution of problem* (P), *then* x^0 *is also an optimal solution of problem* (PC);

b) *if* x^0 *is an optimal solution of problem* (PC), *then there is an optimal solution* $z^0 \in X$ *of problem* (P) *such that*

$$(11.5.5) \qquad f\,(z^0) = h\,(x^0).$$

Proof. a) Let x^0 be an optimal solution of (P). Suppose that x^0 is not an optimal solution of (PC). Then there is $x \in$ conv (X) such that

$$(11.5.6) \qquad h\,(x^0) < h\,(x).$$

Because $(x, h(x)) \in \text{conv}(\text{gr } f)$ then there are: a natural non null number m, $m \leq n+2$, m elements $(x^k, f(x^k)) \in \text{gr } f$, $k \in \{1, ..., m\}$, and m non negative real numbers $t_1, ..., t_m$, with $t_1 + ... + t_m = 1$, such that

$$(x, h(x)) = \sum_{j=1}^{m} t_j \cdot (x^j, f(x^j)) = (\sum_{j=1}^{m} t_j \cdot x^j \ , \ \sum_{j=1}^{m} t_j \cdot f(x^j)\ ).$$

Then

$$(11.5.7) \qquad h(x) = \sum_{j=1}^{m} t_j \cdot f(x^j) \leq \max \{f(x^k) \mid k \in \{1, ..., m\}\}.$$

From (11.5.6) and (11.5.7) we get that

$$h(x^0) < \max \{f(x^k) \mid k \in \{1, ..., m\}\}.$$

But $f(x^0) \leq h(x^0)$ and then $f(x^0) < \max \{f(x^k) \mid k \in \{1, ..., m\}\}$, which contradicts the hypothesis that x^0 is a maximum point of f on X. It follows that the assumption that x^0 is not a maximum point of h on conv (X) is false.

b) Let x^0 be an optimal solution of (PC). Since $(x^0, h(x^0)) \in \text{conv}(\text{gr } f)$, there are a natural non null number $m \leq n + 2$, m elements $(x^1, f(x^1)), ..., (x^m, f(x^m))$ of gr f, and m non negative real numbers $t_1, ..., t_m$, with $t_1 + ... + t_m = 1$, such that $(x^0, h(x^0)) = t_1 \cdot (x^1, f(x^1)) + ... + t_m \cdot (x^m, f(x^m))$. Then

$$h(x^0) = t_1 \cdot f(x^1) + ... + t_m \cdot f(x^m).$$

As h is concave, it follows that

$$h(x^0) \leq \max \{f(x^k) \mid k \in \{1, ..., m\}\}.$$

Without restricting the generality, we suppose that

$$f(x^1) = \max \{f(x^k) \mid k \in \{1, ..., m\}\} \text{ and } h(x^0) \leq f(x^1).$$

From lemma 11.5.2, we have $f(x^1) \leq h(x^1)$. Hence $h(x^0) \leq h(x^1)$. Because x^0 is a maximum point of h on conv (X), it results that $h(x^0) = h(x^1)$. Then $h(x^0) = f(x^1) = h(x^1)$. Therefore x^1 is also a maximum point of h on conv (X). As $X \subseteq \text{conv}(X)$ and $x^1 \in X$, it follows that x^1 is a maximum point of f on X.

Let $D = \{d \in \mathbf{R}^m \mid \text{the set } \{x \in \mathbf{R}^n \mid x \cdot A^T \leq d\} \text{ is bounded}\}$, and let FSC = $\{F: D \to \mathbf{R} \mid F \text{ is strongly concave with respect to } D \times \mathbf{R} \text{ and non-increasing}\}$.

From lemma 11.5.1 it results that $b \in D$. For every $x^0 \in \text{conv}(X)$ we have $x^0 \cdot A^T \in D$, because $b \in D$ and

$$\{x \in \mathbf{R}^n \mid x \cdot A^T \leq x^0 \cdot A^T\} \subseteq \{x \in \mathbf{R}^n \mid x \cdot A^T \leq b\}.$$

In what follows, we consider the problem

$$(\mathrm{D})\qquad \begin{cases}\inf F(b)\\ F(x\cdot A^T)\geqq h(x)\\ F\in FSC\end{cases}$$

Theorem 11.5.2. *If* $X\subseteq \mathbf{Z}^n$ *is a finite set,* x^0 *is a feasible solution of* (P) *and* F^0 *is a feasible solution of* (D), *then*

(11.5.8) $\quad f(x^0)\leqq F^0(b).$

Proof. From assertion b) of lemma 11.5.2 it follows that

(11.5.9) $\quad f(x^0)\leqq h(x^0).$

Because F^0 is feasible for (D), we have

(11.5.10) $\quad h(x^0)\leqq F^0(Ax^0).$

But $x^0\in X\subseteq$ conv (X), hence $Ax^0\leq b$. Since F^0 is non-decreasing, we have

(11.5.11) $\quad F^0(Ax^0)\leq F^0.$

From (11.5.9) – (11.5.11) it results (11.5.8).

Lemma 11.5.3. *If set* X *is finite then the perturbation function* $G{:}D\to\mathbf{R}\cup\{-\infty\}$, *given by*

$$G(d)=\begin{cases}\sup\{h(x)\mid x\cdot A^T, \text{ if } \{x\in R^n\mid x\cdot A^T\leqq d\}\neq\varnothing\\ -\infty, \text{ if } \{x\in R^n\mid x\cdot A^T\leqq d\}=\varnothing\end{cases},$$

for all $d\in D$, *is strongly concave with respect to* $D\times\mathbf{R}$ *and non-decreasing.*

Proof. As X is a finite set, it follows conv (epi (- G)) $\cap D\times\mathbf{R}\neq\varnothing$. Let $z=(d, y)\in$ conv (epi (- G)) $\cap D\times\mathbf{R}$. There are: a natural non null number p, $p\leqq n+2$, p elements $z^1=(d^1, y_1), \ldots, z^p=(d^p, y_p)$ of epi (-G), and p non negative real numbers $t_1, \ldots, t_p$, with $t_1+\ldots+t_p=1$, such that $z=\sum_{j=1}^{p}t_jz^j$.

Then $d=\sum_{j=1}^{p}t_jd^j$, and $y=\sum_{j=1}^{p}t_jy_j\in\mathbf{R}$. This implies

(11.5.12) $\quad y_j\in\mathbf{R}$, for each $j\in\{1, \ldots, p\}$.

As $(d^k, y_k)\in$ epi (- G), $k\in\{1, \ldots, p\}$, we get

(11.5.13) $\quad G(d^j)\geq y_j\in\mathbf{R}$, for each $j\in\{1, \ldots, p\}$.

Then, from (11.5.12) and (11.5.13) it results $G(d^j)>-\infty$, for all $j\in\{1, \ldots, p\}$. It follows that, for each $j\in\{1, \ldots, p\}$, we have $\{x\in\mathbf{R}^n\mid x\cdot A^T\leqq d^j\}\neq\varnothing$, and because $d^i \in D$, the set $\{x \in \mathbf{R}^n \mid x \cdot A^T \leqq d^j\}$ is bounded, for all $j \in \{1, ..., p\}$. Hence $\{x \in \mathbf{R}^n \mid x \cdot A^T \leqq d^j\}$ is a nonempty compact set, for all $j \in \{1, ..., p\}$. Thus, $h(x^j) = G(d^j)$ for all $j \in \{1, ..., p\}$. Let $x^0 = t_1x^1 + ... + t_px^p$. Obviously $x^0 \in \text{conv}(X)$, and

$$x^0 \cdot A^T = \sum_{j=1}^{p} t_j \cdot x^j \cdot A^T \leqq \sum_{j=1}^{p} t_j d^j = d.$$

Then $x^0 \in \{ x \in \mathbf{R}^n \mid x \cdot A^T \leqq d\}$, and therefore,

(11.5.14) $\quad g(d) \geqq h(x^0)$.

Because h is a concave function, we have

(11.5.15) $\quad h(x^0) \geq t_1 h(x^1) + ... + t_p h(x^p)$.

From (11.5.14) and (11.5.15) we get

(11.5.16) $\quad -g(d) \leq t_1(-G(d^1)) + ... + t_p(-G(d^p))$.

From (11.5.13) and (11.5.16) it follows

$$-G(d) \leq t_1 \cdot y^1 + ... + t_p \cdot y^p = y.$$

This inequality implies $z = (d, y) \in \text{epi}(-G)$. Hence $-G$ is strongly convex with respect to $D \times \mathbf{R}$. Thus G is strongly concave with respect to $D \times \mathbf{R}$.

Lemma 11.5.4. *The perturbation function* G *is feasible for* (D).

Proof. From lemma 11.5.4 we get that $G \in FSC$. Let $z \in \text{conv}(X)$. Because $z \in \{x \in \mathbf{R}^n \mid x \cdot A^T \leq z \cdot A^T\}$, we have

$$G(z \cdot A^T) = \sup\{h(x) \mid x \cdot A^T \leq z \cdot A^T\} \geq h(z).$$

Hence $G(z \cdot A^T) \geq h(z)$, for every $z \in \text{conv}(X)$. It follows that G is feasible for (D).

Theorem 11.5.3. *If* $X \subseteq \mathbf{Z}^n$ *is a finite set, then if* x^0 *is an optimal solution of* (P) *and* F^0 *is an optimal solution for* (D), *then* $f(x^0) = F^0(b)$.

Proof. Obviously x^0 and F^0 are feasible solutions for (P), respectively for (D). By theorem 11.5.2 we get

(11.5.17) $\quad f(x^0) \leq F^0(b)$.

But, by lemma 11.5.4, it follows that

(11.5.18) $\quad F^0(b) \leq G(b) = \sup\{h(x) \mid x \cdot A^T \leq b\}$.

In view of theorem 11.5.1 we have

(11.5.19) $\quad f(x^0) = h(x^0) = \sup\{h(x) \mid x \cdot A^T \leq b\}$.

Then, from (11.5.17) – (11.5.19) we obtain $f(x^0) = F^0(b)$.

11.6 Divided differences and generalized convex functionals in metric spaces

L. Lupşa (1997) published the notion of divided difference of a real function defined on a metric space, for the first time. In our generalization we shall follow the same methods of thinking as in 11.3.

Let (X, d) be a metric space, $E: X \to X$ an injective function and $A \subset X$ an E-convex set, according to definition 1.1. Consider the distinct points x_1, x_2, ..., x_{n+1} belonging to A and a functional $f: A \to \mathbf{R}$.

Definition 11.6.1. *The* (E)*-divided difference of order* n *of functional* f *on the points* $x_1, x_2, \ldots, x_{n+1}$ *is the real number obtained by the following recurrence:*

(11.6.1)

$$\begin{cases} [x_1;f]_{(E)} = f(E(x_1)) \\ [x_1,x_2,\ldots,x_{n+1};f]_{(E)} = \dfrac{[x_2,x_3,\ldots,x_{n+1};f]_{(E)} - [x_1,x_2,\ldots,x_n;f]_{(E)}}{e(E(x_1),E(x_{n+1}))d(E(x_{n+1}),E(x_1))} \end{cases}$$

Number e (a, b) is defined by (11.3.3). The injectivity condition for function E is necessary to avoid the situation when $d(E(x_{n+1}), E(x_1)) = 0$.

Proposition 11.6.1. *If* $[x_1, x_2, \ldots, x_{n+1};f]$ *means the usual divided difference of order* n *of the functional* f *on* $x_1, x_2, \ldots, x_{n+1}$, *then*

(11.6.2) $\quad [x_1, x_2, \ldots, x_{n+1};f]_{(E)} = [E(x_1), E(x_2), \ldots, E(x_{n+1}); f]$.

Proof. We use the induction. First, we notice that

$$[x_1; f]_{(E)} = f(E(x_1)) = [E(x_1); f].$$

Supposing now that the property is valid for the E-divided differences of order n-1, from the formula (11.6.1) we get

$$[x_1,x_2,\ldots,x_{n+1};f]_{(E)} = \frac{[x_2,x_3,\ldots,x_{n+1};f]_{(E)} - [x_1,x_2,\ldots,x_n;f]_{(E)}}{e(E(x_1),E(x_{n+1}))d(E(x_{n+1}),E(x_1))} =$$

$$= \frac{[E(x_2),E(x_3),\ldots,E(x_{n+1});f] - [E(x_1),E(x_2),\ldots,E(x_n);f]}{e(E(x_1),E(x_{n+1}))d(E(x_{n+1}),E(x_1))} =$$

$$= [E(x_1),E(x_2),\ldots,E(x_{n+1});f]$$

as required.

Corollary 11.6.1. *The* E*-divided difference of orders* 1, 2 *and* 3 *of function* f *can be computed by:*

i) $[x_1, x_2; f]_{(E)} = \dfrac{f(E(x_2)) - f(E(x_1))}{d(E(x_2), E(x_1))}$;

ii)
$$[x_1, x_2, x_3; f]_{(E)} = \frac{1}{d(E(x_1),E(x_2))\cdot d(E(x_1),E(x_3))\cdot d(E(x_2),E(x_3))}\{f(E(x_1))d(E(x_2),E(x_3)) - f(E(x_2))[d(E(x_1),E(x_2)) + d(E(x_2),E(x_3))] + f(E(x_3))d(E(x_1),E(x_2))\}.$$

iii)
$$\begin{aligned}
[x_1, x_2, x_3, x_4; f]_{(E)} &= \\
&= \frac{1}{\prod\limits_{1\le i<j\le 4} d(E(x_i), E(x_j))}\{-f(E(x_1))d(E(x_2), E(x_3)) \\
&\quad d(E(x_2), E(x_4))d(E(x_3), E(x_4)) + \\
&\quad + f(E(x_2))d(E(x_3), E(x_4))[d(E(x_1), E(x_2))d(E(x_1), E(x_3)) + \\
&\quad + d(E(x_1), E(x_2))d(E(x_2), E(x_4)) + \\
&\quad + d(E(x_2), E(x_3))d(E(x_2), E(x_4))] - \\
&\quad - f(E(x_3))d(E(x_1), E(x_2))[d(E(x_1), E(x_3))d(E(x_2), E(x_3)) + \\
&\quad + d(E(x_1), E(x_3))d(E(x_1), E(x_4)) + \\
&\quad + d(E(x_2), E(x_4))d(E(x_3), E(x_4))] + \\
&\quad + f(E(x_4))d(E(x_1), E(x_2))d(E(x_1), E(x_3))d(E(x_2), E(x_3))\}.
\end{aligned}$$

Corollary 11.6.1 is obvious if we take into account proposition 11.6.1 and lemma 11.3.1.

Definition 11.6.2. *The metric Vandermonde determinant of points* x_1, x_2, ..., x_{n+1} *is the real number given by*

$$(11.6.3)\qquad \begin{cases} VM(x_1) = 1 \\ VM(x_1, x_2, \ldots, x_{n+1}) = \prod\limits_{1\le i<j\le n+1} d(x_i, x_j)\cdot \end{cases}$$

We denote by F (X, **R**) the set of all the functionals f: X → **R**. For each distinct points $x_1, \ldots, x_{n+1} \in X$ we define the functional UM: F (X, **R**) → **R**,

$$(11.6.4)\qquad \begin{aligned} &UM(x_1, x_2, \ldots, x_{n+1}; f)_{(E)} = \\ &= \sum_{k=1}^{n+1} f(E(x_k))\cdot VM(E(x_1), E(x_2), \ldots, E(x_{k-1}), E(x_{k+1}), \ldots, E(x_{n+1})) \end{aligned}$$

and this implies the following result:

Corollary 11.6.2.

$$UM(x_1, x_2, \ldots, x_{n+1}; f)_{(E)} =$$

(11.6.5) $$= \sum_{k=1}^{n+1} (-1)^{n+k-1} f(E(x_k)) \cdot \frac{\prod\limits_{\substack{1 \le i < j \le n+1 \\ i \ne k}} d(E(x_i), E(x_j))}{\prod\limits_{\substack{i=1 \\ i \ne k}}^{n+1} d(E(x_k), E(x_i))}.$$

Then, corollary 11.6.1 may be partially written as

Corollary 11.6.3. $[x_1, x_2; f]_{(E)} = \dfrac{UM(x_1, x_2; f)_{(E)}}{VM(E(x_1), E(x_2))}$.

Generally, a similar equality does not fulfil for the E-divided differences of order greater then 2 of the functional f. This fact is due to the triangle inequality, which is strict if the points, are not collinear. To correct it we need collinearity and order conditions between the considered points. The best framework is provided by the d-convex structures in the sense of Menger (1928) – Soltan (1984). We recall this notion. Let (X, d) be a metric space and x, y, z three points of X. The point y metrically lies between x and z if and only if the points x, y, z are distinct and d (x, y) + d (y, z) = d (x, z). If y metrically lies between x and z then we shall denote it by (xyz). A subset $Y \subset X$ is said to be d - convex if for each two points $x \in Y$ and $z \in Y$ there is a point $y \in Y$ that metrically lies between x and z. If the whole set X satisfies this condition then the metric space (X, d) is said to be convex.

In what follows, (X, d) will be supposed to be a convex metric space.

Definition 11.6.3. *The mapping* E: $X \to X$ *preserves the betweenness if and only if for every* $x \in X$, $y \in X$ *and* $z \in X$ *such that* (xyz) *it follows that* (E(x)E(y)E(z)).

In a convex metric space points $x_1, x_2, \ldots, x_n$ are said to be *metrically collinear* if and only if for each $j \in \{2, 3, \ldots, n\}$, we have $(x_i x_j x_k)$ for all $i \in \{1, 2, \ldots, j-1\}$ and $k \in \{j+1, j+2, \ldots, n\}$.

Corollary 11.6.4. *If the mapping* E: $X \to X$ *preserves the betweenness and if points* $x_1, x_2, \ldots, x_{n+1}$ *are metrically collinear then*

(11.6.6) $$[x_1, x_2, \ldots, x_{n+1}; f]_{(E)} = \frac{UM(x_1, x_2, \ldots, x_{n+1}; f)_{(E)}}{VM(E(x_1), E(x_2), \ldots, E(x_{n+1}))}.$$

The collinearity condition might be avoided if metric space (X, d) is endowed with a total order relation. This is easy to be obtained following the idea of G. Cristescu (2000 [73]) that uses an injective functional $\varphi: X \to \mathbf{R}$. This functional induces a total order relation on X as it follows: if $x, y \in X$ then

(11.6.7) $$x \leq_\varphi y \quad \Leftrightarrow \quad \varphi(x) \leq \varphi(y).$$

Therefore, $(X, d, \leq_\varphi)$ becomes a totally ordered metric space. We denote by $\sigma(x, y) = \varphi(x) - \varphi(y)$, a real number that will appear in various sign conditions.

Let $(X, d, \leq_\varphi)$ be a totally ordered metric space, $Y \subseteq X$ an E-convex set and $n \in \mathbf{Z}$, $n \geq -1$.

Definition 11.6.4. *The functional* $f: Y \to \mathbf{R}$ *is said to be* E-convex *(*E-*non-concave,* E-*polynomial,* E-*non-convex,* E-*concave) of order n if for each* n+2 *distinct points of* Y, $x_1, x_2, \ldots, x_{n+2}$, *the following inequality occurs:*

i) $[x_1; f]_{(E)} > (\geq, =, \leq, <)\ 0$, *for* $n = -1$;

ii) $\sigma(x_1, x_2)\,[x_1, x_2; f]_{(E)} > (\geq, =, \leq, <)\ 0$, *for* $n = 0$;

iii) $[x_1, x_2, \ldots, x_{n+2}; f]_{(E)} > (\geq, =, \leq, <)\ 0$, *for* $n \geq -1$.

If a functional satisfies one of the five inequalities of definition 11.6.4 on Y then it is called of order n with respect to E on Y. The case $n = -1$ describes sign properties and case $n = 0$ contains monotony properties of functional f.

Definition 11.6.5. *The functional* $f: Y \to \mathbf{R}$ *is said to be* E-*pseudo-convex of order* n *if for every* n+3 *distinct points of* Y, $x_1, x_2, \ldots, x_{n+3}$, *we have*

(11.6.8) $$[x_2, x_3, \ldots, x_{n+2}; f]_{(E)} \leq$$
$$\leq \max\{[x_1, x_2, \ldots, x_{n+1}; f]_{(E)}, [x_3, x_4, \ldots, x_{n+3}; f]_{(E)}\}.$$

Theorem 11.6.1. *The functional* $f: Y \to \mathbf{R}$ *is* E-*convex (*E-*non-concave,* E-*polynomial,* E-*non-convex,* E-*concave) of order* n+1 *if for every* n+3 *distinct points of* Y, $x_1, x_2, \ldots, x_{n+3}$, *the following inequality is validated:*

(11.6.9)
$$[x_2, x_3, \ldots, x_{n+2}; f]_{(E)} < (\leq, =, \geq, >)$$
$$< (\leq, =, \geq, >) \frac{d(E(x_2), E(x_{n+3}))}{d(E(x_1), E(x_{n+2})) + d(E(x_2), E(x_{n+3}))}[x_1, x_2, \ldots, x_{n+1}; f]_{(E)} +$$
$$+ \frac{d(E(x_1), E(x_{n+2}))}{d(E(x_1), E(x_{n+2})) + d(E(x_2), E(x_{n+3}))}[x_3, x_4, \ldots, x_{n+3}; f]_{(E)}.$$

Proof. According to (11.6.1), the inequality from definition 11.6.4.iii) becomes a monotony condition for the *E*-divided differences of order n. If we apply (11.6.1) in this monotony property, after an elementary calculus we obtain (11.6.9).

Remark 11.6.1. Let us take $X = \mathbf{R}^n$ together with the Euclidean distance, E: $\mathbf{R}^n \to \mathbf{R}^n$ a linear function and $Y \subseteq \mathbf{R}^n$ an E-convex set. Then the functional f: $Y \to \mathbf{R}$ is E-convex of order 1 if the inequality

$$f(t\,E(x) + (1-t)\,E(y)) < t\,f(E(x)) + (1-t)\,f(E(y)) \tag{11.6.10}$$

holds for each $x, y \in Y$ and $t \in [0, 1]$. Indeed, denoting by $z = t\,x + (1-t)\,y$ then, according to (11.6.9), we obtain

$$f(E(z)) < \frac{d(E(y),E(z))}{d(E(x),E(z)) + d(E(z),E(y))} f(E(x)) +$$
$$+ \frac{d(E(x),E(z))}{d(E(x),E(z)) + d(E(z),E(y))} f(E(y)).$$

Using the linearity of E we obtain

$$d(E(z), E(y)) = t\,d(E(x), E(y)),$$
$$d(E(x), E(z)) = (1-t)\,d(E(x), E(y))$$

and, therefore,

$$f(tE(x) + (1-t)E(y)) <$$
$$< \frac{td(E(x),E(y))}{(1-t)d(E(x),E(y)) + td(E(x),E(y))} f(E(x)) +$$
$$+ \frac{(1-t)d(E(x),E(y))}{(1-t)d(E(x),E(y)) + td(E(x),E(y))} f(E(y)).$$

After the simplification (11.6.10) fulfils.

A similar inequality might be derived for every type of functional of order 1 with respect to E on Y. Condition (11.6.10) means that the E-convexity of order 1 of the functional f is equivalent to the E-convex functions defined by E. A. Youness (1999).

Coming back to the general case, we can establish connections between E-convexity properties of order n and E-pseudo-convexity.

Theorem 11.6.2. *If the functional* f: $Y \to \mathbf{R}$ *is* E-*convex of order* n+1 *then it is* E-*pseudo-convex of order* n.

Proof. From theorem 11.6.1 it follows that if f is E-convex of order n+1, then for each distinct points of Y, $x_1, x_2, \ldots, x_{n+3}$, we have

$$[x_2, x_3, \ldots, x_{n+2}; f]_{(E)} <$$

$$< \frac{d(E(x_2), E(x_{n+3}))}{d(E(x_1), E(x_{n+2})) + d(E(x_2), E(x_{n+3}))} [x_1, x_2, \ldots, x_{n+1}; f]_{(E)} +$$

$$+ \frac{d(E(x_1), E(x_{n+2}))}{d(E(x_1), E(x_{n+2})) + d(E(x_2), E(x_{n+3}))} [x_3, x_4, \ldots, x_{n+3}; f]_{(E)} \leq$$

$$\left(\frac{d(E(x_2), E(x_{n+3}))}{d(E(x_1), E(x_{n+2})) + d(E(x_2), E(x_{n+3}))} + \right. \\ \left. + \frac{d(E(x_1), E(x_{n+2}))}{d(E(x_1), E(x_{n+2})) + d(E(x_2), E(x_{n+3}))} \right) \cdot$$

$$\cdot \max\{[x_1, x_2, \ldots, x_{n+1}; f]_{(E)}, [x_3, x_4, \ldots, x_{n+3}; f]_{(E)}\} =$$

$$= \max\{[x_1, x_2, \ldots, x_{n+1}; f]_{(E)}, [x_3, x_4, \ldots, x_{n+3}; f]_{(E)}\}.$$

Similar convexity properties might be defined by the generalization of the convexities of functions corresponding to the convexities of sets from the definitions given in 7.1.

Chapter 12

Applications in optimisation

The notion of d-base has a very important role in the theoretical foundation of some methods of solving some integer programming problems with one or more objective functions. L. Lupşa (1982[153]) introduced the notion of d-base. It permits the spotlight of necessary conditions or sufficient conditions or necessary and sufficient conditions for a feasible solution of mathematical programming problem to be an optimal solution. Also it allows the elaboration of some duality theorems for discrete programming problem, where the feasible solution is a discrete polyhedral set, integer convex, of n dimension.

12.1 d-bases

Let $m, n \in \mathbf{N}^*$, the integers $a_{ij} \in \mathbf{Z}$, $(i \in \{1, ..., m\}; j \in \{1, ..., n\})$, $b_i \in \mathbf{Z}$, $i \in \{1, ..., m\}$ and $c_j \in \mathbf{Z}, j \in \{1, ..., n\}$.

For future reference we define the following sets:

$$I = \{1, ..., m\}, \quad J = \{1, ..., n\}, \quad K = \{1, ..., n+m\},$$

$$S = \{x = (x_1, \ldots, x_n) \in \mathbf{R}^n \mid a_{i1} x_1 + ... + a_{in} x_n \le b_i, \forall\, i \in I,\ x_j \ge 0, \forall\, j \in J\}$$

and $SI = S \cap \mathbf{Z}^n$. Set SI is a discrete polyhedral set. If SI is finite, then SI is a discrete polytope.

Let $Y = \{y_1, ..., y_{n+m}\}$ be the set of the real functions defined on $\mathbf{R}^n$ by

(12.1.1) $\quad y_k(x) = x_k$, for all $x = (x_1, ..., x_n) \in \mathbf{Z}^n$, if $k \in \{1, ..., n\}$,

and by

(12.1.2) $y_k(x) = b_{k-n} - (a_{k-n,1}x_1+...+a_{k-n,n}x_n)$, for all $x = (x_1, ..., x_n) \in \mathbf{Z}^n$, if $k \in \{n+1, ..., n+m\}$.

Let Y be the set of all real affine functions $y: \mathbf{R}^n \to \mathbf{R}$ satisfying the following conditions: there is $\alpha = (\alpha_1, ..., \alpha_n) \in \mathbf{Z}^n$ and $\beta \in \mathbf{Z}$ such that

(12.1.3) $y(x) = \beta - (\alpha_1 x_1 + ... + \alpha_n x_n)$, for each $x = (x_1, ..., x_n) \in \mathbf{R}^n$

and

(12.1.4) $y(x) \geq 0$, for all $x \in SI$.

Obviously we have $Y \subseteq Y$.

Definition 12.1.1. *The set* $W = \{w_1, ..., w_n\}$, *where* $w_j \in Y$, *for all* $j \in J$, *is said to be a* d-*base for* SI *if the following conditions are satisfied*:

a) *The linear functions* $\hat{w}_j$, $j \in J$, *corresponding to affine functions* w_j, *for* $j \in J$, *are linearly independent*;

and

b) *For each* $k \in K$, *there are* $n + 1$ *integer numbers* d_{kj}, $j \in \{0, 1, ..., n\}$, *such that*

(12.1.5) $y_k = -(d_{k1} w_1 + ... + d_{kn} w_n) + d_{k0}$.

It is easy to see that numbers d_{kj}, $(k \in K, j \in \{0\} \cup J)$, are unique, if they exist. For this reason *the vector* $d = (d_{10}, d_{20}, ..., d_{n+m,0})$ *is called the associated vector of the* d-*base* W, *and the matrix* $D = [d_{kj}]$ $(k \in K, j \in J)$ *is called the associated matrix of the* d-*base* W. If $W = \{w_1, ..., w_n\}$ is a d-base of SI, then the system

$$(12.1.6) \qquad \begin{cases} w_1(x) = 0 \\ \quad ... \\ w_n(x) = 0 \end{cases}$$

has a unique solution $z = (z_1, ..., z_n)$. For this reason *the point* z *is said to be generated by the* d-*base* W.

Example 12.1.1. From the definition of the functions $y_1, ..., y_n$, we get that the functions $\hat{y}_1, ..., \hat{y}_n$, are linearly independent. Because $a_{ij} \in \mathbf{Z}$ for all $i \in I$, $j \in J$, and $b_i \in \mathbf{Z}$, for all $i \in I$, we get that the set $W = \{y_1, ..., y_n\}$ is a d-base for SI with associated matrix $D = [d_{kj}]$ $(k \in K, j \in J)$ given by :

$d_{jj} = -1$, for all $j \in \{1, ..., n\}$,

$d_{kj} = 0$, for all $k \in \{1, ..., m\}$, and all $j \in \{1, ..., n\}$, $k \neq j$,

$d_{kj} = a_{k-n,j}$, for all $k \in \{n+1, ..., n+m\}$, and all $j \in \{1, ..., n\}$.

The associated vector of this d-base is

$d = (0, 0, ..., 0, b_1, ..., b_n)$.

The point generated by the d-base Y is $z^0 = (0, ..., 0)$.

Theorem 12.1.1. *Let* $W = \{w_1, ..., w_n\}$ *a* d-*base of* SI, *with the associated matrix* $D = [d_{kj}]$ $(k \in K, j \in J)$ *and the associated vector* $d = (d_{10}, ..., d_{n+m,0})$. *If* t *is a real number,* $t \neq 0$, *and* $s \in \{1, ..., n+m\}$, *then*

$$(12.1.7) \qquad \left[\frac{d_{s0}}{t}\right] - \sum_{j=1}^{n} \left[\frac{d_{sj}}{t}\right] \cdot w_j(x) \geq 0, \text{ for all } x \in SI,$$

where [a] *denotes the integer part of* a.

Proof. We have

$d_{s0} - (d_{s1}\, w_1(x) + ... + d_{sn}\, w_n(x)) = y_s(x) \geq 0$, for all $x \in SI$.

Dividing by t, we obtain

$$\left[\frac{d_{s0}}{t}\right] - \sum_{j=1}^{n} \left[\frac{d_{sj}}{t}\right] w_j(x) + \left]\frac{d_{s0}}{t}\right[\geq \sum_{j=1}^{n} \left]\frac{d_{sj}}{t}\right[w_j(x), \text{ for all } x \in SI,$$

where]a[means the fractional part of a,]a[= a - [a]. The right member of this inequality is a nonnegative number. It follows that, for all $x \in SI$, we have

$$(12.1.8) \qquad \left[\frac{d_{s0}}{t}\right] - \sum_{j=1}^{n} \left[\frac{d_{sj}}{t}\right] w_j(x) \geq -\left]\frac{d_{s0}}{t}\right[.$$

Because the left member of (12.1.8) is an integer number and the right member of (12.1.8) is a number belonging to]-1, 0], then for all $x \in SI$, we obtain

$$(12.1.9) \qquad \left[\frac{d_{s0}}{t}\right] - \sum_{j=1}^{n} \left[\frac{d_{sj}}{t}\right] w_j(x) \geq 0.$$

Remark 12.1.1. Let $t \in \mathbf{R} \setminus \{0\}$ and $s \in K$. If we put

$$(12.1.10) \qquad \alpha_j = \left[\frac{d_{sj}}{t}\right], \text{ for each } j \in \{1, ..., n\},$$

and

(12.1.11) $\beta = \left[\frac{d_{s0}}{t}\right]$,

then, by theorem 12.1.1, it results that function f: **R** → **R**,

(12.1.12) $f = \beta - (\alpha_1 w_1 + ... + \alpha_n w_n)$

is an element of set Y.

Theorem 12.1.2. *Let* $W = \{w_1, ..., w_n\}$, *a* d-*base of* SI, *with the associated matrix* $D = [d_{kj}]$ ($k \in K$, $j \in J$) *and the associated vector* $d = (d_{10}, ..., d_{n+m,0})$, *and let* $s \in \{1, ..., n+m\}$ *with* $y_s \notin W$. *If* t *is a real number,* $t \neq 0$, *and if there is* $r \in J$ *such that number* α_r, *given by* (12.1.10) *has the property that* $|\alpha_r| = 1$, *then set* $W' = (W \setminus \{w_r\}) \cup \{f\}$, *where function* f *is given by* (12.1.12), *is a new* d-*base for the set* SI.

Proof. First we prove that functions $\hat{f}$ and $\hat{w}_j$, $j \in J \setminus \{r\}$, are linearly independent. If these functions are not linearly independent, then there are n real numbers $\gamma_1, ..., \gamma_n$, with $|\gamma_1| + ... + |\gamma_n| \neq 0$, such that

(12.1.13) $\gamma_1 w_1(x) + ... + \gamma_{r-1} w_{r-1}(x) + \gamma_r f(x) + \gamma_{r+1} w_{r+1}(x) + ... + \gamma_n w_n(x) = 0$,

for all $x \in \mathbf{R}^n$. Because functions $\hat{w}_j$, $j \in J \setminus \{r\}$, are linearly independent, we get that $\gamma_r \neq 0$. From (12.1.13) and (12.1.12) we have

(12.1.14) $$\sum_{\substack{j=1 \\ j\neq r}}^{n} (\gamma_j - \gamma_r \alpha_r)\, \hat{w}_j(x) - \gamma_r \alpha_r\, \hat{w}_r(x) = 0, \text{ for all } x \in \mathbf{R}.$$

Taking $u_j = \gamma_j - \gamma_r \alpha_j$, $j \in J \setminus \{r\}$, $u_r = -\gamma_r \alpha_r$, we have n real numbers, not all null ($u_r \neq 0$), such that $\sum_{j=1}^{n} u_j \hat{w}_j(x) = 0$, for all $x \in \mathbf{R}^n$. That contradicts the linearly independence of functions $\hat{w}_j$, $j \in J$. Therefore functions $\hat{f}$ and $\hat{w}_j$, $j \in J \setminus \{r\}$, are linearly independent.

Now we show that condition b) from definition of d-base is satisfied. From (12.1.12) we get

(12.1.15) $$w_r = \frac{1}{\alpha_r}(\beta - \alpha_1 w_1 - ... - \alpha_{r-1} w_{r-1} - f - \alpha_{r+1} w_{r+1} - ... - \alpha_n w_n).$$

If we replace w_r from (12.1.15) in (12.1.5), we obtain

$$(12.1.16)\quad y_i = d_{i0} - \frac{d_{ir}\beta}{\alpha_r} - \sum_{\substack{j=1\\ j\neq r}}^{n} (d_{ij} - \frac{d_{ir}\alpha_j}{\alpha_r})\, w_j - \frac{-d_{ir}}{\alpha_r}\, f,$$

for each $i \in K$. In view of $|\alpha_r| = 1$, if we put

$$(12.1.17)\qquad \delta_{i0} = d_{i0} - \frac{d_{ir}\beta}{\alpha_r}, \text{ for all } i \in K,$$

$$(12.1.18)\qquad \delta_{ij} = d_{ij} - \frac{d_{ir}\alpha_j}{\alpha_r}, \text{ for all } i \in K, j \in J \setminus \{r\},$$

$$(12.1.19)\qquad \delta_{ir} = \frac{-d_{ir}}{\alpha_r}, \text{ for all } i \in K,$$

it is easy to see that

$$(12.1.20)\qquad \delta_{ij} \in \mathbf{Z}, \text{ for all } i \in K, j \in J \cup \{0\}, \text{ and}$$

$$(12.1.21)\qquad y_i = \delta_{i0} - (\sum_{\substack{j=1\\ j\neq r}}^{n} \delta_j w_j + \delta_r f), \text{ for each } i \in K.$$

Then, from (12.1.20) and (12.1.21), it follows that b) condition of definition of the d-base is satisfyied. Therefore set W' is a d-base of SI.

Definition 12.1.2. *A point* $z \in \mathbf{Z}^n$ *is said to be a* p-vertex *of set* SI *if there is a* d-base W *of* SI *such that* z *is generated by* W.

Theorem 12.1.3. *If* $z \in \mathbf{Z}^n$ *is a* p-vertex *of* SI *generated by the* d-base $W = \{w_1, ..., w_n\}$, *with the associated matrix* $D = [d_{kj}]$ $(k \in K, j \in J)$ *and the associated vector* $d = (d_{10}, ..., d_{n+m,0})$, *then the coordinates of point* z *are given by*

$$(12.1.22)\qquad z_j = d_{j0}, \text{ for each } j \in \{1, ..., n\}.$$

Proof. Because z is generated by $W = \{w_1, ..., w_n\}$, we have $w_j(z) = 0$, for all $j \in \{1, ..., n\}$. Let be $j \in \{1, ..., n\}$. From (12.1.1) and (12.1.5) we have

$$z_j = y_j(z) = -(d_{j1} w_1(z) + ... + d_{jn} w_n(z)) + d_{j0} = d_{j0}.$$

Corollary 12.1.1. *If* $z \in \mathbf{Z}^n$ *is a* p-vertex *of* SI, *generated by the* d-base W, *then for any* $j \in \{1, ..., n\}$, *such that* $y_j \in W$, *we have* $z_j = 0$.

Definition 12.1.3. *A point* $z \in \mathbf{Z}^n$ *is said to be an* a-vertex *of* SI *if* z *is a* p-vertex *of* SI *and* $z \in$ SI.

We remark that the notion of p-vertex differs from the notion of a-vertex.

Example 12.1.2. Let

$S = \{(x_1, x_2) \in \mathbf{R}^2 \mid -x_1 \le -1, -x_2 \le -1, x_1 + x_2 \le 1\}$, and $SI = S \cap \mathbf{Z}^2$.

$W = \{y_1, y_2\}$ is a d-base of SI. The point generated by W is $x^0 = (0, 0)$. x^0 is a p-vertex of SI, but it is not an a-vertex of SI because $x^0 \notin SI = \varnothing$.

Theorem 12.1.4. *If* $SI \subseteq \mathbf{N}^n$ *is a finite integer set and* $x \in \mathbf{Z}^n$, *then the following sentences are equivalent:*

i) x^0 *is an* a-*vertex of* SI;

ii) x^0 *is a vertex of* conv (SI);

iii) x^0 *is an extremal point of* SI *with respect to* $\mathbf{Z}^n$.

Proof. First we shall prove that i) => ii). If $x^0 \in SI$ is an a-vertex of SI, then $x^0 \in SI$ and there is a d-base $W = (w_1, ..., w_n)$, such that x^0 is generated by W. As $w_j \in Y$, there are $\beta_j \in Z$, for each $j \in J$, and $\alpha_{ij} \in \mathbf{Z}^n$, for all $i, j \in J$ such that $w_i(x) = \beta_i - \sum_{j=1}^{n} \alpha_{ij} x_j$, for all $x \in R^n$, and all $i \in J$. Then the set

$$M = \{x \in \mathbf{R}^n \mid w_j(x) \ge 0, j \in J\}$$

is an integer polyhedral set and x^0 is one of its vertices (as W is a d-base, functions $\hat{w}_j$, $j \in \{1, ..., n\}$, are linearly independent; then the rank of square matrix $[\alpha_{ij}]$ $(i, j \in \{1, ..., n\})$ equals to n; because x^0 is the solution of the system $w_j(x) = 0, j \in \{1, ..., n\}$, it follows that x^0 is a vertex of M).

From (12.1.4) it results that we have $w_j(x) \ge 0, j \in \{1, ..., n\}$, for $x \in SI$. Hence $SI \subseteq M$. Further more, we show that conv $(SI) \subseteq M$. Let us choose point $z \in$ conv (SI). There is a non null natural number p, $p \le n+1$, p elements $x^k \in SI$, $k \in \{1, ..., p\}$, and p non negative real numbers t_k, $k \in \{1, ..., p\}$, with $t_1 + ... + t_p = 1$, such that $z = t_1 x^1 + ... + t_p x^p$. As $SI \subseteq M$, for each $k \in \{1, ..., p\}$ and each $i \in \{1, ..., n\}$, we have $w_i(x^k) = \beta_i - \sum_{j=1}^{n} \alpha_{ij} x_j^k \ge 0$. It follows that $w_i(z) = \beta_i - \sum_{j=1}^{n} \alpha_{ij} (\sum_{k=1}^{p} t_k x_j^k) = \sum_{k=1}^{p} t_k (\beta_i - \sum_{j=1}^{n} \alpha_{ij} x_j^k) \ge 0$, for all $i \in \{1, ..., n\}$ Hence $z \in M$. Therefore conv $(SI) \subseteq M$.

It is known that if X and Y are two polyhedral sets with $X \subseteq Y$ and the point $x \in X$ is a vertex of Y, then x is also a vertex of X. Because x^0 is a vertex of M, conv $(SI) \subseteq M$, and $x^0 \in SI \subseteq$ conv (SI), applying this property we get that x^0 is a vertex of conv (SI).

Now we prove ii) => iii). As SI is a finite integer set, the set conv (SI) is, in virtue of theorem 2.7.16, an integer polyhedral set. It is known that the set of all vertices of a polyhedral set equals to the set of all its extremal points. As x^0 is a vertex of conv (SI), applying this property, we get that x^0 is an extremal point of conv (SI). Then $x^0 \notin \text{conv}(\text{conv}(SI) \setminus \{x^0\})$. But

$$\text{conv}((SI \setminus \{x^0\}) \cap \mathbf{Z}^n) = \text{conv}(SI \setminus \{x^0\}) \subseteq \text{conv}(\text{conv}(SI) \setminus \{x^0\}).$$

Hence $x^0 \notin \text{conv}((SI \setminus \{x^0\}) \cap \mathbf{Z}^n)$. Therefore x^0 is an extremal point of SI with respect to $\mathbf{Z}^n$.

It remains to prove iii) => i). Let x^0 an extremal point of SI with respect to $\mathbf{Z}^n$. Then $x^0 \in SI \cap \mathbf{Z}^n = SI$ and $x^0 \notin \text{conv}(SI \setminus \{x^0\} \cap \mathbf{Z}^n) = \text{conv}(SI \setminus \{x^0\})$. Suppose that x^0 is not an extremal point of conv (SI). Then

$$x^0 \in \text{conv}(\text{conv}(SI) \setminus \{x^0\}).$$

It follows that there is a natural number q, $1 \le q \le n+1$, q elements

$$y^j \in (\text{conv}(SI) \setminus \{x^0\}), j \in \{1, ..., q\},$$

and q strictly positive real numbers $t_1, ..., t_q$, with $t_1 + ... + t_q = 1$, such that we have $x^0 = t_1 y^1 + ... + t_q y^q$. Since $y^j \in (\text{conv}(SI) \setminus \{x^0\})$, for each $j \in \{1, ..., q\}$, there is a natural number m and m elements $x^h \in SI \setminus \{x^0\}$, $h \in \{1, ..., m\}$, such that $y^j \in \text{conv}\{x^1, ..., x^m\}$, $j \in \{1, ..., q\}$. Then, for each $j \in \{1, ..., q\}$, there are m nonnegative real numbers g_{jk}, $k \in \{1, ..., m\}$ with $\sum_{k=1}^{m} g_{jk} = 1$, such that $y^j = \sum_{k=1}^{m} g_{jk} x^k$, for each $j \in \{1, ..., q\}$. Hence

$$x^0 = \sum_{j=1}^{q} t_j (\sum_{k=1}^{m} g_{jk} x^k) = \sum_{k=1}^{m} (\sum_{j=1}^{q} t_j g_{jk}) x^k .$$

Because $\sum_{k=1}^{m} (\sum_{j=1}^{q} t_j g_{jk}) = \sum_{j=1}^{q} (t_j \sum_{k=1}^{m} g_{jk}) = 1$, and $\sum_{j=1}^{q} t_j g_{jk} \ge 0$, for each $k \in \{1,...,m\}$, we get that $x^0 \in \text{conv}(\{x^k \mid k \in \{1, ..., m\}\})$. Two cases may obtain: $x^k \in SI \setminus \{x^0\}$ for all $k \in \{1, ..., q\}$ or it is $k \in \{1, ..., q\}$ such that $x^k = x^0$.

In the first case it follows that $x^0 \in \text{conv}(SI \setminus \{x^0\})$; this contradicts the fact that $x^0 \notin \text{conv}(SI \setminus \{x^0\})$.

In the second case, without restriction of the generality, we can suppose that $x^m = x^0$. We remark that we have $g_{jm} \neq 1$, for each $j \in \{1, ..., q\}$, because if it is $j \in \{1, ..., q\}$ such that $g_{jm} = 1$, then it follows that $y^j = x^0$, what contradicts $y^j \in \text{conv (SI)} \setminus \{x^0\}$. If we put $u = \sum_{j=1}^{q} t_j g_{jm}$, then we have

$0 < u < 1$, and we can write $(1 - u)\cdot x^0 = \sum_{k=1}^{m-1}(\sum_{j=1}^{q} t_j g_{jk})x^k$. Dividing by (1-u), we obtain

$$x^0 = \sum_{k=1}^{m-1}(\sum_{j=1}^{q}\frac{1}{1-u}t_j g_{jk})x^k .$$

As

$$\sum_{k=1}^{m-1}(\sum_{j=1}^{q}\frac{1}{1-u}t_j g_{jk}) = \frac{1}{1-u}\sum_{j=1}^{q}t_j(\sum_{k=1}^{m-1}g_{jk}) = \frac{1}{1-u}\sum_{j=1}^{q}t_j(1-g_{jm}) =$$

$$= \frac{\sum_{j=1}^{q}t_j - \sum_{j=1}^{q}t_j g_{jm}}{1-u} = \frac{1-u}{1-u} = 1,$$

and $\sum_{j=1}^{q}\frac{1}{1-u}t_j g_{jk} \geq 0$, for each $k \in \{1, ..., m\}$, we get that

$$x^0 \in \text{conv}(\{x^1, ..., x^{m-1}\}) \subseteq \text{conv}\{SI \setminus \{x^0\}).$$

This contradicts the hypothesis that $x^0 \notin \text{conv}(SI \setminus \{x^0\})$. Because in both cases we obtain a contradiction, the supposition that x^0 is not an a-vertx of SI is false. Hence x^0 is an a- vertex of SI.

Theorem 12.1.5. *If* $W = \{w_1, ..., w_n\}$ *is a d-base of* SI, *with the associated matrix* $D = [d_{kj}]$ ($k \in K, j \in J$) *and the associated vector* $d = (d_{10}, ..., d_{n+m,0})$, *then the following propositions are true*:

i) *If* x^0 *is generated by* W, *then* $x^0 \in SI$ *if and only if* $d_{k0} \geq 0$, *for all* $k \in K$.

ii) *If it is* $k \in K$ *such that* $d_{k0} < 0$ *and* $d_{kj} \geq 0$, *for all* $j \in J$, *then* $SI = \Phi$.

iii) *If* $z \in \mathbf{R}^n$, *then we have* $z \in SI$ *if and only if*

(12.1.23) $\quad -(d_{k1}w_1(x) + ... + d_{kn}w_n(x)) + d_{k0} \in \mathbf{N}$, *for all* $k \in K$.

Proof i) Because x^0 is generated by $W = \{w_1, ..., w_n)\}$, we have

(12.1.24) $\quad w_j(x^0) = 0$, for each $j \in J$.

Necessity. Let $x^0 \in SI$. It follows that $y_k(x^0) \geq 0$, for all $k \in K$. Then, in view of (12.1.5), we get that

(12.1.25) $0 \leq y_k(x^0) = -(d_{k1}w_1(x^0)+...+d_{kn}w_n(x^0)) + d_{k0} = d_{k0}$,

for all $k \in K$. Hence $d_{k0} \geq 0$, for all $k \in K$.

Sufficiency. Let now x^0 generated by the d-base W, such that its associated vector satisfies $d_{k0} \geq 0$, for all $k \in K$. Then, from theorem 12.1.3, we get

(12.1.26) $x_k^0 = d_{k0} \geq 0$, for all $k \in \{1, ..., n\}$.

On the other hand, from (12.1.5) we have

(12.1.27) $b_{k-n} - \sum_{j=1}^{n} a_{k-n,j}x_j^0 = y_k(x^0) = -\sum_{j=1}^{n} d_{k,j}\, w_j(x^0) + d_{k0} \geq 0$,

$k \in \{n+1, ..., n+m\}$. From (12.1.26) and (12.1.27) it results that $x^0 \in SI$.

ii) Suppose that $SI \neq \Phi$ and let be $x \in SI$. Then

(12.1.28) $y_k(x) \geq 0$, for all $k \in K$,

and

(12.1.29) $w_j(x) \geq 0$, for each $j \in J$.

On the other hand, for each $k \in K$, we have

$$y_k(x) = -(d_{k-n,1}\, w_1(x)+...+d_{k-n,n}\, w_n(x)) + d_{k-n} \leq d_{k-n,0} < 0,$$

that contradicts (12.1.28). Hence $SI = \varnothing$.

iii) **Necessity.** Let $x \in SI$. Then $w_j(x) \in \mathbf{Z}$ for all $j \in \{1, \ldots, n\}$ and, in view of definition of the d-base, we get that

$$-(d_{k1}\, w_1(x) + ... + d_{kn}w_n(x)) + d_{k0} \in \mathbf{Z},\ k \in K.$$

On the other hand, because $x \in SI$, we have

$$-(d_{k1}w_1(x) +...+ d_{kn}w_n(x)) + d_{k0} = y_k(x) \geq 0.$$

Hence $-(d_{k1}w_1(x) +...+ d_{kn}w_n(x)) + d_{k0} \in \mathbf{N}$.

Sufficiency. Now let $x \in \mathbf{R}^n$. We suppose that (12.1.23) is true. In view of theorem 12.1.3, we have $x_j = d_{j0} \in \mathbf{N}$, $j \in \{1, \ldots, n\}$. Hence $x \in \mathbf{N}^n$. Also, from $y_{k-n}(x) = -(d_{k1}\, w_1(x) + ... + d_{kn}\, w_n(x)) + d_{k0} \in \mathbf{N}$, $k \in \{n+1, \ldots, n+m\}$, we get that $y_i(x) \geq 0$, for all $i \in \{1, \ldots, m\}$. Therefore $x \in SI$.

Theorem 12.1.6. *If* z *is an* a-*vertex of* SI *and* $W = \{w_1, ..., w_n\}$ *is a* d-*base which generates the point* z, *then*

(12.1.30) $SI \setminus \{z\} = \{x \in SI \mid w_1(x) + \ldots + w_n(x) \geq 1\}$.

Proof. z being generated by W, we have $w_j(z) = 0$, for each $j \in J$. Then

$$z \notin \{ x \in SI \mid w_1(x) + \ldots + w_n(x) \geq 1 \}.$$

Hence

(12.1.31) $\{x \in SI \mid w_1(x) + \ldots + w_n(x) \geq 1 \} \subseteq SI \setminus \{z\}$.

Let $x \in SI \setminus \{z\}$. Then it is $k \in J$ such that $w_k(x) \neq 0$. Because $w_k \in Y$ and $x \in SI \subseteq \mathbf{Z}^n$, we have $w_k(x) \in \mathbf{Z} \cap [0, +\infty[$. Hence $w_k(x) \geq 1$. On the other hand, from (12.1.4), we have $w_j(x) \geq 0$, for each $j \in J$. It follows that $w_1(x) + \ldots + w_n(x) \geq w_k(x) \geq 1$. Hence $x \in \{x \in SI \mid w_1(x) + \ldots + w_n(x) \geq 1\}$. Therefore

(12.1.32) $SI \setminus \{z\} \subseteq \{x \in SI \mid w_1(x) + \ldots + w_n(x) \geq 1\}$

From (12.1.31) and (12.1.32) it results (12.1.30).

Let $W = \{w_1, \ldots, w_n\}$ be a d-base of SI, with the associated matrix $D = [d_{kj}]$ ($k \in K$, $j \in J$) and the associated vector $d = (d_{10}, \ldots, d_{n+m,0})$. For each index $j \in \{1, \ldots, n\}$ we denote

(12.1.33) $I_j^+ = \{i \in \{1, \ldots, n + m\} \mid y_i \notin W \text{ and } d_{ij} > 0\}$,

(12.1.34) $t_j = \min\left\{\dfrac{d_{i0}}{d_{ij}} \mid i \in I_j^+\right\}$, if $I_j^+ \neq \varnothing$, and $t_j = +\infty$, if $I_j^+ = \varnothing$.

We put

(12.1.35) $J^1 = \{j \in \{1, \ldots, n\} \mid t_j \geq 1, t_j < +\infty\}$,

and, for each $j \in J$, we denote

(12.1.36) $\bar{t}_j = \begin{cases} [t_j], & \text{if } t_j \geq 1 \\ 1, & \text{if } t_j < 1 \end{cases}$.

Theorem 12.1.7. *If* $W = \{w_j \mid j \in \{1, \ldots, n\}\}$ *is a* d-*base of* SI, $D = (d_{kj})$ ($k \in K$; $j \in J$) *its associated matrix,* $d = (d_{10}, \ldots, d_{m0})$ *its associated vector,* x^0, *the point generated by W, is an* a-*vertex of* SI, *and* $J^1 \neq \varnothing$, *then the solution* x^r *(for each* $r \in J^1$*) of the system*

(12.1.37) $\begin{cases} w_h(x) = 0, & h \in \{1, \ldots, n\} \setminus \{r\} \\ w_r(x) = \bar{t}_r \end{cases}$

is also an a-*vertex of* SI.

Proof. Let $r \in J^1$, with $t_r < \infty$. First we show that x^r is a p-vertex of SI. We choose the index $s \in K$, such that

(12.1.38) $\quad d_{sr} > 0,$

and

(12.1.39) $\quad \dfrac{d_{s0}}{d_{sr}} = t_r.$

From (12.1.36) and (12.1.39) it follows that $\dfrac{d_{s0}}{d_{sr}} \geq 1$. Two cases are possible: $d_{sr} = 1$, and $d_{sr} \neq 1$.

Case 1: $d_{sr} = 1$. We prove that the set

$$W' = \{w_1, \ldots, w_{r-1}, y_s, w_{r+1}, \ldots, w_n\}$$

is a d-base for SI. In an analogous manner to that used in the proof of theorem 12.1.2 we can show that the functions $\hat{w}_1, \ldots, \hat{w}_{r-1}, \hat{y}_s, \hat{w}_{r+1}, \ldots, \hat{w}_n$ are linearly independent. It remains to prove that condition b) of definition of d-base is satisfied. From (12.1.5), we obtain

(12.1.40) $w_r = d_{s0} - (d_{s1} w_1 + \ldots + d_{s,r-1} w_{r-1} + y_s + d_{s,r+1} w_{r+1} + \ldots + d_{sn} w_n).$

If we replace w_r from (12.1.40) in (12.1.5), we obtain

(12.1.41) $y_i = d_{i0} - d_{ir} d_{s0} - \sum_{\substack{j=1 \\ j \neq r}}^{n} (d_{ij} - d_{ir} d_{sj}) w_j + d_{ir} y_s,$

for each $i \in K$. Because $d_{ij} \in \mathbf{Z}$, for all $i \in K, j \in J \cup \{0\}$, if we put

(12.1.42) $\quad \delta_{i0} = d_{i0} - d_{ir} d_{s0}$, for all $i \in K$,

(12.1.43) $\quad \delta_{ij} = d_{ij} - d_{ir} d_{sj}$, for all $i \in K, j \in J \setminus \{r\}$,

(12.1.44) $\quad \delta_{ir} = -d_{ir}$, for all $i \in K$,

it is easy to see that

(12.1.45) $\quad \delta_{ij} \in \mathbf{Z}$, for all $i \in K, j \in J \cup \{0\}$, and

(12.1.46) $\quad y_i = \delta_{i0} - (\sum_{\substack{j=1 \\ j \neq r}}^{n} \delta_{ij} w_j + \delta_{ir} y_s)$, for each $i \in K$.

Then, from (12.1.45) and (12.1.46), it follows that the b) condition of definition of the d-base is satisfied. Therefore the set W' is a d-base of SI. The vertex z, generated by W', is the unique solution of the sistem

$$\begin{cases} w_j(x) = 0, & j \in \{1,\dots,n\} \setminus \{r\} \\ y_s(x) = 0 \end{cases}.$$

Then we have $w_j(z) = 0$, for all $j \in \{1, \dots, n\} \setminus \{r\}$ and $y_s(z) = 0$. From 12.1.39 we get that $w_r(z) = d_{s0} = \frac{d_{s0}}{1} = \frac{d_{s0}}{d_{sr}} = t_r$. As $d_{s0} \in \mathbf{Z}$ and $r \in J^1$, it results that $\bar{t}_r = [t_r] = w_r(z)$. It follows that z is a solution of the system (12.1.37). Because the functions $\hat{w}_1,\dots,\hat{w}_{r-1},\hat{y}_s,\hat{w}_{r+1},\dots,\hat{w}_n$ are linearly independent, the system (12.1.37) has a unique solution, denoted by x^r. Then $z = x^r$. Therefore, x^r is a p-vertex of SI.

Now we prove that x^r is an a-vertex of SI. For this we apply theorem 12.1.6. We remark that:

- the associated vector of the d-base W' is the vector $(\delta_{10}, \dots, \delta_{1,n+m})$;

- as W is a d-base for SI and x^0, the point generated by W, is an a-vertex of SI, we have

(12.1.47) $\quad d_{i0} \geq 0$, for all $i \in K$.

Let $i \in K$. We have

(12.1.48) $\quad \delta_{i0} = d_{i0} - d_{ir}\, d_{s0}$.

Two cases may appear either $d_{ir} \leq 0$ or $d_{ir} > 0$.

If $d_{ir} \leq 0$, from (12.1.47) and (12.1.48), it results

(12.1.49) $\quad \delta_{i0} \geq d_{i0} \geq 0$.

If $d_{ir} > 0$, also two cases are possible: $y_i \notin W$, or $y_i \in W$.

If $y_i \notin W$, then $\frac{d_{i0}}{d_{ir}} \geq t_r = d_{s0}$. This implies that $\delta_{i0} = d_{i0} - d_{ir}d_{s0} \geq 0$

If $y_i = w_h \in W$, then we have $d_{ij} = 0$, for all $j \in (\{1, \dots, n\} \cup \{0\}) \setminus \{h\}$. On the other hand, as $y_i = w_h \in W$, we get $r \neq h$. It follows that

$$\delta_{i0} = 0 - 0 \cdot d_{s0} = 0.$$

Therefore, in all cases, $\delta_{i0} \geq 0$. In view of theorem 12.1.5, point x^r, generated by the d-base W', is an a-vertex of SI.

Case 2: $d_{sr} \neq 1$. From $d_{sr} > 0$, $d_{sr} \in \mathbf{Z}$, and $d_{sr} \neq 1$, we get that $d_{sr} > 1$. If we take $t = d_{sr}$, the hypotheses of theorem 12.1.2 are satisfied. Then it results that set $W' = (W \setminus \{w_r\}) \cup \{f\}$, where function f is given by (12.1.12), is a new d-base for set SI. Then the point z generated by the d-base W' is a p-vertex of SI. The associate vector of the d-base W' is the vector $(\delta_{10}, \ldots, \delta_{1,n+m})$, where δ_{i0} is given by (12.1.17) for all $i \in K$.

Let $i \in K$. We have

$$(12.1.50) \qquad \delta_{i0} = d_{i0} - d_{ir} \cdot \left[\frac{d_{s0}}{d_{sr}}\right], \text{ for all } i \in K,$$

where,

$$(12.1.51) \qquad d_{i0} \geq 0,$$

because W is a d-base for SI and x^0, the point generated by W, is an a-vertex of SI.

Two cases may appear: $d_{ir} \leq 0$, or $d_{ir} > 0$.

If $d_{ir} \leq 0$, from (12.1.50) and (12.1.51), it results

$$\delta_{i0} \geq d_{i0} \geq 0.$$

If $d_{ir} > 0$, also two cases are possible: $y_i \notin W$, or $y_i \in W$.

If $y_i \notin W$, then $\dfrac{d_{i0}}{d_{ir}} \geq t_r \geq [t_r] = \left[\dfrac{d_{s0}}{d_{sr}}\right]$. This implies that

$$\delta_{i0} = d_{i0} - d_{ir} \cdot \left[\frac{d_{s0}}{d_{sr}}\right] \geq 0.$$

If $y_i = w_h \in W$, then we have $d_{ij} = 0$, for all $j \in (\{1, \ldots, n\} \cup \{0\}) \setminus \{h\}$. On the other hand, as $y_i = w_h \in W$, we get $r \neq h$. It follows that

$$\delta_{i0} = 0 - 0 \cdot \left[\frac{d_{s0}}{d_{sr}}\right] = 0.$$

Therefore, in all cases, $\delta_{i0} \geq 0$. In view of theorem 12.1.5, point x^r, generated by the d-base W', is an a-vertex of SI.

Theorem 12.1.8. *If* $W = \{w_1, \ldots, w_n\}$ *is a d-base of* SI, *with the associated matrix* $D = [d_{kj}]$ $(k \in K, j \in J)$ *and the associated vector* $d = (d_{10}, \ldots, d_{n+m,0})$, *satisfies the conditions:*

i) *There is* $r \in J$ *such that* $d_{kr} \leq 0$ *for all* $k \in K$, *with at last one strict inequality;*

ii) $d_{k0} \geq 0$, *for all* $k \in K$,

then SI *is a non bounded set.*

Proof. Let m be a natural number, $m \geq 1$, and z^m the unique solution of the system

$$(12.1.52) \qquad \begin{cases} w_h(x) = 0, & h \in \{1,\dots,n\} \setminus \{r\} \\ w_r(x) = m \end{cases}.$$

Because we have

$$y_i(z^m) = d_{i0} - \sum_{j=1}^{n} d_{ij} w_j(z^m) = d_{i0} - m \cdot d_{ir} \geq d_{i0}, \text{ for all } i \in K,$$

and

$$y_i(z^m) = d_{i0} - m \cdot d_{ir} \in \mathbf{N}, \text{ for all } i \in K,$$

in view of the theorem 12.1.5, we obtain $z^m \in$ SI. But from condition i), it results that there is $k \in K$ such that $d_{kr} < 0$. As

$$\lim_{m \to \infty} y_k(z^m) = +\infty, \text{ and } z^m \in \text{SI}$$

for all natural numbers m, we get that SI is non bounded.

Corollary 12.1.2. *If* SI *is a bounded set,* $W = \{w_j \mid j \in \{1, \dots, n\}\}$ *is a* d-*base of* SI, $D = [d_{kj}]$ $(k \in K; j \in J)$ *its associated matrix,* $d = (d_{10}, \dots, d_{m+n,0})$ *its associated vector, and* x^0, *the point generated by* W, *is an a vertex of* SI, *then we have* $I_j^+ \neq 0$, *for all* $j \in J$, *where* I_j^+ *is given by* (12.1.30).

Proof. Suppose that there is $j \in J$ such that, $I_j^+ = \varnothing$. Let $i \in K$. Two cases may arise:

If $y_i \notin W$; then from definition of I_j^+, we have $d_{ij} \leq 0$;

If $y_i \in W$, then there is $h \in J$ such that $y_i = w_h = 0 - (-1)\, w_h$. It implies that we have also $d_{ij} \leq 0$.

In both cases we obtain $d_{ij} \leq 0$. We cannot have $d_{ij} = 0$, for all $i \in K$; this fact contradicts that the rank of matrix D is equal to $n = \max\{n, m\}$. It follows that there is $i \in K$ such that $d_{ij} < 0$. Applying theorem 12.1.8, we get that SI is a non-bounded set. This contradicts the hypothesis. Therefore we have $I_j^+ \neq 0$, for all $j \in J^1$.

12.2 Simplexes and p - vertices

Let SI be a discrete polyhedral set, $W = \{w_j \mid j \in \{1, ..., n\}\}$ a d-base of SI, $D = [d_{kj}]$ ($k \in K$; $j \in J$) the associated matrix of W, $d = (d_{10}, ..., d_{n+m,0})$ the associated vector of W and P^0 the point generated by W. There are n^2 natural numbers α_{jk}, ($k \in J, j \in J$), with

$$\text{(12.2.1)} \qquad \operatorname{rank}\begin{pmatrix} \alpha_{11} & \cdots & \alpha_{1n} \\ \cdots & \cdots & \cdots \\ \alpha_{n1} & \cdots & \alpha_{nn} \end{pmatrix} = n,$$

and there are n natural numbers β_j, ($j \in J$) such that, for all $j \in J$, we have

$$\text{(12.2.2)} \qquad w_j(x) = \beta_j - (\alpha_{j1} x_1 + ... + \alpha_{jn} x_n), \text{ for each } x = (x_1, ..., x_n) \in \mathbf{R}^n.$$

Let $\lambda = (\lambda_1, ..., \lambda_n) \in (\mathbf{R}^*)^n$, let $h_\lambda: \mathbf{R}^n \to \mathbf{R}$ be the function given by

$$\text{(12.2.3)} \qquad h_\lambda(x) = \sum_{j=1}^{n} \frac{w_j(x)}{\lambda_j} - 1,$$

for each $x \in \mathbf{R}^n$, and let

$$\text{(12.2.4)} \qquad Y_\lambda = \{ x \in \mathbf{R} \mid w_j(x) \geq 0, j \in J, \sum_{j \in J} \frac{w_j(x)}{\lambda_j} \leq 1 \}.$$

Obviously we have

$$\text{(12.2.5)} \qquad Y_\lambda = \{x \in \mathbf{R} \mid \sum_{j=1}^{n} \alpha_{kj} x_j \leq \beta_k , k \in J,$$

$$\sum_{j=1}^{n} (\sum_{k=1}^{n} \frac{\alpha_{kj}}{\lambda_k}) \cdot x_j \leq 1 - \sum_{k=1}^{n} \frac{\beta_k}{\lambda_k} \}.$$

Remark 12.2.1. From (12.2.1) it results that, for each $j \in J$, we have

$$(12.2.6)\qquad \text{rank}\begin{pmatrix} \alpha_{11} & \dots & \alpha_{1n} \\ \dots & \dots & \dots \\ \alpha_{j-1,1} & \dots & \alpha_{j-1,n} \\ \alpha_{j+1,1} & \dots & \alpha_{j+1,n} \\ \dots & \dots & \dots \\ \alpha_{n1} & \dots & \alpha_{nn} \\ \sum_{k=1}^{n}\frac{\alpha_{k1}}{\lambda_k} & \dots & \sum_{k=1}^{n}\frac{\alpha_{kn}}{\lambda_k} \end{pmatrix} = n.$$

Theorem 12.2.1. *If* SI *is a discrete polyhedral set,* $W = \{w_j \mid j \in \{1, ..., n\}\}$ *is a* d- *base of* SI *and* $\lambda = (\lambda_1, ..., \lambda_n) \in (\mathbf{R}^*)^n$, *then the set* Y_λ *is a simplex and its vertices are the points* $P^0 = (d_{10}, ..., d_{n0})$, *and* $P^j(\lambda), j \in J$, *where*

$$(12.2.7)\qquad P^j(\lambda) = (p_1^j(\lambda), ..., p_n^j(\lambda)) = (-d_{1j}\lambda_j + d_{10}, ..., -d_{nj}\lambda_j + d_{n0}).$$

Proof. Obviously, the set Y_λ is a polyhedral set. From (12.2.5) and (12.2.6) we get that Y_λ has n+1 vertices. A vertex, denoted by P^0, is the solution of the system

$$(12.2.8)\qquad \begin{cases} \sum_{j=1}^{n}\alpha_{1j}x_j = \beta_1 \\ \dots \\ \sum_{j=1}^{n}\alpha_{nj}x_j = \beta_n \end{cases},$$

that can be written as

$$(12.2.9)\qquad \begin{cases} w_1(x) = 0 \\ \dots \\ w_n(x) = 0 \end{cases}.$$

Hence P^0 is the p-vertex generated by d-base W. Therefore, from theorem 12.1.3 we have $P^0 = (d_{10}, ..., d_{n0})$.

Let now $j \in \{1, ..., n\}$. By $P^j(\lambda)$ we denote the vertex of Y_λ which is the solution of system

$$(12.2.10)\quad \begin{cases} \sum_{h=1}^{n} \alpha_{1h} x_h = \beta_1 \\ \qquad \ldots \\ \sum_{h=1}^{n} \alpha_{j-1,h} x_h = \beta_{j-1} \\ \sum_{h=1}^{n} \alpha_{j+1,h} x_h = \beta_{j+1} \\ \qquad \ldots \\ \sum_{h=1}^{n} \alpha_{nh} x_h = \beta_n \\ \sum_{h=1}^{n} (\sum_{k=1}^{n} \frac{\alpha_{kh}}{\lambda_k}) x_h = 1 - \sum_{k=1}^{n} \frac{\beta_k}{\lambda_k} \end{cases},$$

which can be written as

$$(12.2.11)\quad \begin{cases} w_1(x) = 0 \\ \qquad \ldots \\ w_{j-1}(x) = 0 \\ w_{j+1}(x) = 0 \\ \qquad \ldots \\ w_n(x) = 0 \\ \sum_{k=1}^{n} \frac{w_k(x)}{\lambda_k} = 1 \end{cases}.$$

From (12.2.11) we get that $w_i\,(P^j(\lambda)) = 0$, for each $i \in J \setminus \{j\}$, and $w_j(P^j(\lambda)) = \lambda_j$. Then, from (12.1.1) and (12.1.5), we have

$$p_k^j = y_k(P^j(\lambda)) = -\,d_{kj}\,\lambda_j + d_{k0}, \text{ for each } k \in \{1, \ldots, n\}.$$

Therefore, $P^j(\lambda) = (-d_{1j}\,\lambda_j + d_{10}, \ldots, -d_{nj}\,\lambda_j + d_{n0})$. We remark that

$$P^1(\lambda) - P^0 = (-d_{11}\,\lambda_1, \ldots, -d_{n1}\,\lambda_1),$$

$$\ldots,$$

$$P^n(\lambda) - P^0 = (-d_{1n}\,\lambda_n, \ldots, -d_{nn}\,\lambda_n),$$

Then, as W is a d- base and $\lambda_j \neq 0$ for each $j \in \{1, \ldots, n\}$, we get

$$(12.2.12)\quad \operatorname{rank}\begin{pmatrix} -d_{11}\cdot\lambda_1 & \dots & -d_{1n}\cdot\lambda_n \\ \dots & \dots & \dots \\ -d_{n1}\cdot\lambda_1 & \dots & -d_{nn}\cdot\lambda_n \end{pmatrix} = \operatorname{rank}\begin{pmatrix} d_{11} & \dots & d_{1n} \\ \dots & \dots & \dots \\ d_{n1} & \dots & d_{nn} \end{pmatrix} = n.$$

It follows that points $P^1(\lambda) - P^0, \dots, P^n(\lambda) - P^0$ are linearly independent. Therefore Y_λ is a simplex.

Let Ω be a convex, closed, nonempty subset of $\mathbf{R}^n$, with $SI \subseteq \Omega$. Suppose that x^0 is an a-vertex of SI, generated by d-base $W = \{w_j \mid j \in \{1, \dots, n\}\}$, the matrix $D = [d_{kj}]$ $(k \in K; j \in J)$ is the associated matrix of W, and $d = (d_{10}, \dots, d_{n+m,0})$ is the associated vector of W.

Let $t \in \mathbf{R}_+^*$, and $\lambda = (t, \dots, t)$. In what follows, for each $j \in J$, we consider point $P^j(t) = (p_1^j(t), \dots, p_n^j(t))$, where p_k^j is given by (12.2.7), for each $k \in \{1, \dots, n\}$. Obviously, we have

$$(12.2.13)\quad p_k^j(t) = d_{k0} - d_{kj}\, t, \text{ for each } k \in \{1, \dots, n\}.$$

For all $j \in J$, we denote

$$(12.2.14)\quad m_j = \sup\{t \geq 0 \mid P^j(t) \in \Omega\}.$$

Let

$$(12.2.15)\quad J^0 = \{j \in \{1, \dots, n\} \mid m_j = 0\},$$

and

$$(12.2.16)\quad J^\infty = \{j \in J \mid m_j = +\infty\}.$$

Theorem 12.2.2. *If* Ω *is a convex, closed, nonempty subset of* $\mathbf{R}^n$, $SI \subseteq \Omega$ *is a discrete polyhedral set,* $W = \{w_j \mid j \in \{1, \dots, n\}\}$ *is a* d- *base of* SI *and* $J^0 = \varnothing$, *then the set*

$$(12.2.17)\quad Y_\lambda = \{x \in \mathbf{R} \mid w_j(x) \geq 0, j \in J, \sum_{j\in J} \frac{w_j(x)}{\lambda_j} \leq 1\},$$

where $0 < \lambda_j \leq m_j$, *for all* $j \in \{1, \dots, n\}$, *is a subset of* Ω.

Proof. From theorem 12.2.1 it results that Y_λ is a simplex and its vertices are the points P^0, solution of system (12.2.9), and $P^j(\lambda)$, $j \in \{1, \dots, n\}$, where $P^j(\lambda)$ is the solution of system (12.2.11). Then it is easy to see that $P^0 \in Y_\lambda$, and $P^j(\lambda) \in Y_\lambda$, for each $j \in J$. Then, because Ω is a convex set, it follows

$$(12.2.18)\quad \operatorname{conv}(\{P^0, P^1(\lambda), \dots, P^n(\lambda)\}) \subseteq \Omega.$$

On the other hand we have

$$(12.2.19) \qquad Y_\lambda = \text{conv}(\{P^0, P^1(\lambda), ..., P^n(\lambda)\}).$$

From (12.2.18) and (12.2.19) we get $Y_\lambda \subseteq \Omega$.

Corollary 12.2.1. *If* Ω *is a convex, closed, nonempty subset of* $\mathbf{R}^n$, $SI \subseteq \Omega$ *is a discrete polyhedral set,* $W = \{w_j \mid j \in \{1, ..., n\}\}$ *is a* d*-base of* SI, *and* $J^0 = J^\infty = \varnothing$, *then*

$$(12.2.20) \qquad Y = \{x \in \mathbf{R} \mid w_j(x) \geq 0, j \in J, \sum_{j\in J} \frac{w_j(x)}{m_j} \leq 1\} \subseteq \Omega.$$

Proof. As $J^0 = J^\infty = \varnothing$, we get that $0 < m_j < +\infty$, for all $j \in J$. Applying theorem 12.2.2 with $\lambda = (m_1, ..., m_n)$, we obtain (12.2.20).

Corollary 12.2.2. *If* Ω *is a convex, closed, nonempty subset of* $\mathbf{R}^n$, $SI \subseteq \Omega$ *is a discrete polyhedral set,* $W = \{w_j \mid j \in \{1, ..., n\}\}$ *is a* d*-base of* SI, $J^0 = \varnothing$ *and* $J \setminus J^* \neq \varnothing$, *then*

$$(12.2.21) \; Y = \{x \in \mathbf{R} \mid w_j(x) \geq 0, j \in J, \sum_{j\in J\setminus J^*} \frac{w_j(x)}{m_j} < 1\} \subseteq \Omega.$$

Proof. Let t be a non-null natural number. We take $\lambda_j = t$, for each $j \in J^*$, and $\lambda_j = m_j$, for each $j \in J \setminus J^*$. In view of theorem 12.2.2, we have

$$(12.2.22) \qquad Y_\lambda = \{x \in \mathbf{R} \mid w_j(x) \geq 0, j \in J, \sum_{j\in J} \frac{w_j(x)}{\lambda_j} \leq 1\} \subseteq \Omega.$$

Let $x \in Y$, let $w = \max\{w_j(x) \mid j \in J^*\}$ and let $p = \text{card}(J^*)$. If we take

$$t = \left[\frac{p \cdot w}{1 - \sum_{j\in J^*} w_j(x)} \right] + 1,$$

then we have $x \in Y_\lambda$. From (12.2.22) we get that $x \in \Omega$. Hence $Y \subseteq \Omega$.

Corollary 12.2.3. *If* Ω *is a convex, closed, nonempty subset of* $\mathbf{R}^n$, $SI \subseteq \Omega$ *is a discrete polyhedral set,* $W = \{w_j \mid j \in \{1,...,n\}\}$ *is a* d*-base of* SI, $J^1 = J$, *then*

$$(12.2.23) \qquad \{x \in \mathbf{R}^n \mid w_j \geq 0, j \in J, \text{ and } \sum_{j=1}^{n} \frac{w_j(x)}{\bar{t}_j} \leq 1\} \subseteq \Omega$$

Proof. From theorem 12.1.7 we get that, for each $j \in J$, the solution of the system (12.1.37) is an element of SI. Then $x^j \in \Omega$, for each $j \in J$.

Therefore $\bar{t}_j \le m_j$, for each $j \in J$. Applying theorem 12.2.1 we get that (12.2.23) is true.

Let $W = \{w_1, ..., w_n\}$ a d-base of SI, with the associated matrix $D = [d_{kj}]$ ($k \in K$, $j \in J$) and the associated vector $d = (d_{10}, ..., d_{n+m,0})$, and let x^0 be the vertex generated by W, an a- vertex of SI. If (W, D, d) is irreducible, we put

(12.2.24) $J^- = \{j \in J \mid f(x^0) < f(x^j)\}$

and

(12.2.25) $J^+ = \{j \in J \mid f(x^0) > f(x^j)\}$,

where x^j is the solution of system (12.1.37).

Definition 12.2.1. *We say that the* a-vertex x^0 *has the* (MINL) *property if* $J^+ = \emptyset$ *and it has the* (MAXL) *property if* $J^- = \emptyset$.

Remark 12.2.2. If x^0 is a minimum point of f on SI, then x^0 has the (MINL) property; if x^0 is a maximum point of f on SI, then x^0 has the (MAXL) property.

Remark 12.2.3. From theorem 12.1.5 we have

(12.2.26) $x^0 = (d_{10}, ..., d_{n0})$, and $x^j = (d_{10} - d_{1j}\,\bar{t}_j, ..., d_{n0} - d_{nj}\,\bar{t}_j)$.

12.3 Characterisation of minimum (maximum) points of linear functions using d-bases

Let $f: \mathbf{R}^n \to \mathbf{R}$ be a linear function given by

$$f(x) = (c \mid x), \text{ for each } x \in \mathbf{R}^n,$$

where $c = (c_1, ..., c_n) \in \mathbf{Z}^n$, and let SI be the set of all solutions of the system

$$\begin{cases} \sum_{j=1}^{n} a_{ij}x_j \le b_i, & i \in \{1,...,m\} \\ x_j \in N, & j \in \{1,...,n\} \end{cases}.$$

If $W = \{w_1, ..., w_n\}$ is a d-base of SI with the associated matrix $D = [d_{kj}]$ ($k \in K$, $j \in J$) and the associated vector $d = (d_{10}, ..., d_{n+m,0})$, then, using (12.1.3), we can write

$$f(x) = \sum_{k=1}^{n} c_k y_k(x) = \sum_{k=1}^{n} c_k (d_{k0} - \sum_{j=1}^{n} d_{kj} w_j(x)) =$$

$$= \sum_{k=1}^{n} c_k d_{k0} - \sum_{j=1}^{n} \left(\sum_{k=1}^{n} c_k d_{kj} \right) w_j(x), \text{ for each } x \in \mathbf{R}^n.$$

We denote

$$\text{(12.3.1)} \qquad d_{0j} = \sum_{k=1}^{n} c_k d_{kj}, \text{ for each } j \in \{1, ..., n\}$$

and

$$\text{(12.3.2)} \qquad d_{00} = \sum_{k=1}^{n} c_k d_{k0}.$$

Then

$$\text{(12.3.3)} \qquad f(x) = d_{00} - \sum_{j=1}^{n} d_{0j} w_j(x), \text{ for each } x \in \mathbf{R}^n.$$

It is easy to see that numbers d_{0j}, $j \in \{0\} \cup J$, are unique. *For this reason vector* $d_f = (d_{01}, ..., d_{0n})$ *is said to be the associated vector of* d- *base* W *and of function* f. We remark that number d_{00} is the value of function f on the p-vertex generated by d-base W.

Remark 12.3.1. If W is a d-base of SI then, for each $j \in J$ we have $m_j = +\infty$, where m_j is given by (12.2.14) and $\Omega = \mathbf{R}^n$.

Theorem 12.3.1. *If* SI *is a discrete polytope,* x^0 *is an* a-*vertex of* SI, *generated by d-base* $W = \{w_1,...,w_n\}$, *such that* (W, D, d) *is irreducible, then* x^0 *is a minimum point of a linear function* f *on* SI *if and only if*

$$\text{(12.3.4)} \qquad d_{0j} \le 0, \textit{ for each } j \in J,$$

where $(d_{01}, ..., d_{0n})$ *is the associated vector of* d- *base* W *and of function* f.

Proof. We remark that we have $J^1 = J$, because SI is a finite set and (W, D, d) is irreducible.

Necessity. Let x^0 be a minimum point of f on SI. Then x^0 has the (MINL) property, i.e.

$$\text{(12.3.5)} \qquad f(x^0) \le f(x^k), \text{ for each } k \in J,$$

where $x^k = (d_{10},...,d_{k-1,0}, d_{k0} - \bar{t}_k d_{kk}, d_{k+1,0},...,d_{n0})$.

From (12.3.3.) and (12.3.5) it follows

$$\text{(12.3.6)} \qquad d_{00} - \sum_{j=1}^{n} d_{0j} w_j(x^0) \le d_{00} - \sum_{j=1}^{n} d_{0j} w_j(x^k), \text{ for each } k \in J.$$

Because $w_j(x^0) = 0$, and $w_j(x^k) = \bar{t}_k$, if $k = j$, and $w_j(x^k) = 0$, if $k \in J \setminus \{j\}$, from (12.3.6) we obtain

(12.3.7) $\quad 0 \leq d_{0k}\bar{t}_k$, for each $k \in J$.

Since x^0 is an a-vertex of SI, $\bar{t}_k \geq 0$, for each $k \in J$. Then, from (12.3.7), it results (12.3.4).

Sufficiency. Let (12.3.4) be true. Then (12.3.7) is also true. This implies that

(12.3.8) $\quad f(x^0) \leq f(x^k)$, for each $k \in J$.

Then $J^+ = \emptyset$. On the other hand, since SI is a finite set, it follows that there are n real numbers α_j, with $\alpha_j \geq \bar{t}_j$, $j \in J$, such that $SI \subseteq Y_\alpha$, where Y_α is given by (12.2.6). If $j \in J$, because $J^+ = \emptyset$, two cases may arise:

If $f(x^0) < f(x^k)$ then applying proposition 12.2.2, in view of remark 12.3.1, we obtain $f(P^k(\alpha_k)) \geq f(x^0)$.

If $f(x^0) = f(x^k)$ then, by linearity of f it follows that $f((P^k(\alpha_k)) = f(x^0)$. Therefore,

(12.3.9) $\quad f(P^0) \leq f(P^j(\alpha_j))$, for each $j \in J$.

Let $x^* \in SI$. Then $x^* \in Y_\alpha$. As Y_α is a simplex with vertices x^0, $P^j(\alpha)$, $j \in J$, in view of linearity of f, we get

(12.3.10) $\quad f(x^*) \geq \min \{f(x^0), f(P^j(\alpha), j \in J\}$

Then, from (12.3.9) and (12.3.10) it results

(12.3.11) $\quad f(x^*) \geq f(x^0)$.

As x^* was arbitrarily chosen in SI, it follows that x^0 is a minimum point of f on SI.

12.4 Characterisation of minimum (maximum) points of quasi-monotonic functions using d-bases

In this paragraph we give necessary, sufficient, and necessary and sufficient conditions for a point $z \in SI$ to be a local minimum point (or maximum) point of f on SI. We suppose that:

i) $\quad SI \subseteq \mathbf{R}^n$ is a discrete polytope;

ii) $\Omega \subseteq \mathbf{R}_+^n$ is a convex set with $SI \subseteq \Omega$;

f: $\Omega \to \mathbf{R}$ is a quasi-monotonic function, i.e. for all $x, y \in \Omega$ we have

$$\text{(12.4.1)} \qquad \min \{f(x), f(y)\} \le f(tx + (1-t)y) \le \max \{f(x), f(y)\},$$

for each $t \in [0, 1]$. We remark that, if we denote the convex hull of SI by S, i. e. S = conv (SI), then, because Ω is a convex set, we have $S \subseteq \Omega$.

We point that the notations used in this paragraph are taken from paragraphs 12.1 and 12.2.

Proposition 12.4.1. *If* SI *is a discrete polytope and function* f: $\Omega \to R$ *is quasi-monotonic, then there are two a-vertices* x *and* y *of* SI *such that*

$$f(x) = \min \{f(z) \mid z \in SI\} \le \max \{f(z) \mid z \in SI\} = f(y).$$

Proof. Because SI is a discrete polytope, set S = conv (SI) is a polytope. Therefore the minimum, respectively the maximum, of f on S is taken on a vertex x, respectively y, of S. As SI is a bounded set, in view of theorem 12.1.4, each vertex of S is an a-vertex of SI. Then x and y are vertices of SI and we have

$$f(x) = \min \{f(z) \mid z \in S\} \le \min \{f(x) \mid z \in SI\} \le \max \{f(z) \mid z \in SI\} \le$$

$$\le \max \{f(z) \mid z \in S\} = f(y).$$

But $x \in SI$ implies $f(x) \ge \min \{f(z) \mid z \in SI\}$ and $y \in SI$ implies

$$f(y) \le \max \{f(z) \mid z \in SI\}.$$

Hence $f(x) = \min \{f(z) \mid z \in SI\}$ and $f(y) = \max \{f(z) \mid z \in SI\}$.

Proposition 12.4.2. *If* $j \in \{1, ..., n\} \setminus J^0$, *then the following sentences are true*:

a) *If there is a* $t^0 > 0$ *such that* $f(P^0) < f(P^j(t^0))$, *then we have*

$$f(P^0) = \min \{f(P^j(t) \mid 0 \le t \le m_j\};$$

b) *If there is a* $t^0 > 0$ *such that* $f(P^0) > f(P^j(t^0))$, *then we have*

$$f(P^0) = \max \{f(P^j(t)) \mid 0 \le t \le m_j\}.$$

Proof. a) Suppose that $f(P^0) \ne \min \{f(P^j(t)) \mid 0 \le t \le m_j\}$. Then there is t, $0 < t \le m_j$ such that $f(P^j(t)) < f(P^0)$.

Two cases may obtain: $t \in]0, t^0[$ or $t \in]t^0, m_j[$.

If $0 < t < t^0$, then we have $P^j(t) \in [P^0, P^j(t^0)]$, and

$$f(P^j(t)) < f(P^0) = \min \{f(P^0), f(P^j(t^0))\}.$$

This contradicts the fact that f is a quasi-monotonic function.

If $0 < t^0 < t \leq m_j$, then we have $P^j(t^0) \in [P^0, P^j(t)]$, and

$$f(P^j(t^0)) > f(P^0) = \max\{f(P^0), f(P^j(t))\}.$$

This also contradicts the fact that f is a quasi-monotonic function. Hence

$$f(P^0) = \min\{f(P^j(t) \mid 0 \leq t \leq m_j\}.$$

Analogously we can prove statement (b).

Suppose that SI is a discrete polytope. Let α_j, $j \in J$, be real numbers with $0 < \alpha_j$, for each $j \in J$, and let

$$\text{(12.4.3)} \qquad Y_\alpha = \{x \in \mathbf{R}^n_+ \mid w_j(x) \geq 0, j \in J, \sum_{j \in J} \frac{w_j(x)}{\alpha_j} \leqq 1\}.$$

Proposition 12.4.3. *If* Ω *is a convex, closed, nonempty subset of* $\mathbf{R}^n$, *with* SI $\subseteq Y_\alpha$, f: $\Omega \to \mathbf{R}$ *is a quasi-monotone function,* $J^+ = \varnothing$, *and* x^0 *is an a-vertex of* SI *generated by* W, *with* (W, D, d) *irreducible, then* x^0 *is a minimum point of* f *in* SI.

Proof. Let $x^* \in$ SI. As SI $\subseteq Y_\alpha$, from remark 12.2.3 it results that there are n+1 nonnegative real numbers t_j, $j \in \{0, 1, ..., n\}$, with $t_0 + t_1 + ... + t_n = 1$, such that $x^* = \sum_{j=0}^{n} t_j P^j(\alpha_j)$. Since f is a quasi-monotone function, we get

$$\text{(12.4.3)} \qquad f(x^*) \geq \min\{f(P^0), f(P^j(\alpha_j)), j \in J\}.$$

But, as $J^+ = \varnothing$. Then

$$\text{(12.4.4)} \qquad f(P^0) \leq f(P^j(\alpha_j)), \text{ for each } j \in J.$$

From (12.4.3) and (12.4.4) it follows that $x^0 = P^0$ is a minimum point of f on SI.

12.5 Efficient points and a-vertices

Let SI be a discrete polytope and let $f = (f_1, ..., f_p)$: $\mathbf{R}^n \to \mathbf{R}^p$ a linear function given, for each $h \in \{1, ..., p\}$, by

$$\text{(12.5.1)} \qquad f_h(x) = (c^h \mid x), \text{ for each } x \in \mathbf{R}^n,$$

where $c^h = (c_1^h, ..., c_n^h) \in \mathbf{Z}^n$.

In this paragraph we study the following vectorial programming problem, denoted by (MIP):

$$\text{(MIP)} \qquad \begin{cases} f(x) \to v - \min \\ x \in SI \end{cases}$$

Definition 12.5.1. *We said that a point* z ∈ SI *is a solution of problem* (MIP) *it it is a min-efficient point of* f *on* SI, *i.e. there is no a point* y ∈ SI *such that*

$$(12.5.2) \qquad f(y) \leq f(z).$$

In what follows we denote by H = {1, ..., p} and we consider function $s:\mathbf{R}^n \to \mathbf{R}$, given by

$$(12.5.3) \qquad s(x) = \sum_{h=1}^{p} f_h(x), \text{ for each } x \in \mathbf{R}^n.$$

Theorem 12.5.1. *A point* z ∈ SI *is a min efficient point of* f *on* SI *if and only if the system*

$$(12.5.4) \qquad \begin{cases} f_h(x) \leqq f_h(z), & h \in H \\ s(x) \leqq s(z) - 1 \\ x \in SI \end{cases}$$

is inconsistent.

Proof. Necessity. Let y ∈ SI. Two cases may arise.

I) There is h ∈ H such that $f_h(z) < f_h(y)$; then y is not a solution of system (12.5.4);

II) We have

$$(12.5.5) \qquad f_h(y) \leqq f_h(z), \text{ for each } h \in H.$$

As z is a min-efficient point of f on SI, we get

$$(12.5.6) \qquad f_h(z) = f_h(y), \text{ for each } h \in H.$$

Then, from (12.5.3), it follows s(y) = s(z). Hence z is not a solution of system (12.5.4).

Since in both cases, y is not a solution of system (12.5.4) and because y was arbitrarily chosen in SI, we get that system (12.5.4) is inconsistent.

Sufficiency. Suppose that z is not a min-efficient point of f on SI. Then there is a point x ∈ SI such that

$$(12.5.7) \qquad f_h(x) \leqq f_h(z), \text{ for each } h \in H,$$

and there is i ∈ H such that

(12.5.8) $f_i(x) < f_i(z)$.

Then

(12.5.9) $$\sum_{h=1}^{p} f_h(x) < \sum_{h=1}^{p} f_h(z).$$

As x, z $\in \mathbf{Z}^n$, we get $f_h(x), f_h(z) \in \mathbf{Z}$, for each h $\in$ H. Then, from (12.5.9) it follows that

(12.5.10) $$s(x) = \sum_{h=1}^{p} f_h(x) \leq \sum_{h=1}^{p} f_h(z) - 1 = s(z) - 1.$$

Hence x is a solution of system (12.5.4). This contradicts the fact that system (12.5.4) is inconsistent. Therefore z is a min-efficient point for f on SI.

If W = $\{w_1, ..., w_n\}$ is a d-base of SI with the associated matrix D = $[d_{kj}]$ (k $\in$ K, j $\in$ J) and the associated vector d = $(d_{10}, ..., d_{n+m,0})$, then using (12.1.3) and a reasoning similar to that from paragraph 12.3, for each h $\in$ H, we can write

$$f_h(x) = \sum_{k=1}^{n} c_k^h d_{k0} - \sum_{j=1}^{n}\left(\sum_{k=1}^{n} c_k^h d_{kj}\right) w_j(x), \text{ for each } x \in \mathbf{R}^n.$$

We denote

(12.5.11) $$g_{hj} = \sum_{k=1}^{n} c_k^h d_{kj}, \text{ for each } j \in \{1, ..., n\}$$

and

(12.5.12) $$g_{h0} = \sum_{k=1}^{n} c_k^h d_{k0}.$$

Then

(12.5.13) $$f_h(x) = g_{h0} - \sum_{j=1}^{n} g_{hj} w_j(x), \text{ for each } x \in \mathbf{R}^n.$$

From (12.5.3) and (12.5.13) we obtain

(12.5.14) $$s(x) = \sum_{h=1}^{p} g_{h0} - \sum_{j=1}^{n}\left(\sum_{h=1}^{p} g_{hj}\right) \cdot w_j(x), \text{ for each } x \in \mathbf{R}^n.$$

If we denote

$$(12.5.15) \qquad e_0 = \sum_{h=1}^{p} g_{h0}$$

and

$$(12.5.16) \qquad e_j = \sum_{h=1}^{p} g_{hj}\ , \text{ for all } j \in J,$$

then we can write

$$(12.5.17) \qquad s(x) = e_0 - \sum_{j=1}^{n} e_j w_j(x)\ , \text{ for each } x \in \mathbf{R}^n.$$

Theorem 12.5.2. *If* z *is an* a-*vertex of* SI *generated by* d-*base* W, *then it is a solution of problem* (MIP) *if and only if the system*

$$(12.5.18) \qquad \begin{cases} -\sum_{j=1}^{n} d_{kj} \cdot t_j + d_{k0} \in N, \quad k \in K \\ -\sum_{j=1}^{n} g_{hj} \cdot t_j \leqq 0 \\ -\sum_{j=1}^{n} e_j \cdot t_j \leqq -1 \\ t_1, \ldots, t_n \in N \end{cases}$$

is inconsistent.

Proof. Necessity. If z is a solution of problem (MIP), z is a min-efficient point of f on SI. In virtue of theorem 12.5.1, it results that system (12.5.4) is inconsistent. Suppose that system (12.5.18) is consistent and let $t_1^0, \ldots, t_n^0$ a solution of it. As $W = \{w_1, \ldots, w_n\}$ is a d – base, the system

$$(12.5.19) \qquad \begin{cases} w_1(x) = t_1^0 \\ \ldots \\ w_n(x) = t_n^0 \end{cases}$$

has a solution x^0 and only one. Obviously

$$(12.5.20) \qquad w_j(x^0) = t_j^0\ , \text{ for each } j \in J.$$

From (12.1.1), (12.1.5), (12.5.19) and (12.5.20) we obtain

$$y_k(x^0) = - \sum_{j=1}^{n} d_{kj} w_j(x^0) + d_{k0} = - \sum_{j=1}^{n} d_{kj} t_j^0 + d_{k0} \in \mathbf{N},$$

for each $k \in K$. Then, in view of theorem 12.1.5, $x^0 \in SI$. Also, as $w_j(z) = 0$ for all $j \in J$, from (12.5.13), (12.5.17) and (12.5.18), it results

$$f_h(x^0) - f_h(z) = -\sum_{j=1}^{n} g_{hj}(w_j(x^0) - w_j(z)) = -\sum_{j=1}^{n} g_{hj} \cdot t_j^0 \leqq 0,$$

for all $h \in H$, and $s(x^0) - s(z) = -\sum_{j=1}^{n} e_j(w_j(x^0) - w_j(z)) = -\sum_{j=1}^{n} e_j t_j^0 \leqq -1.$

It follows that x^0 is a solution of system (12.5.4), contradiction with the inconsistency of (12.5.4). Therefore system (12.5.18) is inconsistent.

Sufficiency. Let (12.5.18) be inconsistent. Suppose that z is not a solution of the problem (MIP). Then z is not a min-efficient point of f on SI, and, in virtue of theorem 12.5.1, we get that system 12.5.4 is compatible. Let x^0 be a solution of system (12.5.4). Then, because z is generated by d-base W, from (12.5.13) and (12.5.17) we get

$$\text{(12.5.21)} \quad -\sum_{j=1}^{n} g_{hj} w_j(x^0) = -\sum_{j=1}^{n} g_{hj}(w_j(x^0) - w_j(z)) =$$

$$= f_h(x^0) - f_h(z) \leqq 0, \text{ for each } h \in H,$$

and

$$\text{(12.5.22)} \quad -\sum_{j=1}^{n} e_j w_j(x^0) = -\sum_{j=1}^{n} e_j(w_j(x^0) - w_j(z)) =$$

$$= s(x^0) - s(z) \leq -1.$$

Let

$$\text{(12.5.23)} \quad t_j^0 = w_j(x^0), j \in J.$$

As $x^0 \in SI$, $w_j(x^0) \in \mathbf{N}$, for each $j \in J$. Hence

$$\text{(12.5.24)} \quad t_j^0 \in \mathbf{N}, \text{ for each } j \in J.$$

Also, in view of theorem 12.1.5, because $x^0 \in SI$, we have

$$\text{(12.5.25)} \quad -\sum_{j=1}^{n} d_{kj} t_j^0 + d_{k0} = -\sum_{j=1}^{n} d_{kj} w_j(x^0) + d_{k0} \in \mathbf{N},$$

for each $k \in K$. On the other hand, from (12.5.21) and (12.5.22) we obtain

(13.5.26) $$- \sum_{j=1}^{n} g_{hj} t_j^0 = - \sum_{j=1}^{n} g_{hj} w_j(x^0) \leqq 0, \text{ for each } h \in H,$$

and

(12.5.27) $$- \sum_{j=1}^{n} e_j t_j^0 = - \sum_{j=1}^{n} e_j w_j(x^0) \leqq -1.$$

Then (12.5.24) – (12.5.27) imply that $t_1^0, \ldots, t_n^0$ is a solution of system (12.5.18). This contradicts the inconsistency of (12.5.18). Therefore z is a solution of problem (MIP).

Remark 12.5.1. As d_{kj}, $g_{hj} \in \mathbf{Z}$, for all $k \in K$ and all $j \in J \cup \{0\}$, and $e_j \in \mathbf{Z}$ for each $j \in J$, the system (12.5.18) is equivalent to

$$\begin{cases} -\sum_{j=1}^{n} d_{kj} \cdot t_j + d_{k0} \geqq 0, \quad k \in K \\ -\sum_{j=1}^{n} g_{hj} \cdot t_j \leqq 0 \\ -\sum_{j=1}^{n} e_j \cdot t_j \leqq -1 \\ t_1, \ldots, t_n \in N \end{cases}$$

Corollary 12.5.1. *If* z *is an a-vertex of* SI, *generated by* d-*base* W, *and there is* $h \in H$ *such that*

(12.5.28) $g_{hj} < 0$, *for each* $j \in J$,

where g_{hj}, $j \in J$, *is given by* (12.5.11), *then* x^0 *is a solution of problem* (MIP).

Proof. Suppose that x^0 is not a solution of problem (MIP). Applying theorem 12.5.2 we get that system (12.5.18) is consistent. Let $t_1^0, \ldots, t_n^0$ a solution of (12.5.18). Then

(12.5.29) $$- \sum_{j=1}^{n} g_{1j} t_j^0 \leqq 0.$$

As $g_{hj} < 0$ and $t_j^0 \geqq 0$, for each $j \in J$, (12.5.29) imply $t_j^0 = 0$, for each $j \in J$. Then $\sum_{j=1}^{n} e_j t_j^0 = 0$. This contradicts the fact that $\sum_{j=1}^{n} e_j t_j^0 \leqq -1$. Therefore x^0 is a solution of (MIP).

In what follows we set

(12.5.30) $J_0 = J$, and $J_h = \{j \in J_{h-1} \mid g_{hj} = 0\}$, for each $h \in H$.

Corollary 12.5.2. *If* x^0 *is an* a-*vertex of* SI, *generated by* d-*base* W, *and if for each* $h \in H$ *we have* $g_{hj} < 0$, *for all* $j \in (J_{h-1} \setminus J_h)$ *then* x^0 *is a solution of problem* (MIP).

Proof. First we remark that we have

(12.5.31) $g_{hj} \leq 0$, for each $j \in J$ and each $h \in H$.

Let be $h \in H$. Two cases may obtain

a) $J_1 = \varnothing$. Then $g_{1j} < 0$, for each $j \in J$. By corollary 12.5.1, for $h = 1$, it follows that x^0 is a solution of (MIP).

b) $J_1 \neq \varnothing$. Then we have

(12.5.32) $g_{1j} = 0$, if $j \in J_1$, and $g_{1j} < 0$, if $j \in J \setminus J_1$.

Suppose that x^0 is not a solution of (MIP). In view of theorem 12.5.2, system (12.5.18) is consistent. Let $t_1^0, \ldots, t_n^0$ a solution of (12.5.18). Then

$$- \sum_{j=1}^{n} g_{1j} t_j^0 \leq 0,$$

and, from (12.5.31) it follows

(12.5.33) $t_j^0 = 0$, for each $j \in J \setminus J_1$.

Then

$$(12.5.34) \qquad - \sum_{j \in J_1} g_{hj} \cdot t_j^0 = \sum_{j=1}^{n} g_{hj} \cdot t_j^0 \leq 0, \text{ for all } h \in H,$$

and

$$(12.5.35) \qquad - \sum_{j \in J_1} e_j \cdot t_j^0 = \sum_{j=1}^{n} e_j \cdot t_j^0 \leq -1.$$

Let now $q \in J$. Two cases may arise.

b1) There is $h \in \{1, \ldots, p-1\}$ such that $q \in J_h$ and $q \notin J_{h+1}$. Then

(12.5.36) $g_{h+1,q} < 0$.

From (12.5.31), (12.5.34) and (12.3.36), we obtain

$$0 \geq - \sum_{j \in J_1} g_{hj} \cdot t_j^0 \geq - g_{h+1,q} \cdot t_q^0 \geq 0.$$

This implies that

$$(12.5.37) \qquad t_q^0 = 0.$$

b2) $q \in J_h$, for all $h \in \{2, ..., p\}$. Then $g_{hq} = 0$, for each $h \in H$. It follows that $e_q = 0$.

We conclude that, if $q \in J$, then $t_q^0 = 0$, or $e_q = 0$. Therefore $e_q t_q^0 = 0$, for each $q \in J$. This implies that

$$(12.5.38) \qquad \sum_{j=1}^{n} e_j \cdot t_j^0 = 0.$$

But (12.5.38) contradicts the fact that $t_1^0,...,t_n^0$ is a solution of (12.5.18). Then the hypothesis that x^0 is not a solution of (MIP) is false.

Therefore, in both cases a) and b) we conclude that x^0 is a solution of problem (MIP).

Theorem 12.5.3. *If* $W = (w_1, ..., w_n)$ *is a* d-*base of* SI, *point* x^0, *generated by* W, *is an* a-*vertex of* SI *and if there is* $q \in J$ *such that*

$$(12.5.39) \qquad d_{kq} \leq 0, \textit{for all } k \in K,$$

$$(12.5.40) \qquad g_{hq} \geq 0, \textit{for all } h \in H$$

and

$$(12.5.41) \qquad e_q > 0,$$

where $D = [d_{kj}]$ $(k \in K, j \in J)$ is *the associated matrix* of W, *the numbers* g_{hk}, $(h \in H, j \in J)$ *are given by* (12.5.11) *and* e_j $(j \in J)$, *by* (12.5.16), *then the problem* (MIP) *has not solution, i.e. there are no min-efficient points of* f *on* SI.

Proof. Let $m \in N^*$. In the proof of theorem 12.1.8 we showed that solution z^m of system

$$(12.5.42) \qquad \begin{cases} w_j(x) = 0, & j \in J \setminus \{q\} \\ w_q(x) = m \end{cases}$$

is an element of SI. Since

$$(12.5.43) \qquad f_h(z^m) = - \sum_{j=1}^{n} g_{hj} \cdot w_j(z^m) + g_{h0} = - g_{hq} \cdot m + g_{h0} \leq g_{h0} = f_h(x^0),$$

for each $h \in H$, and

$$(12.5.44) \quad s(z^m) = -\sum_{j=1}^{n} e_j w_j(z^m) + e_0 = -e_q \cdot m + e_0 < e_0 = s(x^0).$$

As $s(z^m) \in \mathbf{Z}$ and $s(x^0) \in \mathbf{Z}$, the inequality (12.5.44) is equivalent to

$$(12.5.45) \quad s(z^m) \leq s(x^0) - 1.$$

From (12.5.44) and (12.5.45) we get that the point z^m is a solution of system (12.5.4). In view of theorem 12.5.1 it follows that x^0 is not a min-efficient point of f on SI.

Let $z \in$ SI. For each $k \in K$, in view of linearity of function w_j, $j \in J$, and the definition of point z^m, we can write

$$y_k(z + z^m) = -\sum_{j=1}^{n} d_{kj} w_j(z + z^m) + d_{k0} =$$

$$= -\sum_{j=1}^{n} d_{kj} w_j(z) + d_{k0} - d_{kq}\, m = y_k(z) - d_{kq}\, m.$$

Let $k \in K$. As $z \in$ SI, in view of theorem 12.1.5, we have $y_k(z) \in \mathbf{N}$. From definition of d-base, we get that $d_{kq} \in \mathbf{Z}$, and from hypothesis of theorem we have $d_{kq} \leq 0$. Then $d_{kq} \in \mathbf{N}$, for all $k \in K$. It follows that $y_k(z + z^m) \in \mathbf{N}$, for each $k \in K$. Applying theorem 12.1.5, we get that $z + z^m \in$ SI. We have

$$(12.5.46) \quad f_h(z + z^m) = -\sum_{j=1}^{n} g_{hj} \cdot w_j(z + z^m) + g_{h0} =$$

$$= -\sum_{j=1}^{g} g_{hj} w_j(z) + g_{h0} - g_{hq}\, m = f_h(z) - g_{hq}\, m \leq f_h(z), \text{ for each } h \in H,$$

As $-e_q \cdot m \leq -1$, it follows

$$(12.5.47) \quad s(z + z^m) = -\sum_{j=1}^{n} e_j w_j(z + z^m) + e_0 =$$

$$= -\sum_{j=1}^{n} e_j w_j(z) + e_0 - e_q\, m = s(z) - e_q\, m \leq s(z) - 1.$$

It results that $z + z^m$ is a solution of system (12.5.42). Then, in view of theorem 12.5.1, we get that z is not a min-efficient point of f on SI. Because $z \in$ SI was arbitrarily chosen, it follows that there is no min-efficient point of f on SI. Therefore the problem (MIP) has not solution.

Chapter 13

Applications in pharmaco-economics

In the previous chapter we studied the optimization problems in which the set of the admissible solutions is integer convex. In this chapter we intend to present some concrete applications of the same type of multi-criteria optimization problems in a very recent research domain, called pharmaco-economics.

Pharmaco-economic research identifies, measures and compares the costs (resources consumed) and the outcome (effectiveness, quality of life, utility, benefit, mortality, morbidity) of pharmaceutical products. Pharmaco-economics studies consider a treatment in terms of its results and its costs by means of several analyses such as: cost–effectiveness analysis, cost–benefit analysis, cost–efficacy analysis, cost minimization analysis, cost–utility analysis. The most frequent analysis is cost–efficacy analysis and cost–benefit analysis. In both cases we calculate the value of the ratio

$$r = (C_2 - C_1)/(Q_2 - Q_1)$$

where C_1, C_2 are the costs of two treatments expressed in the same monetary units, and Q_1, Q_2 represent a certain type of medical result obtained after the treatment. The result can be expressed either in natural units or in monetary units. We notice that the cost–effectiveness and the cost–utility ratios enable us to compare two treatments taking into account only one aspect of the treatment at a certain point.

The method presented in this chapter aims to construct a medico-economic index for the characterization of the medical treatments from different points of view. Also, we present a method to generate a preference relation between drugs belonging to the same group. From a mathematical point of view, the method deals with splitting a set of points in integer convex subsets. Each integer convex subset contains points, which are dominated with respect to the preference relation generating it.

The methods described in this chapter hide an ethical debate on the methodology of approaching a practical problem within an interdisciplinary research. The method offering a min-efficient drug from a class of drugs is a dictatorial one, leading a very limited information on the best drug, according to some criteria. A preference relation in a set of drugs allows an enlargement of the set of criteria within an interactive approach, taking into account the particularities of the patient, that might not be able to tolerate the min-efficient drug, but another one, which is more preferred that others.

The same mathematical methods might be used in sociology, to solve problems related to efficiency evaluation of a social policy within a population.

13.1 Algorithm for determining min-efficient points

Let A be a non-void set and let $f = (f_1, f_2, \ldots, f_p)$: $A \to \mathbf{R}^p$ be a vectorial function. A point $a \in A$ is said to be a min–efficient point of f on A if there is no $x \in A$ such that

(13.1.1) $$f_i(x) \le f_i(a) \text{ for all } i \in \{1, 2, \ldots, p\}$$

and

(13.1.2) $$\sum_{i=1}^{p} f_i(x) < \sum_{i=1}^{p} f_i(a).$$

It is known that (see E.A. Galperin (1992))

Theorem 13.1.1. *If* $\lambda_1 > 0, \ldots, \lambda_p > 0$ *are* p *given real numbers, then every minimum point of function* $F: A \to \mathbf{R}$ *defined by* $F(x) = \sum_{j=1}^{p} \lambda_j f_j(x)$ *for all* $x \in A$, *is a min–efficient point of* f *on* A.

In what follows we suppose that function f is bounded on A. We denote by $m_i = \inf\{f_i(x) \mid x \in A\}$ for all $i \in \{1, 2, \ldots, p\}$. Let $s = (s_1, \ldots, s_p) \in \mathbf{R}^p$. As Luciana Lupşa (1999), we shall use the following terminology.

Definition 13.1.1. *A point* $x \in A$ *is called a* s - *balanced point of* f *on* A *if it is a solution of the system*

(13.1.3) $f_i(x) \leq m_i + s_i \quad i = 1, 2, \ldots p$.

Consider $u = (u_1, u_2, \ldots, u_p)$ and $v = (v_1, v_2, \ldots, v_p)$ be two points of $\mathbf{R}^p$ and the order relation $u \leq v$ if $u_i \leq v_i$ for all $i \in \{1, 2, \ldots, p\}$ and $\sum_{i=1}^{p} u_i < \sum_{i=1}^{p} v_i$. Let $v = (v_1, v_2, \ldots, v_p) \in \mathbf{R}^p$ be such that (13.1.3) is compatible for $s = v$.

Theorem 13.1.2: *If there is no* $r \in \mathbf{R}^p$, $r \leq v$, *so that the system* (13.1.3) *is compatible for* $s = r$, *then every solution of the system* (13.1.3) *is a min–efficient point of* f *on* A.

Proof.: Let x^0 be a solution of system (13.1.3). Suppose that x^0 is not a min–efficient solution of f on A. Then there is $x \in A$ such that $f_i(x) \leq f_i(a)$ for all $i \in \{1, 2, \ldots, p\}$ and $\sum_{i=1}^{p} f_i(x) < \sum_{i=1}^{p} f_i(x^0)$. If we take $r = v - f(x^0) + f(x)$, then $r \leq v$. As x^0 is a solution of system (13.1.3) for $s = v$, we obtain

$$f_i(x) = f_i(x^0) + r_i - v_i \leq m_i + v_i + r_i - v_i = m_i + r_i$$

for all $i \in \{1, 2, \ldots, p\}$. Hence system (13.1.3) is compatible for $r \in \mathbf{R}^p$, with $r \leq v$, which is a contradiction with the hypothesis. Hence x^0 is a min–efficient solution.

For the case of a finite set A, the previous theorem leads to a simple algorithm to generate all the min-efficient points. Suppose that $A = \{a_1, a_2, \ldots, a_p\}$ be a nonempty finite set and let $f = (f_1, f_2, \ldots, f_p): A \to \mathbf{R}^p$. For all $i \in \{1, 2, \ldots, p\}$ we sort the numbers $f_i(a_1), f_i(a_2), \ldots, f_i(a_m)$ as an increasing sequence. Let i_k be the position of element $f_i(a_k)$ in the sorted sequence. We establish the convention that if a_k and a_h are distinct elements then

i) if $f_i(a_k) \leq f_i(a_h)$ for all $i \in \{1, 2, \ldots, p\}$ and $\sum_{i=1}^{p} f_i(a_k) < \sum_{i=1}^{p} f_i(a_h)$, then $i_k < i_h$;

ii) if $f_i(a_k) \leq f_i(a_h)$ for all $i \in \{1, 2, \ldots, p\}$ then we have $i_k < i_h$ for all $i \in \{1, 2, \ldots, p\}$, or $i_k > i_h$ for all $i \in \{1, 2, \ldots, p\}$.

Let be $j_0 = \min \{j \in \{1, 2, \ldots, m\} \mid \exists\, k \in \{1, 2, \ldots, m\}, i_k \leq j, i \in \{1, 2, \ldots, p\}\}$ and let $\gamma \in \{1, 2, \ldots, m\}$ be so that $i_\gamma \leq j_0$ for all $i \in \{1, 2, \ldots, p\}$.

Corollary 13.1.1. *The point* a_γ *is a min–efficient point of* f *on* A.

Corollary 13.1.2. *If* h $\in$ {1, 2, ..., m} *has the property that* $i_h \geq j_0$ *for all* i $\in$ {1, 2, ..., p} *and at least one inequality is strict, then* a_h *is not a min–efficient point of* f *on* A.

We shall give an algorithm based on theorem 13.1.2 and corollaries 13.1.1 and 13.1.2 for the determination of all min-efficient elements of the set A.

Algorithm 13.1.1. Let A = $\{a_1, ..., a_m\}$ and K = {1, ..., m}.

Step 0: We put E = $\emptyset$.

Step 1: We take $M_1 = \emptyset, ..., M_p = \emptyset$.

For all i $\in$ {1, 2, ..., p} we sort the numbers $f_i(a_k)$, k $\in$ K, as an increasing sequence. Let i_k be the position of the element $f_i(a_k)$ in the sorted sequence respecting the convention.

Step 2: j := 1

Step 3: i := 1

Step 4: $M_i := M_i \cup \{k\}$ where k $\in$ K and $i_k = j$.

Step 5: If $\bigcap_{i=1}^{p} M_i = \varphi$, then we continue with step 6, otherwise we continue with step 8.

Step 6: i := i + 1.

If $i \leq p$, then return to step 4.

Step 7: j: = j + 1 and return to step 3.

Step 8: $E := E \cup \{a_k\}$, $j_0 := j$, $\gamma := k$.

Step 9: For every i $\in$ {1, 2, ..., p} we construct the set

$$S_i := \{k \in K \mid i_k \geq j_0\}.$$

Step 10: We put $S := \bigcap_{i=1}^{p} S_i$ and for each i $\in$ {1, 2, ..., p} we take

$$M_i := M_i \setminus (S \cup \{\gamma\}).$$

Step 11: If $M_i = \emptyset$ for all i $\in$ {1, 2, ..., p} then we stop. Otherwise, we put $A := A \setminus \{a_k \mid k \in S \cup \{\gamma\}\}$, $K := K \setminus (S \cup \{\gamma\})$ and return to step 1.

Remark: All elements in E are min–efficient points.

13.2 Application of min-efficient points to construct a medico–economic effectiveness index which characterises a vaccine

For testing a vaccine two groups of people are used. One group is vaccinated, the other is not vaccinated. It is known that out of the vaccinated people:

p_{VN} % have no disease.

Out of the vaccinated people who have the disease:

p_{VD} % died;

p_{VR} % have complete recovery;

p_{VC} % have complications.

Out of the people who are in the control group:

p_N % have no disease;

p_D % died;

p_R % have the disease with complete recovery;

p_C % have the disease with complications.

In the case of vaccinated people p_{VA} % have negative reactions to the vaccine.

The following costs are known:

c_V for the vaccine/ person;

c_R for the treatment/ person in the case of complete recovery;

c_C for the treatment/ person in the case of complications or death;

c_A for the treatment/ person in the case of negative reactions.

Our purpose is to have disease and death cases as less as it is possible and to minimize the total amount of money spent. Knowing this data we must establish whether the vaccine is pharmaco-economically efficient. A method to answer this question is to attach a mathematical model to this problem – a multi-criteria programming problem with variables 0 and 1. Let us introduce the variables x_1 and x_2 that can be only 0 and 1:

$x_1 = 1$ means that the vaccine is used;

$x_1 = 0$ means that the vaccine is not used;

$x_2 = 1$ means that no vaccination is preferred;

$x_2 = 0$ means that the vaccination is preferred.

We have $x_1 + x_2 = 1$, because we can accept or reject the vaccination. The objective functions will be

$$f_1 : \{0,1\} \to R, \quad f_1(x_1, x_2) = (1 - p_{VN})x_1 + (1 - p_N)x_2 ;$$

$$f_2 : \{0,1\} \to R, \quad f_2(x_1, x_2) = p_{VD}x_1 + p_D x_2 ;$$

$$f_3 : \{0,1\} \to R, \quad f_3(x_1, x_2) = x_1 + \frac{\alpha_N}{\alpha_V} x_2$$

where $\alpha_N = p_R c_R + (p_D + p_C)c_C$ and

$$\alpha_V = c_V 100 + p_{VR}c_R + p_{VA}c_A + (p_{VD} + p_{VC})c_C .$$

For solving this problem we can apply the weight method, and the corresponding problem is:

$$f(x_1, x_2) = \lambda_1(-p_{VN}x_1 - p_N x_2) +$$
$$+ \lambda_2(p_{VD} + p_D x_2) + \lambda_3(x_1 + \frac{\alpha_N}{\alpha_V} x_2) \to min$$

subject to

$$x_1 + x_2 = 1, \; x_1, x_2 \in \{0,1\} .$$

We have

$$f(0,1) = -\lambda_1 p_N + \lambda_2 p_D + \frac{\alpha_N}{\alpha_V}\lambda_3 ,$$

$$f(1,0) = -\lambda_1 p_{VN} + \lambda_2 p_{VD} + \lambda_3 ,$$

$$f(1,0) - f(0,1) = \lambda_1(p_N - p_{VN}) + \lambda_2(p_{VD} - p_D) + \lambda_3(1 - \frac{\alpha_N}{\alpha_V}) .$$

If $f(1,0) - f(0,1) \geq 0$ then the vaccination is profitable. This result allows us to define a medico–economic effectiveness index:

$$imev = \lambda_1(p_N - p_V) + \lambda_2(p_{VD} - p_{ND}) + \lambda_3(1 - \frac{\alpha_N}{\alpha_V}).$$

The *imev* index that we introduced above can be used to compare more vaccines that are equivalent for the same disease, the vaccine with the greatest *imev* being the best.

13.3 Application of min-efficient points to choose the best medico-economic drug

For the treatment of the osteoartritis, one or two of the drugs from the next table are used:

Drug no	Drug name	p_a	p_u	c_t
1	Ibuprofen	14.55	12.9	200.031
2	Nabumetone	9.23	0.8	217.098
3	Piroxicam Gel	1.0	0.1	223.748
4	Indometacin	27.54	17.3	295.430
5	Naproxen	11.22	8.8	327.372
6	Piroxicam	3.78	12.0	343.380
7	Ibuprofen + Misoprostol	35.48	2.0	378.922
8	Diclofenac	24.05	7.5	388.459
9	Fenoprofen	44.25	17.6	405.288
10	Naproxen + Helidac	34.56	4.0	408.091
11	Sulindac	35.40	15.3	433.906
12	Aspirin	8.93	36.7	457.389
13	Etodolac	12.20	14.7	484.315
14	Diclofenac + Misoprostol	33.65	1.9	582.776
15	Ketoprofen	38.74	28	656.836
16	Flurbiprofen	18.37	26.5	651.145

Table 13.3.1. Performance parameters of drugs for osteoartritis (according to J.L. Shreve, J. Van Den Bos, K. Pedlow (2000))

The use of any drug has also secondary effects, the most frequent in our case being the ulcer. A statistical study shows that for any of the 16 medicaments is known the percentage p_u of the people who had ulcer and p_a

the percentage of the people with other side effects than ulcer. We also know the medium cost c_u of the treatment/ patient for an ulcer patient, c_a the medium cost of the treatment/ patient for a patient with other side effects than ulcer, and the cost c_m of the drug administrated to a patient. With these data the problem is to choose the treatment which should be applied so that the cost for side effects is as low as possible, the chance to have ulcer be minimum and the total cost $c_t = c_m + p_u c_u + p_a c_a$ is the lowest.

According to the reasoning used in the first case, we introduce the variables $x_1,\dots,x_{16}$ which can take only the values 0 and 1, with the meaning $x_i=1$ if the drug i is chosen and $x_i=0$ in the contrary case. We obtain the following vectorial programming problem.

$$\text{(PV)} \qquad f(x) = \begin{pmatrix} f_1(x) = \sum_{i=1}^{16} (p_{ui} + p_{ai}) x_c \\ f_2(x) = \sum_{i=1}^{16} p_{ui} x_i \\ f_3(x) = \sum_{i=1}^{16} c_{ti} x_i \end{pmatrix} \to v-\min$$

subject to

$$x_1 + \dots + x_{16} = 1,$$
$$x_i \in \{0,1\}, \quad i \in \{1,2,\dots,16\}.$$

For solving this problem we can apply the weight method, and the corresponding problem is:

$$\lambda_1 \sum_{i=1}^{16} (p_{ai} + p_{ui}) x_i + \lambda_2 \sum_{i=1}^{16} p_{ui} x_i + \lambda_3 \sum_{i=1}^{16} c_{ti} x_i \to \min$$

subject to

$$x_1 + \dots + x_{16} = 1,$$
$$x_i \in \{0,1\}, \quad i \in \{1,\dots,16\}.$$

The objective function can be written as

$$\sum_{i=1}^{16} (\lambda_1 (p_{ai} + p_{ui}) + \lambda_2 p_{ui} + \lambda_3 c_{ti}) x_i .$$

Comparing the coefficients p_{ai}, p_{ui} and c_{ti}, i=1,2,..., 16 we can easily see that the minimum of the function is obtained if $x_4=...=x_{16}=0$; so we can reduce the problem to the following problem:

$$\text{(PP)} \quad \begin{aligned} &(27.45\lambda_1 + 12.9\lambda_2 + 200.031\lambda_3)x_1 + \\ &+(10.03\lambda_1 + 0.8\lambda_2 + 217.098\lambda_3)x_2 + \\ &(1.1\lambda_1 + 0.1\lambda_2 + 223.748\lambda_3)x_3 \to min \\ &\text{subject to} \\ &x_1 + x_2 + x_3 = 1, \\ &x_1, x_2, x_3 \in \{0,1\}. \end{aligned}$$

For different choices of the weights $\lambda_1 > 0$, $\lambda_2 > 0$, and $\lambda_3 > 0$ we obtain ($x_1=1$, $x_2=0$, $x_3=0$) or ($x_1=0$, $x_2=0$, $x_3=0$) or ($x_1=0$, $x_2=0$, $x_3=1$) as solution of the problem (PP). Therefore, depending on the weights $\lambda_1, \lambda_2, \lambda_3$ (the importance allotted to each criterion), Ibuprofen, Nabumetone or Piroxicam Gel will be the preferred drug.

The vectorial programming problem (PV) can be solved using the algorithm from paragraph 13.1.1.

l_k	k	$f_1(a_k)$
1	3	1.10
2	2	10.30
3	6	15.78
4	5	20.02
5	13	26.90
6	1	27.45
7	8	31.55
8	14	35.55
9	7	37.48
10	10	38.56
11	4	44.84
12	16	44.87
13	12	45.63
14	11	50.70
15	9	61.85
16	15	66.74

Table 13.3.2. Sorted sequence ($f_1(a_k)$)

2_k	k	$f_2(a_k)$
1	3	0.1
2	2	0.8
3	14	1.9
4	7	2.0
5	10	4.0
6	8	7.5
7	5	8.8
8	6	12.0
9	1	12.9
10	13	14.7
11	11	15.3
12	4	17.3
13	9	17.6
14	16	26.5
15	15	28.0
16	12	36.7

Table 13.1.3. Sorted sequence ($f_2(a_k)$)

3_k	k	$f_3(a_k)$
1	1	200.031
2	2	217.098
3	3	223.748
4	4	295.430
5	5	327.372
6	6	343.380
7	7	378.922
8	8	388.459
9	9	405.288
10	10	408.091
11	11	433.906
12	12	457.389
13	13	484.315
14	14	582.776
15	15	651.145
16	16	656.836

Table 13.1.3. Sorted sequence ($f_3(a_k)$)

We obtain $E = \{a_2, a_1, a_3\}$. So the min-efficient drugs are also Nabumetone, Ibuprofen, and Piroxicam Gel.

13.4 Algorithm for description of a preference relation between drugs

The previous algorithm provides the user with a dictatorial solution, choosing only the best drugs, according to certain criteria. But, the diversity of patients requires a less restrictive point of view on drug quality, a preference relation between drugs, according to certain criteria, being more useful. To present in front of a doctor a preference relation instead of a dictatorial solution is an ethical decision for a researcher in pharmaco-economics. A preference relation between drugs lets an open window to doctor to choose the best treatment according to the particularities of the patient's body. The following algorithm also based on theorem 13.1.2 and corollaries 13.1.1 and 13.1.2 for the determination of all min-efficient elements of the set A, allows a description of the preference relation between more drugs that are used for curing the same disease. The algorithm is used to determine the sets:

$B(a_i)$ of the drugs which are better than the drug a_i,

$W(a_i)$ of the drugs which are worse than the drug a_i.

The drugs that are neither better nor worse than the drug a_i are not comparable from this point of view with this drug.

Algorithm 13.4.1. Let be $A = \{a_1,...,a_m\}$ and $K = \{1,...,m\}$.

Step 1. Put $E = \varnothing$ and for all $i \in \{1, 2, \ldots, p\}$ sort the numbers $f_i(a_k)$ for $k \in K$, as an increasing sequence. Denote by i_k the position of $f_i(a_k)$ in the sorted sequence.

Step 2. Take $t = 1$.

Step 3. Construct the set

$EE(t) = \{j \in \{1, 2, \ldots, m\} \mid f_i(a_t) = f_i(a_j), \text{ for all } i \in \{1, 2, \ldots, p\}\}$.

Step 4. Determine the number

$\alpha_t = \min \{j \in \{1, 2, \ldots, m\} \mid i_t \le j, \text{ for all } i \in \{1, 2, \ldots, p\}\}$.

Step 5. Construct the sets

$B(t) = \{k \in \{1, 2, \ldots, m\} \mid i_k \le \alpha_t, \text{ for all } i \in \{1, 2, \ldots, p\}\}$,

$W(t) = \{k \in \{1, 2, \ldots, m\} \mid i_k \ge \alpha_t, \text{ for all } i \in \{1, 2, \ldots, p\}\}$

and

$NC(t) = \{1, 2, \ldots, m\} \setminus (EE(t) \cup B(t) \cup W(t))$.

Step 6. $t := t + 1$. If $t \le m$, then go to Step 3.

Step 7. Put E = {t ∈ {1, 2, ..., m} | B (t) = ∅} and stop.

Algorithm 13.4.1 has the advantage of providing both the set of min-efficient points and the whole preference relation.

Now we apply algorithm 13.4.1 in order to determine the sets $B(a_t)$, of the drugs which are better than the drug a_t, and the set $W(a_t)$ of the drugs which are worse than the drug a_t. We also get E = {7, 11, 15} and the preference relation is described in table 13.4.1.

Drug no.	B (k)	W (k)
1	{2, 7, 8, 9, 11, 12, 13, 14, 15}	∅
2	{11, 15}	{1, 5, 6, 9, 10, 16}
3	{11, 15}	{6, 10}
4	{11, 12, 14, 15}	{6, 10}
5	{2, 7, 8, 9, 11, 12, 14, 15}	{10}
6	{2, 3, 4, 7, 8, 9, 11, 12, 13, 14, 15}	∅
7	∅	{1, 5, 6, 9, 10, 16}
8	{3, 11, 15}	{1, 5, 6, 10, 13, 16}
9	{7, 11, 15}	{1, 5, 6, 10}
10	{2, 3, 4, 5, 7, 8, 9, 11, 12, 13, 14, 15, 16}	∅
11	∅	{1, 2, 3, 4, 5, 6, 7, 8, 9, 10, 12, 13, 14, 16}
12	{11, 15}	{1, 4, 5, 6, 10, 16}
13	{8, 11, 15}	{1, 6, 10, 16}
14	{11, 15}	{1, 4, 5, 6, 10, 16}
15	∅	{1, 2, 3, 4, 5, 6, 8, 9, 10, 12, 13, 14, 16}
16	{2, 7, 8, 11, 12, 13, 14, 15}	{10}

Table 13.4.1. The preference relation

13.5 Multiple criteria programming used in medico-economic analysis of treatment protocols

The increasing number of cancer patients and the high cost of treatment protocols enforce a serious study of these protocols including the outcomes, the quality of life and their economic cost. For this analysis we construct an identifier which, in the sequel, we shall call medico-economic index (m.e.i). It is based on results of multiple criteria programming. The construction of m.e.i. is made using the Markov tree (decision tree) attached to the protocol, as in Luciana Lupşa (1999).

In this paragraph we consider two types of Markov tree, a simple one and another a little bit complicated. This types of trees result from real data. Using the constructed m.e.i., we shall perform an analysis of two real cases.

Case I.

The cancer patients, having the same disease, are treated with two protocols of treatment, let they be T_1 and T_2. By analysing the outcomes after a period of time, the same period for all patients, we establish the following percentages:

In case of protocol T_1

- p_v^1 % of the treated patients are still alive;
- p_d^1 % of the treated patients are dead;
- p_a^1 % of the treated patients are lost from records;

In case of protocol T_2

- p_v^2 % of the treated patients are still alive;
- p_d^2 % of the treated patients are dead;
- p_a^2 % of the treated patients are lost from records.

The cost/patient ratio being known, denoted by c_1 monetary units for the first protocol and c_2 monetary units for the second treatment, the problem is to establish which of the two protocols is to be applied.

If $p_v^1 \geq p_v^2$ and $c_1 \leq c_2$, then it is obvious that the first protocol deserves to be applied. The discussion appears when $p_v^1 \geq p_v^2$ and $c_1 > c_2$. For this real situation we use the results of the multiple criteria programming.

Let us introduce the variables x_1 and x_2, which can be only 0 and 1:

$x_1 := 1$ means that the first treatment is chosen;

$x_1 := 0$ means that the first is not chosen.

$x_2 := 1$ means that the second treatment is chosen;

$x_2 := 0$ means that the second is not chosen.

As a patient cannot have both treatments at the same time, we must obviously have $x_1 + x_2 = 1$.

The objective functions are defined by:

$f_1 : \{0, 1\}^2 \to \mathbf{R}$, $f_1(x_1, x_2) = p_v^1\ x_1 + p_v^2\ x_2$, for each $(x_1, x_2) \in \{0, 1\}^2$,

$f_2 : \{0, 1\}^2 \to \mathbf{R}$, $f_2(x_1,x_2) = p_d^1\ x_1 + p_d^2\ x_2$, for each $(x_1, x_2) \in \{0, 1\}^2$,

$f_3 : \{0, 1\}^2 \to \mathbf{R}$, $f_3(x_1, x_2) = \frac{c_1}{c} x_1 + \frac{c_2}{c} x_2$, for each $(x_1, x_2) \in \{0, 1\}^2$.

Our aim is to maximise function f_1, and minimise function f_2 and f_3. Since a function is to be maximised and another to be minimised we shall operate the following transformation: the function that is to be minimised will be replaced by functions that are to be maximised. This is allowed knowing very well that any minimum point of function g is a maximum point for function -g.

The mathematical model for the medical problem will consist in a multiple criteria optimisation problem with the objective function

$f: \{0, 1\}^2 \to \mathbf{R}^3$,

(13.5.1) $f(x_1, x_2) = (f_1(x), f_2(x), f_3(x))$, for each $x = (x_1, x_2) \in \{0, 1\}^2$,

and the restriction set

(13.5.2) $X = \{(x_1, x_2) \in \{0, 1\}^2 \mid x_1 + x_2 = 1\}$.

Using the pounds method with the real pounds λ_1, λ_2, λ_3, with $\lambda_1 > 0$, $\lambda_2 > 0$, $\lambda_3 > 0$, and $\lambda_1 + \lambda_2 + \lambda_3 = 1$, any solution of the problem

$$(P)\quad \begin{cases} g(x) \to \max \\ x \in X \end{cases},$$

where $g: \{0, 1\}^2 \to \mathbf{R}$, is given by

$g(x_1, x_2) = \lambda_1\ f_1(x_1, x_2) + \lambda_2\ f_2(x_1, x_2) + \lambda_3\ f_3(x_1, x_2)$, for each $(x_1,x_2) \in \{0,1\}^2$,

can be considered as a solution of the multiple criteria problem. The simplicity of restrictions allows us to consider that if

$$\lambda_1 \cdot p_v^1 - \lambda_2 \cdot p_d^1 - \lambda_3 \cdot \frac{c_1}{c} \geq \lambda_1 \cdot p_v^2 - \lambda_2 \cdot p_d^2 - \lambda_3 \cdot \frac{c_2}{c},$$

then $x_1 = 1$, $x_2 = 0$, is a solution. Conversely, the solution is $x_1 = 0$, $x_2 = 1$.

This mathematical result leads to the introduction of the m.e.i. for that protocols T_i, which have a Markov tree corresponding to case I, by

$$(13.5.3)\quad \text{m. e. i}(T_i) = \lambda_1\ p_v^i - \lambda_2 \cdot p_d^i - (1 - \lambda_1 - \lambda_2)\ \frac{c_i}{c}.$$

In (13.5.3), p_v^i is the percentage of the living patients, to whom the T_i protocol has been applied, p_d^i is the percentage of dead patients, c_i is the cost/patient ratio of protocol T_i, $c = \frac{\sum_{i=1}^{n} c_i}{n}$, λ_1 is the pound for survival, and λ_2 is the pound for the money spent.

Let's remark that if $p_a^1 = 0$, and $p_a^2 = 0$, then $p_d^1 = 1 - p_v^1$, and $p_d^2 = 1 - p_v^2$.

Therefore

$$\lambda_1\, p_v^i - \lambda_2\, p_d^i - \lambda_3 \frac{c_i}{c} = (\lambda_1 + \lambda_2)\, p_v^i - \lambda_3 \frac{c_i}{c} - \lambda_2, \text{ for } i \in \{1, 2\}.$$

Denoting by $\lambda = \lambda_1 + \lambda_2$, it results $\lambda_3 = 1 - \lambda$. Then the m.e.i is given by

$$\text{m. e. i}\,(T_i) = \lambda \cdot p_v^i + (\lambda - 1) \cdot \frac{c_i}{c}, \text{ where } c = \frac{\sum_{i=1}^{n} c_i}{n}.$$

Case II.

The most branched out (more data) the Markov tree, the most pertinent is the analysis.

Let's suppose that for a treatment protocol T we know that p_v % patients are cured and p_r % patients have recurrences. Among p_v % cured patients

- p_{vc} % have complications during treatment, the others have not;

- c_{vc} is the cost for the patients with complications;

- c_{vf} for the others.

Among p_r % of the patients who have recurrence, p_{rc} %, have complications during the treatment, the others, don't. The patients with recurrences follow one protocol of the two possible. From these:

- p_{rf1} % patients with recurrences presenting no complications during the first phase of protocol 1;

- p_{rf2} % patients with recurrences presenting no complications during the first phase of protocol 2;

- p_{rc1} % patients with recurrences presenting complications during the first phase of protocol 1;

- p_{rc2} % patients with recurrences presenting complications during the first phase of protocol 2.

We know the average cost/patient of the entire treatment:

- c_{rf1} for the patients with recurrences presenting no complications during the first phase of protocol 1;

- c_{rf2} for the patients with recurrences presenting no complications during the first phase of protocol 2;

- c_{rc1} for the patients with recurrences presenting complications during the first phase of protocol 1;

- c_{rc2} for the patients with recurrences presenting complications during the first phase of protocol 2.

Let us do the medico-economic analysis of the treatment protocols, with the initial protocols $T_1, ..., T_n$, given (knowing for each the above data). The purpose is to choose the protocol that maximizes the percentage of cured patients, minimizes the percentage of patients with complications, minimizes the ratio of the total amount spent for patients with complications the total amount spent for patients without complications and minimizes the total cost of the protocols.

For the two protocols T_1 and T_2 we shall construct a multiple criteria programming problem.

Let us introduce the variables

x_1, x_2, that can be only 0 and 1:

$x_1 := 1$, means that the first treatment is chosen;

$x_1 := 0$, means that the first treatment is not chosen.

$x_2 := 1$, means that the second treatment is chosen;

$x_2 := 0$, means that the second treatment is not chosen.

As a patient cannot have both treatments at the same time, we obviously have $x_1 + x_2 = 1$.

The objectives functions are:

f_1: $\{0, 1\}^2 \to \mathbf{R}$, $f_1(x_1, x_2) = p_v^1 \cdot x_1 + p_v^2 \cdot x_2$, for each $(x_1, x_2) \in \{0, 1\}^2$,

f_2: $\{0, 1\}^2 \to \mathbf{R}$, $f_2(x_1, x_2) = p_{vc}^1 \cdot x_1 + p_{vc}^2 \cdot x_2$, for each $(x_1, x_2) \in \{0, 1\}^2$,

f_3: $\{0, 1\}^2 \to \mathbf{R}$, $f_3(x_1, x_2) = (p_{rc1}^1 + p_{rc2}^1) \cdot x_1 + (p_{rc1}^2 + p_{rc2}^2) \cdot x_2$, for each $(x_1, x_2) \in \{0, 1\}^2$,

$f_4: \{0, 1\}^2 \to \mathbf{R}$, $f_4(x_1, x_2) = \frac{c_{vc}^1 \cdot p_{vc}^1}{c_{vf}^1 \cdot p_{vf}^1} \cdot x_1 + \frac{c_{vc}^2 \cdot p_{vc}^2}{c_{vf}^2 \cdot p_{vf}^2} \cdot x_2$, for each $(x_1, x_2) \in \{0, 1\}^2$,

$f_5: \{0, 1\}^2 \to \mathbf{R}$, $f_5(x_1, x_2) = (\frac{c_{rc1}^1 \cdot p_{rc1}^1}{c_{rf1}^1 \cdot p_{rf1}^1} + \frac{c_{rc2}^1 \cdot p_{rc2}^1}{c_{rf2}^1 \cdot p_{rf2}^1}) \cdot x_1 + (\frac{c_{rc1}^2 \cdot p_{rc1}^2}{c_{rf1}^2 \cdot p_{rf1}^2} + \frac{c_{rc2}^2 \cdot p_{rc2}^2}{c_{rf2}^2 \cdot p_{rf2}^2}) \cdot x_2$, for each $(x_1, x_2) \in \{0, 1\}^2$,

$f_6: \{0, 1\}^2 \to \mathbf{R}$, $f_6(x_1, x_2) = \frac{c_1}{c} x_1 + \frac{c_2}{c} x_2$, for each $(x_1, x_2) \in \{0, 1\}^2$,

where $c_i = c_{vf}^i \cdot p_{vf}^i + c_{vc}^i \cdot p_{vc}^i + c_{rf1}^i \cdot p_{rf1}^i + c_{rf2}^i \cdot p_{rf2}^i + c_{rc1}^i \cdot p_{rc1}^i + c_{rc2}^i \cdot p_{rc2}^i$, for $i \in \{1, 2\}$, and $c = \frac{c_1 + c_2}{2}$.

Our aim is to maximise function f_1, and to minimise functions f_2, f_3, f_4, f_5, and f_6. Since a function is to be maximised and other to be minimised we shall operate the following transformation: the function to be minimised will be replaced by the functions to be maximised. This is possible because any min point of function g is a max point of function -g. The mathematical model for the medical problem will consist of a multiple criteria optimisation problem with the objective function f

(15.5.4) $f: \{0, 1\}^2 \to R^6$, $f(x_1, x_2) = (f_1(x), f_2(x), f_3(x), f_4(x), f_5(x), f_6(x))$,

for all $x = (x_1, x_2) \in \{0, 1\}^2$, and the restriction set

(13.5.5) $X = \{(x_1, x_2) \in \{0, 1\}^2 \mid x_1 + x_2 = 1\}$.

For solving this multiple criteria programming problem we apply the pounds method. According to this method we choose the strictly positive pounds λ_1, λ_2, λ_3, λ_4, λ_5 and λ_6 with $\lambda_1 + \lambda_2 + \lambda_3 + \lambda_4 + \lambda_5 + \lambda_6 = 1$ and we construct the synthesis function $f: \{0, 1\}^2 \to \mathbf{R}$,

$f(x_1, x_2) = \lambda_1 \cdot f_1(x_1, x_2) - \lambda_2 \cdot f_2(x_1, x_2) - \lambda_3 \cdot f_3(x_1, x_2) - \lambda_4 \cdot f_4(x_1, x_2) - \lambda_5 \cdot f_5(x_1, x_2) - \lambda_6 \cdot f_6(x_1, x_2)$,

for each $x = (x_1, x_2) \in \{0, 1\}^2$.

Any optimal solution of the problem

$$(Q) \quad \begin{cases} f(x) \to \max \\ x \in X \end{cases}$$

will be an effective max point, so any solution of problem (Q) is an acceptable answer, if we notice that $x_1 + x_2 = 1$, implies that $x_2 = 1 - x_1$. Consequently the problem (Q) is reduced to solve the problem (QR), which consists in: determine the maximum point of the function $\psi : \{0, 1\} \to \mathbf{R}$,

$\psi(u) = \lambda_1 \cdot f_1(u, 1-u) - \lambda_2 \cdot f_2(u, 1-u) - \lambda_3 \cdot f_3(u, 1-u) - \lambda_4 \cdot f_4(u, 1-u) - \lambda_5 \cdot f_5(u, 1-u) - \lambda_6 \cdot f_6(u, 1-u)$,

for all $u \in \{0, 1\}$. The simplicity of restrictions allows that if

$$\lambda_1 \cdot p_v^1 - \lambda_2 \cdot p_{vc}^1 - \lambda_3 \cdot (p_{rc1}^1 + p_{rc2}^1) - \lambda_4 \cdot \frac{c_{vc}^1 \cdot p_{vc}^1}{c_{vf}^1 \cdot p_{vf}^1} - \lambda_5 \cdot \left(\frac{c_{rc1}^1 \cdot p_{rc1}^1}{c_{rf1}^1 \cdot p_{rf1}^1} + \frac{c_{rc2}^1 \cdot p_{rc2}^1}{c_{rf2}^1 \cdot p_{rf2}^1}\right) - \lambda_6 \cdot \frac{c_1}{c} \geq \lambda_1 \cdot p_v^2 - \lambda_2 \cdot p_{vc}^2 - \lambda_3 \cdot (p_{rc1}^2 + p_{rc2}^2) - \lambda_4 \cdot \frac{c_{vc}^2 \cdot p_{vc}^2}{c_{vf}^2 \cdot p_{vf}^2} - \lambda_5 \cdot \left(\frac{c_{rc1}^2 \cdot p_{rc1}^2}{c_{rf1}^2 \cdot p_{rf1}^2} + \frac{c_{rc2}^2 \cdot p_{rc2}^2}{c_{rf2}^2 \cdot p_{rf2}^2}\right) - \lambda_6 \cdot \frac{c_2}{c},$$

then $u = 1$ is a solution of the problem (QR). Conversely $u = 0$.

This mathematical result leads us to introduce the m.e.i. for a protocol T_i, which has a Markov tree corresponding to case II, defined by

$$(13.5.6)\quad \text{m. e. i}(T_i) = \lambda_1 \cdot p_v^i - \lambda_2 \cdot p_{vc}^i - \lambda_3 \cdot (p_{rc1}^i + p_{rc2}^i) - \lambda_4 \cdot \frac{c_{vc}^i \cdot p_{vc}^i}{c_{vf}^i \cdot p_{vf}^i} - \lambda_5 \cdot \left(\frac{c_{rc1}^i \cdot p_{rc1}^i}{c_{rf1}^i \cdot p_{rf1}^i} + \frac{c_{rc2}^i \cdot p_{rc2}^i}{c_{rf2}^i \cdot p_{rf2}^i}\right) - \lambda_6 \cdot \frac{c_i}{c}.$$

In (13.5.6)

p_v^i is the percentage of the patients which are cured by applying T_i protocol;

p_{vf}^i is the percentage of the patients which are cured without complication;

p_{vc}^i is the percentage of the cured patients which have complication during treatment T_i;

p_{rf1}^i is the percentage of the patients with recurrences presenting no complications during the first phase of protocol 1;

p_{rc1}^i is the percentage of the patients with recurrences presenting complications during the first phase of protocol 1;

p^{i}_{rf2} is the percentage of the patients with recurrences presenting no complications during the first phase of protocol 2;

p^{i}_{rc2} is the percentage of the patients with recurrences presenting complications during the first phase of protocol 2;

c_{vf} is the average cost/patient of the entire treatment for the patients cured without complications;

c_{vc} is the average cost/patient of the entire treatment for the patients cured with complications;

c_{rf1} is the average cost/patient of the entire treatment for the patients with recurrences presenting no complications during protocol 1;

c_{rc1} is the average cost/patient of the entire treatment for the patients with recurrences presenting complications during protocol 1;

c_{rf2} is the average cost/patient of the entire treatment for the patients with recurrences presenting no complications during protocol 2;

c_{rc2} is the average cost/patient of the entire treatment for the patients with recurrences presenting complications during protocol 2;

λ_1 is the survival pound given to cured patients;

λ_2 is the factor which signifies the importance given to complications of the cured patients;

λ_3 is the pound given to the complications of the patients with recurrences;

λ_4 is the pound given to ratio of the total amount spent for patients with complications / the total amount spent for patients without complications for cured patients;

λ_5 is the pound given to ratio of the total amount spent for patients with complications / the total amount spent for patients without complications for patients with recurrences;

λ_6 is the pound given to the all money spent for the protocol.

Remark 13.5.1. c is a number of cost type, which harmonises the amount of the total cost with the other data, that are less than 1.

Example 13.5.1. With the data presented in [Gregoire, F. G. Badía, and J. de la Cal. *Cost minimization analysis of treatment options for T1N0 glottic squamous cell carcinoma: comparison between external radiotherapy, laser microsurgery and partial laryngectomy. Radiotherapy and Oncology* 1999;

V, 53: 1-13], we calculate m.e.i. for protocols presented there for the glottic squamous cell carcinoma.

For this we change the presented tree with a tree of the second type. We obtain the following data. Considering radiotherapy (rt) with total laryngectomy (tl) as protocol 1 and radiotherapy (rt) with partial laryngectomy (pl) as protocol 2, for patients with recurrences, we have:

$p^{rt}_{vf} = 0.8590$ %, and $c^{rt}_{vf} = 4052$, euro/patient;

$p^{rt}_{vc} = 0.0110$ %, and $c^{rt}_{vc} = 12470$, euro/patient;

$p^{rt}_{rfpl} = 0.0440$ % and $c^{rt}_{rfpl} = 13662$, euro/patient;

$p^{rt}_{rftl} = 0.0850$ %, and $c^{rt}_{rftl} = 14366$, euro/patient;

$p^{rt}_{rcpl} = 0.0005$ %, and $c^{rt}_{rcpl} = 30560$, euro/patient;

$p^{rt}_{rctl} = 0.0005$ %, and $c^{rt}_{rctl} = 31270$, euro/patient.

Then the m.e.i. equals to $\text{m.e.i}(T_{rt}) = \lambda_1 \cdot 0.87 - \lambda_2 \cdot 0.011 - \lambda_3 \cdot 0.001 - \lambda_4 \cdot 0.04 - \lambda_5 \cdot 0.0382 - \lambda_6 \cdot \dfrac{5470.14}{c}$

For the patients with recurrences after partial laryngectomy (pl) we can choose between total laryngectomy (tl) and radiotherapy (rt).

We have

$p^{pl}_{vf} = 0.8310$ %, and $c^{pl}_{vf} = 9715$ euro/patient;

$p^{pl}_{vc} = 0.1010$ %, and $c^{pl}_{vc} = 23657$ euro/patient;

$p^{pl}_{rftl} = 0.0545$ %, and $c^{pl}_{rftl} = 20030$ euro/patient;

$p^{pl}_{rfrt} = 0.0067$ %, and $c^{pl}_{rfrt} = 13491$ euro/patient;

$p^{pl}_{rctl} = 0.0067$ %, and $c^{pl}_{rctl} = 27873$ euro/patient;

$p^{pl}_{rcrt} = 0.0001$ %, and $c^{pl}_{rcrt} = 25692$ euro/patient.

Then the m. e. i. equals to

$\text{m.e.i}(T_{pl}) = \lambda_1 \cdot 0.932 - \lambda_2 \cdot 0.1010 - \lambda_3 \cdot 0.0066 - \lambda_4 \cdot 0.296 - \lambda_5 \cdot 0.459 - \lambda_6 \cdot \dfrac{11848.8}{c}$

We take $c = \dfrac{5470.1 + 11848.8}{2} \cong 8660$.

The comparison analysis of the treatment from medico-economic and patient's quality of life point of view, is made by parameters $\lambda_1, \dots, \lambda_6$. Thus we can assign the same importance to complications from the points of view of patient's quality of life as well as the economic costs to both situation: patients cured with complications and patients with recurrences; as well $\lambda_2 = \lambda_3$ and $\lambda_4 = \lambda_5$.

If $\lambda_1 = 0.6$, $\lambda_2 = \lambda_3 = 0.05$, $\lambda_4 = \lambda_5 = 0.05$, and $\lambda_6 = 0.3$, then

$$\text{m. e. i.}\,(T_{pl}) < \text{m. e. i.}\,(T_{rt}),$$

so we consider that the protocol with radiotherapy applied in phase 1 to be more advantageous.

If $\lambda_1 = 0.9$, $\lambda_2 = \lambda_3 = 0.03$, $\lambda_4 = \lambda_5 = 0.02$, and $\lambda_6 = 0.05$,

then m. e. i. (T_{pl}) > m. e. i. (T_{rt}), so we consider that the protocol with partial laryngectomy applied in phase 1 to be more advantageous.

References

1. Aleman, A. On the some generalizations of convex sets and convex functions. Mathematica - Rev. d'Anal. Numér. et de Théor. de l'Approx., Ser. L'Anal. Numér. et la Théor. de l'Approx. 1985; 14: 1-6.
2. Aleksandrov, A.D. *Die innere Geometrie der konvexen Flächen.* Berlin, Akademie-Verlag 1955.
3. Altman, M. Bilinear programming. Bull. Acad. Polon. Sci. Math. Astronom. Phys. 1968; 16: 741-746.
4. Aracelli, C., Levialdi, S., Concavity extraction by parallel processing. IEEE Trans. Systems Man Cybernet. 1971; 1: 394-396.
5. Auslender, A., Coutat, P. Closed convex sets without boundary rays and asymptotes, Set-Valued Anal. 1994; 2:19-33.
6. Avriel, M., Diewert, W.E., Schaible, S., Zang, I. *Generalized Concavity.* New-York, Plenum 1988.
7. Baciu, A., Pascu, A., Puşcaş, E. *Applications of Operational Research.* (in Romanian) Bucharest, Editura Militară 1988.
8. Beck, A. A convexity condition in Banach spaces and the strong law of large numbers. Proc. Amer. Math. Soc. 1962; 13: 329-334.
9. Beer, G. The index of convexity and the visibility function. Pacific J. Math.. 1973; 44, 1: 59-67.
10. Beltz, S.E., Yee, G.E. Pharmacoeconomics of Cancer Therapy, Cancer Control: J.M.C.C.1998; 5(5): 415-424.
11. Bing, R.H. Tame Cantor Sets in $\mathbf{R}^3$. Pacific J. Math. 1961; 11: 435-476.
12. Birkhoff, G. *Lattice theory.* American Mathematical Society Colloquium Publications, vol. 25. New York: AMS, 1948.
13. Bisztriczky, T. Some examples in projective convexity. C.R. Math. Rep. Acad. Sci. Canada 1987; 9(4): 199-204.
14. Bisztriczky, T., Fejes Tóth, G. A generalization of the Erdös-Szekeres convex n-gon theorem. J. Reine Angew. Math. 1989; 395: 167-170.

15. Bisztriczky, T., Fejes Tóth, G. Nine convex sets determine a pentagon with convex sets as vertices. Geom. Dedicata 1989; 31: 89-104.

16. Blaga,L., Kolumbàn, I. "Optimization on closely convex sets." In *Generalized Convexity*. S. Komlósi, T. Rapcsák, S. Schaible eds. Lect. Notes in Economics and Math. Systems, vol. 405, Berlin et al.: Springer-Verlag 1994: 19-34.

17. Blaga, L., Lupşa, L. "Convexity space with respect to a given set." In *Generalized Convexity, Generalized Monotonicity*, J.P.Crouzeix, J.E.Martinez-Legaz, M.Vollè (Eds.) Dordrecht/Boston/London: Kluwer Academic Publishers, 1998: 199-208.

18. Blind, R., Valette, G., Zamfirescu, T. Rectangular convexity. Geom. Dedicata 1980; 9: 317-327.

19. Blumenthal L. M. *Theory and applications of distance geometry*. Oxford: Clarendon Press, 1953.

20. Blumenthal, L.M., Menger, K. *Studies in Geometry*. San Francisco: W.H. Freeman and Company, 1970.

21. Boltyanskii, V.G. Helly's theorem for H-convex sets (Russian). Soviet Math. Dokl. 1976; 17: 78-81; Dokl. Akad. Nauk SSSR 1976; 226: 249-252.

22. Bow, S.T. *Pattern Recognition and Image Preprocessing*. Marcel Dekker Inc., New York - Bassel - Hong Kong, 1992.

23. Boxer, L. Computing deviations from convexity in polygons. Pattern Recognition Letters 1993; 14: 163-167.

24. Breckner, W. W. *Introducere în teoria problemelor de optimizare convexă cu restricţii*. Partea I. Cluj, Ed. Dacia 1974.

25. Breckner, W. W., Kassay, G. A Systematization of Convexity Concepts for Sets and Functions, J. Convex. Anal. 1997; 4: 109-127.

26. Breen, M. Decomposition theorems for 3-convex subsets of the plane. Pacific J. Math. 1974; 53, 1: 43-57.

27. Breen, M. A decomposition Theorem for m-convex sets in $\mathbf{R}^d$. Canad. J. Math. 1976; XXVIII(5): 1051-1057.

28. Breen, M. m-convex sets whose ln c points lie in a hyperplane. J. London Math. Soc. 1977; (2), 16, 3: 529-535.

29. Brondsted, A. *An Introduction to Convex Polytopes*. Berlin, Springer Verlag, 1983.

30. Brunn, H. *Über Ovale und Eiflächen*. Inaugural dissertation. Munich 1887.

31. Brunn, H. ÜberKerneigebiete. Mathematische Annalen 1913; 73: 430-440.

32. Bryant, V.W., Webster, R.J.: Generalization of the theorems of Radon, Helly and Carathéodory, Monath. Math. 1969; 73 ,1: 309-315.

33. Bryant, V.W., Webster, R.J. Convexity Spaces I, The Basic Properties, J. Math. Anal. Appl. 1972; 37, 1: 206-213.

34. Bryant, V.W., Webster, R.J., Convexity Spaces II, Separation, J. Math. Anal. Appl., 1973; 43, 2: 321-327.

35. Bryant, V.W., Webster, R.J. Convexity Spaces III, Dimension, J. Math. Anal. Appl., 1977; 57, 2: 382-392.

36. Bryant, V. *Metric spaces, Iteration and application*. Cambridge University Press. 1994.

37. Calder, J.R. Some elementary properties of interval convexities. J. London Math. Soc. 1971; 3(2): 422-428.

38. Cantor, G. Uber die Ausdehnung eines Satzes aus der Theorie der trigonometrichen Reihen, Mathematische Annalen 1872, 5: 123-132.

39. Chagoya-Guzman, Al., Programmation bilineaire: une approche de résolution par rélaxation. Thèse, l'Université Scientifique et Médicale de Grénoble; 1980.

40. Chassery, J.M. Convexité et enveloppe convexe dans $\mathbf{Z}^2$. Séminaire Analyse Numérique, No. 297, IMAG, Grenoble (1978).

41. Chassery, J.M. Connectivity and consecutivity in digital pictures. Computer Graphics and Image Processing; 1979, 9: 294-300.

42. Chassery, J.M. Discrete convexity: definition, parametrization and comparability with continuous convexity. Computer Vision, Graphics and Image Processing; 1983, 21: 326-344.

43. Chassery, J.M. Représentation discrète, interprétation numérique et description des images: des concepts à l'applications, Thèse, Université Scientifique et Médicale de Grénoble; 1984.

44. Chassery, J.M., Garbay, C. An Iterative Segmentation Method Based on a Contextual Color and Shape Criterion, IEEE Trans. Pattern Anal. Mach. Intel. 1984; PAMI-6, 6: 794-800.

45. Chassery, J.M., Garbay, C., Cinquin, P. Segmentation d'images cytologiques: Etude de méthodes. 4ème congrès Reconnaîssance des Formes et Intelligence Artificielle, AFCET-INRIA; 1984: 51-71.

46. Chaudhuri, B.B. Some Shape definitions in fuzzy geometry of space. Pattern Recognition Letters 1991; 12: 531-535.

47. Chaudhuri, B.B. Fuzzy convex hull determination in 2D-space. Pattern Recognition Letters 1991; 12: 591-594.

48. Chaudhuri, B.B. Concave fuzzy set: a concept complementary to the convex fuzzy set. Pattern Recognition Letters 1992; 13: 103-106.

49. Cobzaş, Ş. Separation Theorems for convex sets in non-archimedean spaces. (Romanian) Revista de Analiză Numerică şi Teoria Aproximaţiei 1974; 3, 1: 61-70.

50. Cochand, M., Duchet, P. "Sous les pavés" Ann. Discrete Math. 1983; 17: 191-202.

51. Coppel, W. A. Axioms for convexity. Bull. Austral. Math. Soc. 1993; 47: 179-197.

52. Corson, H., Klee, V. L. "Topological classification of convex sets." In *Convexity*, Proc. Symp. in Pure Math. VII (American Mathematical Society, Providence, RI) 1963: 37-51.

53. Coutant, P., Volle, M., Martinez-Legaz, J.E. Convex functions with continuous epigraph or continuous level sets. J. Optim. Theory Appl. 1996; 88: 365-379.

54. Cristescu, G. Cones in a convexity space; Ordered Convexity Spaces. Révue d'Analyse Num. Th. Approx , Cluj-Napoca; 1980, 9, 2: 157-162.

55. Cristescu, G. Behaviours. In Proc. Sci. Com. Meeting of the Aurel Vlaicu University, Third Edition, Arad, 16-17 May 1996, Aurel Vlaicu University Publishing House, Arad, 14A (1996) 25-32.

56. Cristescu, G. Convexities with respect to a set and two behaviours. In Proc. Sci. Com. Meeting of the Aurel Vlaicu University, Third Edition, Arad, 16-17 May 1996, Aurel Vlaicu University Publishing House, Arad, 14A (1996) 35-43.

57. Cristescu, G. Convexity close by undetectable concavity. In Proc. Sci. Com. Meeting of the Aurel Vlaicu University, Third Edition, Arad, 16-17 May 1996, Aurel Vlaicu University Publishing House, Arad, 14A (1996) 45–50.

58. Cristescu, G. Sets looking convex – like. In 5th Symposium in Descriptive Geometry, Design, Engineering and Computer Graphics, "Graphics, a scientific language", 17-19th June 1996, Politehnica University Publishing House, Timişoara, III (1996) I, 725 – 730.

59. Cristescu, G. A geometrical characterization of the sets looking convex-like, Analele Universităţii Eftimie Murgu Reşiţa; 1996, I: 121-128.

60. Cristescu, G. Convexity and Concavity Coarseness of some Fractals. In Academia Tehnică Militară, A XXVII-a Sesiune de Comunicări Ştiinţifice cu Participare Internaţională, Secţiunea 18 Matematică şi Fizică, Subsecţiunea 1 Matematică, Bucureşti, 13-14 noiembrie 1997, Military Technical Academy Publishing House, Bucharest (1997) 100 –107.

61. Cristescu, G. Behaviours and Allure. In Proc. Sci. Com. Meeting of the Aurel Vlaicu University, Fourth Edition, Arad, 29-30 October 1997, Aurel Vlaicu University Publishing House, Arad, 1 (1997) 19-28.

62. Cristescu, G. On the Approximation of the classical Convexity. In Proc. Sci. Com. Meeting of the Aurel Vlaicu University, Fourth Edition, Arad, 29-30 October 1997, Aurel Vlaicu University Publishing House, Arad, 1 (1997) 39-46.

63. Cristescu, G. Convexity related to the Coarseness of the Concavity. Révue d'Analyse Num. Th. Approx; 1997, 26, 1-2: 29-37.

64. Cristescu, G. A new route to the Convexity related to the Concavity Coarseness. In Proc. Sci. Com. Meeting of the Aurel Vlaicu University, Fourth Edition, Arad, 29-30 October 1997, Aurel Vlaicu University Publishing House, Arad, 1 (1997) 29-38.

65. Cristescu, G. The Extension of the notions of behaviour and approximation to some particular domains of the research (Romanian). Thesis, Babeş-Bolyai University of Cluj-Napoca, Faculty of Mathematics and Computer Science, 1997.

66. Cristescu, G. Approximation of the classical convexity in totally bounded metric spaces. Bulletins for Applied & Computer Mathematics, Budapest; 1998, LXXXV-B/'98: 191-196.

67. Cristescu, G. On the concavity coarseness of the tame Cantor sets. Bulletins for Applied & Computer Mathematics, Budapest; 1998, LXXXVI-A/'98: 37-46.

68. Cristescu, G. Convexity spaces generated by means of behaviours. Bulletins for Applied & Computer Mathematics, Budapest; 1999, LXXXVIII/'99: 51-58.

69. Cristescu, G. Weak convergence of some convexities to the classical convexity. In Academia Tehnică Militară, A XXVIII-a Sesiune de Comunicări Ştiinţifice cu Participare Internaţională organizată cu ocazia Sărbătoririi a 50 de Ani de la Înfiinţarea Academiei Tehnice Militare, Secţiunea Matematică şi Fizică, Subsecţiunea 1 Matematică, Bucureşti, 21-22 Octombrie 1999, Military Technical Academy Publishing House, Bucharest 1999: 86-91.

70. Cristescu, G. On some approximate convexities. Proc. Eighth Symp. Math. Appl., November 4-7 1999, Timişoara, Politehnica University of Timişoara Publishing House, 1999: 47-54.

71. Cristescu, G. *Behaviours: Implications and Applications in Convexity Recognition and in Topology*, Monographical Booklets in Applied and Computer Mathematics, MB-14, Technical University of Budapest, Budapest, 2000.

72. Cristescu, G. Topological properties of α-convex sets, In E. Popoviciu et al. (Eds.): "Tiberiu Popoviciu" Itinerant Seminar of Functional Equations, Approximation and Convexity, Cluj-Napoca 23-29 May 2000, SRIMA, Cluj-Napoca (2000) 65-68.

73. Cristescu, G. E-Convexity of superior order of functionals on a metric space. Bulletins for Applied & Computer Mathematics, Budapest 2000; BAM – 1775/2000 XCII: 65-72.

74. Cristescu, G., Lupşa, L. On a Class of Convexity Properties applied in Mathematical Programming. Bulletins for Applied & Computer Mathematics, Budapest 2001; BAM – 1792/2001 XCIV: 151-160.

75. Cristescu, G. Generalized convexity and a weak approximation problem. Bulletins for Applied & Computer Mathematics, Budapest 2001; BAM/2001 XCV-C: (to appear).

76. Cristescu, G. Sequences of convexities converging to the classical convexity. Révue d'Analyse Num. Th. Approx., Cluj-Napoca, (to appear).

77. Cristescu, G., Lupşa, L. Classes of (a, e) - ((S, s), R) convexities. Bulletins for Applied & Computer Mathematics, Budapest; 2001, XCVI/'01: (to appear).

78. Cristescu, G., Lupşa, L. Three classes of generalised convexity properties. Proc. Ninth Symp. Math. Appl., November 2-4, 2001, Timişoara, Politehnica University of Timişoara Publishing House (to appear).

79. Cristescu, G., Lupşa, L., Generalised convex sets. In Academia Tehnică Militară, A XXIX-a Sesiune de Comunicări Ştiinţifice cu Participare Internaţională "Tehnologii Moderne în Secolul XXI", Secţiunea Matematică şi Fizică, Subsecţiunea 1 Matematică, Bucureşti, 15-16 Noiembrie 2001, Military Technical Academy Publishing House, Bucharest 2001: 89-94.

80. Cristescu, G., Generalised convex functions and related convex programming. In Academia Tehnică Militară, A XXIX-a Sesiune de Comunicări Ştiinţifice cu Participare Internaţională "Tehnologii Moderne în Secolul XXI", Secţiunea Matematică şi Fizică, Subsecţiunea 1 Matematică, Bucureşti, 15-16 Noiembrie 2001, Military Technical Academy Publishing House, Bucharest 2001: 85-88.

81. Crouzeix, J.P., Ferland, J., Zalinescu, C. α - convex sets and strong quasiconvexity. Mathematics of Operations Research 1997; 22, 4: 998-1022.

82. Danzer, L., Grunbaum, B. Intersection properties of boxes in $\mathbf{R}^d$. Combinatorica 1982; 2: 237-246.

83. Danzer, L., Grunbaum, B., Klee, V. Helly's theorem and its relatives. Proceedings of Symposia in Pure Mathematics, VII, Convexity, 1963:101-177.

84. Das, P.P., Chakrabarti, P.P., Chatterji, B.N. Distance functions in digital geometry. Inform.Sci. 1987; 42: 113-136.

85. Drandell, M. Generalized convex sets in the plane, Duke Math. J. 1952; 19: 537-547.

86. Duca, D., Duca, E., Lupşa, L., Blaga, R. E – Convex Function. Bulletins for Applied & Computer Mathematics, Budapest 2000; BAM - 1804/2000 XCII: 95-101.

87. Duca D., Lupşa, L. (g, h ,M) convex sets. The problem of the best approximation L'Analyse Num. et la Th. de l'Approx (to appear in RGMIA 2001; 4).

88. Duca, D., Duca, E., Lupşa, L., Blaga, R. (φ, ψ) - Convex Sets, Communicated at "Zilele Academice Clujene", Cluj-Napoca, 14th-16th of June 2000 (to appear).

89. Duchet, P. Convexity in combinatorial structures. Rend. Circ. Math. Palermo (2), Suppl. 1987; 14: 261-293.

90. Edelman, P.H., Jamison, R.E. The theory of convex geometries. Geom. Dedicata 1985; 19: 247-270.

91. Edelman, P.H. Abstract convexity and meet-distributive lattices. Contemp. Math. 1986; 57.

92. Eggleston, H.D. *Convexity*. Cambridge: Cambridge Univ. Press, 1959.

93. Evens, J. J. M., van Maaren, H. Duality principles in mathematics and their relations to conjugate functions. Nieuw Arch. Wisk. 1985; 3: 23-68.

94. Fan, K. On Krein-Milman theorem. Proceedings of Symposia in Pure Mathematics, vol. 7, American Mathematical Society, Providence, 1963: 211-220.

95. Fenchel, W. A remark on convex sets and polarity. Comm. Sém. Math. Univ. Lund (Medd. Lunds Univ. Math. Sem.), tome supplém. 1952: 82-89.

96. Forstnerič, F., Stout, E. L. A new class of polynomially convex sets. Ark. Mat. 1991; 29(1):51-62.

97. Franchetti, C., Singer, I. Best approximation by elements of caverns in normed linear spaces. Boll. Unione Mat. Ital., Serie V 1980; 17B:33-43.

98. Franklin, S. P. Some results on order convexity. Amer. Math. Month. 1962; 69(5): 357-359.

99. Freeman, H., Shapira, R. Determining the minimum - area encasing rectangle for an arbitrary closed curve. Commun. ACM 1975; 18: 409-413.

100. Fuks, B. A. *An Introduction to the theory of Analytic Functions of Several Complex Variables* (Russian). Moscow: Fizmatgiz, 1962.

101. Gaafar, M. Convexity verification, block chords and digital straight lines. Comp. Graphics Image Proc. 1977; 5: 360-370.

102. Gale, D., Klee., V. Continuous convex sets. Math. Scand. 1959; 7: 379-391.

103. Gallo, G., Ulkucu, A. Bilinear Programming: an exact algorithm. Math. Programming 1976; 12: 173-194.

104. Galperin, E.A. Nonscalarized Multiobjective Global Optimization, J.O.T.A. 1992; 75,1: 69-85.

105. Ghika, A. Ensembles A-convexes dans les A-modules. Com. Acad. R.P. Române 1952; 2: 669-671.

106. Ghika, A. Séparation des ensembles convexes dans les espaces lignés non vectoriels. Bul. Şti. Acad. R.P. Române, Secţ. Şti. Mat. Fiz. 1955; 7: 287-296.

107. Giles, J. R. *Convex analysis with application in differentiation of convex functions*. Research Notes in Mathematics, vol. 58 Boston: Pitman 1982.

108. Glivenko, V. Géometrie des systemes de choses normées. Am. Jour. 1936; 58:799-828.

109. Glucksman, H. A. A propagation pattern classifier. IEEE Trans. Electron. Comput. 1965; 14: 434-443.

110. Graham, R.L. An efficient algorithm for determining the convex hull of a finite planar set. Inform. Proc. Letters 1972; 1:132-133.

111. Green, J. W., Gustin, W. Quasiconvex sets. Canad. J. Math. 1950; 2:489-507.

112. de Groot, J., de Vries, H. Convex sets in projective space. Compositio Math. 1958; 13: 113-118.

113. Gruber, P.M., Willis, J.M. (eds.) *Handbook of Convex Geometry,* A,B, North-Holland, 1993.

114. Gudder, S., Schroeck, F. *Generalized Convexity*. SIAM J. Math. Anal. 1980; 11: 984-1001.

115. Harary, F., Nieminen, J. Convexity in Graphs. J. Differ. Geom. 1981; 16: 185-190.

116. Hausdorff, F. *Teoria mnojestv*. Gostehizdat, Moskva, 1937.

117. Hausdorff, F. *Mengenlehre*. W. Gruyter & Co., Berlin, 1927.

118. van Heijenoort, J. On locally convex manifolds, Comm. Pure Appl. Math. 1952; 5: 223-242.

119. Held, A., Abe, K. On approximate convexity. Pattern Recognition Letters. 1994; 15: 611-618.

120. Hicks, N.J. *Notes on Differential Geometry*. Princeton, New Jersey: Van Nostrand Publishing Company, 1965.

121. Horn, A., Valentine, F. A. Some properties of L sets in the plane. Duke Math. J. 1949; 16: 131-140.

122. Hőrmander, L. *An introduction to Complex Analysis in Several Variables* (third ed.). Amsterdam: North Holland, 1989.

123. Hőrmander, L. *Notions of Convexity*. Progress in Mathematics, vol. 127. Basel: Birkhäuser, 1994.

124. Hu, Y.D., Sun, E.J. Connectedness of the Efficient Set in strictly Quasiconcave Vector Maximization. J. Optim. Theory Appl. 1993; 78: 613-622.

125. Iacob, M.E. Convexity, Approximation and Optimization on Networks, Thesis, Babeş-Bolyai University of Cluj-Napoca, (1998).

126. Ivan, M. Interpolation methods and their applications (in Romanian). Thesis, University of Cluj-Napoca, 1982.

127. Jamison - Waldner, R. E. A perspective on abstract convexity: Classifying alignments by varieties, In *Convexity and Related Combinatorial Geometry*, D. C. Kay and M. Breen, eds. Lecture Notes in Pure and Applied Mathematics, 76: New York: Marcel Dekker, 1982: 113-150.

128. Janos, L., Rosenfeld, A. Some results on fuzzy (digital) convexity. Pattern Recognition 1982; 15, 5: 379-382.

129. Jensen, J. L. W. V. Om konvexe Funktioner og Uligheder mellem Middelvaerdier. Nyt. Tidsskr. Math. B 1905; 16: 49-69.

130. Jensen, J. L. W. V. Sur les fonctions convexes et les inegalités entre leurs valeurs moyennes. Acta Math. 1906; 30: 175-193.

131. Jeyakumar, V., Gwinner, J. Inequality systems and optimization. J. Math. Anal. Appl. 1991; 159:51-71.

132. Kay, D.C., Womble, E.W. Axiomatic convexity theory and relationships between the Caratheodory, Helly and Radon numbers. Pacific J. Math. 1971; 38: 471-485.

133. Keimel, K., Wieczorek, A. Kakutani property of the polytopes implies Kakutani property of the whole space. J. Math. Anal. Appl. 1988; 130: 97-109.

134. Kim, C.E. A linear time convex hull algorithm for simple polygons. Dept. Computer Sc. Univ. Maryland, TR 956, 1980.

135. Kim, C.E., Rosenfeld, A. On the Convexity of Digital Regions, Proc. 5th. JCPR, (1980) 1010-1015.

136. Kim, C.E. On the cellular convexity of complexes. IEEE Trans. Pattern Anal. Mach. Intel., 1981; PAMI-3: 617-625.

137. Kim, C.E., Sklansky, J. Digital and Cellular Convexity. Pattern Recognition. 1982; 15, 5: 359-367.

138. Klee, V. Convexity of Chebyshev sets. Math. Ann. 1961; 142: 292-304.

139. Klette, R. On the approximation of convex hulls of finite grid point sets. Pattern Recognition Letters 1983; 2: 19-22.

140. von Koch, H. Sur une courbe continue sans tangente obtenue par une construction géometrique élémentaire. Arkiv for Mathematik, Astronomi och Fysik, 1904; 1: 681-704.

141. Kolmogorov, A.N. On some asymptotic characteristics of completely bounded metric spaces. Dokl. Akad. Nauk SSSR, 1956; 108, 3: 385-388.

142. Kuzmin, V. B., Ovchinnikov, S. V. A geometry of preference spaces I (in Russian), Avtomatika i Telemekhanika,1975; 12: 140-145.

143. Lassak, M. On metric b-convexity for which diameters of any set and its hull are equal. Bull. Acad. Polon. Sci. Sér. Sci. Math. Astronom. Phys. 1977; 25: 969-975.

144. Latecki, L., Rosenfeld, A., Silverman, R. Generalized convexity: CP_3 and boundaries of convex sets. Pattern Recognition, 1995; 28(8): 1911-99.

145. Lawrence, J.F., Hare, W.R. Jr., Kenelly, J.W. Finite unions of convex sets. Proc. Am. Math. Soc., 1972; 34(1):225-228.

146. van Lierop, M.L.P., Digitization Functions in Computer Graphics, Ph. D. Thesis, Eindhowen University of Technology, 1987.

147. Levi, F. W. On Helly's theorem and the axioms of convexity. J. Indian Math. Soc., (N.S.), Part A, 1951; 15:65-76.

148. Lowen, R. Convex fuzzy sets, Fuzzy Sets Syst. 1980; 3: 291-310.

149. Lupşa, L. Asupra alurii unor functii hiperbolice. Studia Univ. "Babeş-Bolyai", Mathematica 1978; 2: 66-72.

150. Lupşa, L. On the structure of the essentially efficient solutions of an integer vectorial programming problem (Romanian). In Seminarul itinerant de ecuaţii funcţionale, aproximare şi convexitate, 1980, 16-17 mai; Cluj-Napoca. Cluj-Napoca: Babeş-Bolyai University Publishing House, 1980:77-89.

151. Lupşa, L. On a convexity notion (Romanian). In Seminarul itinerant de ecuaţii funcţionale, aproximare şi convexitate, 1980, 7-8 noiembrie; Timişoara. Timişoara: University of Timişoara Publishing House, 1980:127-135.

152. Lupşa, L. On the strongly convex sets with respect to a given set (Romanian). In Seminarul itinerant de ecuaţii funcţionale, aproximare şi convexitate, 1981, mai; Cluj-Napoca. Cluj-Napoca: Babeş-Bolyai University Publishing House, 1981: 217-227.

153. Lupşa, L., Particular types of linear and nonlinear programming problems (Romanian). Thesis, Babeş-Bolyai University of Cluj-Napoca, 1982.

154. Lupşa, L. On strong convex mappings with respect to a given set (Romanian). In Seminarul itinerant de ecuaţii funcţionale, aproximare şi convexitate, 1982, 20-22 octombrie; Cluj-Napoca. Cluj-Napoca: Babeş-Bolyai University Publishing House, 1982: 181-185.

155. Lupşa, L. Separation of strong convex sets. In Itinerant Seminar on functional equations, approximation and convexity, 1983, Cluj-Napoca. Cluj-Napoca: Babeş-Bolyai University Publishing House, 1983: 85-86.

156. Lupşa, L. Optimum properties of strong convex functions with respect to a given set. In Theodor Angheluta Seminar, 1983, Cluj-Napoca. Cluj-Napoca: Politechnical Institute Publishing House, 1983: 147-152.

157. Lupşa, L. Duality theorems for integer programming problems. In Itinerant Seminar on functional equations, approximation and convexity, 1984, Cluj-Napoca. Cluj-Napoca: Babeş-Bolyai University Publishing House, 1984: 99-102.

158. Lupşa, L. On quasi-convex integer programming problems. In Proceedings of the Colloquium on approximation and optimisation. 1984, 25 – 27 October, Cluj-Napoca: Babeş-Bolyai University Publishing House, 1984: 261 - 268.

159. Lupşa L. On Quasiconvex integer programming problems (II). In Babeş-Bolyai University of Cluj-Napoca, Seminar on Optimization Theory, 1985; Report no. 5: 63-74.

160. Lupşa, L. Slack convexity with respect to a given set. In Itinerant Seminar on functional equations, approximation and convexity, 1985, Cluj-Napoca. Cluj-Napoca: Babeş-Bolyai University Publishing House, 1985: 107-114.

161. Lupşa, L. I-points with respect to a given set. L'Analyse Num. et la Th. de l'Approx., 1986; 15, 2: 145-147.

162. Lupşa, L. An integer theorem of Farkas Lemma type. In "Itinerant Seminar on functional equations, approximation and convexity, 1985, Cluj-Napoca". Cluj-Napoca: Babeş-Bolyai University Publishing House, 1986: 147-150.

163. Lupşa L. Some types of convexity with respect to a given set. In Babeş-Bolyai University of Cluj-Napoca, Seminar on Optimization Theory, 1986; Report no. 8: 89-114.

164. Lupşa, L. Remarks concerning the solvability of integer programming problems. In Itinerant Seminar on functional equations, approximation and convexity, 1987, Cluj-Napoca. Cluj-Napoca: Babeş-Bolyai University Publishing House, 1987: 201-204.

165. Lupşa, L. Duality theorems for rational programming problems. L'Analyse Num. et la Th. de l'Approx., 1987; 16, 1: 51-53.

166. Lupşa, L. On integer bilinear programming problems. "Babeş-Bolyai" University, Cluj-Napoca, Seminar on Optimization theory, Report no. 8, 1987: 51-64.

167. Lupşa, L. Remarks concerning the solvability of integer programming problems. In Itinerant Seminar on functional equations, approximation and convexity, 1987. Pre-print nr. 6, 1987: 201 –204.

168. Lupşa, L. Linear integer inequalities and theorem of the alternative. In: Babeş-Bolyai University, Cluj-Napoca, Seminar on Optimization theory, Report no. 8, 1988, pag. 67 – 80. Publishing House, 1988: 205-208.

169. Lupşa, L. On integer convex sets. I. In Itinerant Seminar on functional equations, approximation and convexity, 1988, Cluj-Napoca. Cluj-Napoca: Babeş-Bolyai University Publishing House, 1988: 205-208.

170. Lupşa, L. On integer convex sets. II. In Itinerant Seminar on functional equations, approximation and convexity, 1989, Cluj-Napoca. Cluj-Napoca: Babeş-Bolyai University Publishing House, 1989: 211-216.

171. Lupşa, L. Linear integer and linear rational inequalities. L'Analyse Num. et la Th. de l'Approx., 1990; 19, 1: 49-52.

172. Lupşa, L. Theorem of the alternative for the linear non-homogeneous integer systems (I). In Babes-Bolyai University, Faculty of Mathematics, Research Seminars, Itinerant Seminar on functional equations, approximation and convexity, Preprint nr. 6, Cluj-Napoca: Babeş-Bolyai University Publishing House, 1990: 181-184.

173. Lupşa, L. On the relationship between efficient points and d-bases. In Babes-Bolyai University, Faculty of Mathematics, Research Seminars, Seminar of Mathematical Analysis. Preprint nr. 7, Cluj-Napoca: Babeş-Bolyai University Publishing House, 1992: 87-100.

174. Lupşa, L., Duca E., Duca D. I., On the structure of the set of points dominated and nondominated in a vectorial optimization problem (I). In Babes-Bolyai University, Faculty of Mathematics, Science Research Seminars, Preprint nr. 6, 1993.

175. Lupşa L. Induced convexity (in Romanian). In Seminarul itinerant "Tiberiu Popoviciu" de ecuaţii funcţionale, aproximare şi convexitate, 16 mai – 21 mai 1995; Cluj-Napoca. Cluj-Napoca: Babeş-Bolyai University Publishing House 1995; 61-64.

176. Lupşa L. On a generalization of the divided differences. In Seminarul itinerant de ecuaţii funcţionale, aproximare şi convexitate, 21 mai – 26 mai 1997; Cluj-Napoca. Cluj-Napoca: Babeş-Bolyai University Publishing House 1997; 51-54.

177. Lupşa, L., Blaga, L.R. Duca, D., Duca, E. "Some types of (g,h,M)-convexity," In *Séminaire de la théorie de la meilleure approximation, convexité et optimisation*, E. Popoviciu, M. Ivan, L. Lupşa, Şt. Ţigan, eds. Cluj-Napoca: SRIMA, 2000: 161-170.

178. Lupşa, L. ((S, s), $(\Lambda_i)_{i\in I}$, M, r) - Convexity, Analele Universităţii Aurel Vlaicu din Arad, Arad: Aurel Vlaicu University Publishing House, 2000: 98-103.

179. Lupşa, L., Cristescu, G. ((S, s), $(\Lambda_i)_{i\in I}$, M, r) convexity at a point and ((S, s), $(\Lambda_i)_{i\in I}$, M, r) starlike sets. Buletinul Ştiinţific al Universităţii din Baia Mare, Seria B 2000; XVI,1: 59-64.

180. Lupşa, L., Cristescu G. About a general notion of convexity. In E. Popoviciu et all. *Proceedings of the Tiberiu Popoviciu Itinerant Seminar on Functional Equations, Approximation and Convexity.* Cluj-Napoca: SRIMA, 2001: 81-88.

181. Lupşa, L., Cristescu G. Topology with respect to a given set. In *Seminar of Best Approximation Theory, Convexity and Optimisation*, E. Popoviciu, M. Ivan, L. Lupşa, Şt. Ţigan, eds. Cluj-Napoca: SRIMA, 2001.

182. Lupşa, L., Cristescu, G. Induced convexity and the problem of the best induced approximation. L'Analyse Num. et la Th. de l'Approx. 2001; 30, 1: 55-59.

183. Lupşa, Luciana. A Criterion for Caracterizing a Medical Treatment that Uses Multicriteria Programming in Pharmacoeconomics, in *Analysis, Functional Equations, Approximation and Convexity*, Cluj-Napoca, Editura Carpatica 1999; 142-146.

184. Lupşa, Luciana. Multicriteria Programming Used in Medico-Economic Analysis of Treatment Protocols, in *Proceedings of the "Tiberiu Popoviciu" Itinerant Seminar of Functional Equations, Approximation and Convexity.* Cluj-Napoca, Editura SRIMA, 2000: 103-111.

185. Lupşa, Luciana. Use of dynamic programming for the supply of a pharmacy. *Proceedings of the "Tiberiu Popoviciu" Itinerant Seminar of Functional Equations, Approximation and Convexity.* Cluj-Napoca, Editura SRIMA, 2001: 237-239.

186. Maes, M. Digitization of Straight Line Segments, Closeness and Convexity. Computer Vision, Graphics and Image Processing, 1990; 52: 297-305.

187. Mandelbrot, B. *Les objets fractals. Forme, hasard et dimension*, Flammarion, Paris, 1975.

188. Mangasarian, O.L. *Nonlinear Programming.* New York, McGraw-Hill Book Co 1968.

189. Martinez-Legaz, J. E. Un concepto generalizado de conjugación. Applicación a las funciones quasiconvexas. Thesis, Univ. of Barcelona 1981.

190. Martinez-Legaz, J. E. "A generalized concept of conjugation." In *Optimization: Theory and Algorithms*, J. B. Hiriart-Urruty, W. Oettli, J. Stoer, eds. Lecture Notes in Pure and Applied Mathematics, vol. 86. New York: Marcel Dekker, 1983:45-49.

191. Martinez-Legaz, J.E., Singer, I. "Multi-order convexity" In Martinez-Legaz, J. E., Singer, I. Two Papers on Order and Convexity, Preprint Series in Mathematics, INCREST, Bucureşti, 6 (1990); printed in DIMACS Series in Discrete Mathematics and Theoretical Computer Science 1991; 4:471-488.

192. Martos, B. *Nonlinear Programming. Theory and Methods*, Budapest, Akademiai Kiado 1975.

193. Marusciac, I. On bilinear programming. L'Analyse Numerique et la Theorie de l'Approximation 1978; 7: 67-79.

194. Menger, K. Untersuchungen über allgemeine Metrik, Math. Ann., 1928; 100: 75-163.

195. Meyer, W., Kay, D.C. A convexity structure admits but one real linearization of dimension greater then one. J. London Math. Soc. 1973; 7: 124-130.

196. Minkowski, H. *Geometrie der Zahlen.* Leipzig and Berlin 1910.

197. Minkowski, H. *Theorie der konvexen Körper, inbesondere Begründung ihres Oberflächenbegriffs.* Gesammelte Abhandlungen II, Leipzig 1911.

198. Minsky, M., Papert, S. *Perceptrons.* Cambridge, Mass.: MIT Press 1969.

199. Mitrinović, D. S., Lacković, I.B., Stanković, M.S. Addenda to the monograph „Analytic inequalities" II. Univ. Beograd. Publik. Elektrotechn. Fak., Ser. Mat. Fiz. 1979; 634-677: 3-24.

200. Moldovan, E. On a generalization of the notion of convexity (Romanian). Stud. Cerc. Sti., 1955; II, 6, 3-4: 65-73.

201. Moldovan, E., On the notion of convex function with respect to an interpolating set of functions (Romanian). Stud. Cerc. Mat. 1958; 9: 161-224.

202. Moldovan, E. Sur une généralisation des fonctions convexes, Mathematica, Cluj, 1959; 1, 24: 49-80.

203. Molho, E. "On a notion of quasiconcave set and some applications to vector optimization". In Castagnoli, E. and Giorgi, G. eds. Scalar and Vector optimization in economic and financial problems. 1995: 113-118.

204. Molho, E., Zaffaroni, A. "Quasiconcavity of sets and connectedness of the efficient frontier in ordered vector spaces." In *Generalized Convexity, Generalized Monotonicity*. Crouzeix, J.P., Martinez-Legaz, J.E., Volle, M. eds. Dordrecht/Boston/London: Kluwer Academic Publishers, 1998: 407-424.

205. Monna, A.F. Séparation d'ensembles convexes dans un espace linéaire topologique sur un corps valué. Proc. Kon. Ned. Akad. v. Wetensch. 1964; 67: 399-421.

206. Monna, A.F. *Analyse non-archimédienne*, Springer Verlag, Berlin, 1967.

207. Moreau, J.J. *Functionnelles convexes*. Séminaire "Equations aux dérivées partielles", Collége de France, Paris 1966-67, no 2.

208. Moreau, J.J. Inf-convolution, sous-additivité, convexité des fonctions numériques. J. Math. Pures Appl. 1970; 9:109-154.

209. Motzkin, T.S. *Linear inequalities*. Mimeographed lecture notes. University of California, Los Angeles, California, 1951.

210. Mot, G., Generalized Convexity in Metric Spaces. In Proc. Sci. Com. Meeting of the Aurel Vlaicu University, Fourth Edition; 1997 October 29-30; Arad. Arad: Aurel Vlaicu University Publishing House, 1997; 1:83-92.

211. Neumann, J. von. On complete topological spaces. Trans. Amer. Math. Soc. 1935; 37: 1-20.

212. Neumann, J. von, Morgernstern, O. *Theory of Games and Economic Behaviour*. Princeton Univ. Press, Princeton NJ. 1944.

213. Nishida, H. Curve description based on directional features and quasi-convexity/ concavity. Pattern Recognition. 1995; 28, 7: 1045-1051.

214. Ovchinnikov S. V. Convexity in subsets of lattices. Stochastica, 1980; IV(2):129-140.

215. Paeck S. Convexlike and concavelike conditions in alternative, minimization theorems. J. Optim. Theory Appl. 1992; 74:317-332.

216. Pavlidis, T. *Structural Pattern Recognition*. Springer, New York, 1977.

217. Perkal, J. Sur les ensembles ε convexes. Colloq. Math. 1956; IV(1):1-10.

218. Pompeiu, D. Sur la continuité des fonctions de variables complexes, Thèses présentées à la Faculté des Sciences de Paris pour obtenir le grade Docteur en Sciences Mathématiques, Paris, Gautier-Villars, 1905.

219. Ponstein, J. Seven kinds of convexity. SIAM Review, 1967; 9: 115-119.

220. Popoviciu, E. Sur la notion de convexité par rapport à un procédé d'interpolation, ISMN, 1969; 10: 321-327.

221. Popoviciu, E. *Mean Theorems in Mathematical Analysis and their Connection with the Interpolation Theory* (Romanian). Dacia Publishing House, Cluj, 1972.

222. Popoviciu, E. Sur certaines proprietes des fonctions quasi-convexes (I). L'Analyse Num. et la Th. de l'Approx., 1983; 12, 2: 175-180.

223. Popoviciu, E. Sur certaines allures rémarquables. In Seminarul itinerant "Tiberiu Popoviciu" de ecuaţii funcţionale, aproximare şi convexitate, 21 mai – 25 mai 1996; Cluj-Napoca. Cluj-Napoca: Babes-Bolyai University Publishing House 1996: 129-134.

224. Popoviciu, E., Essai concernant une analyse quantitative bernsteinienne, In Seminarul itinerant "Tiberiu Popoviciu" de ecuaţii functionale, aproximare si convexitate, 21 mai – 26 mai 1997; Cluj-Napoca. Cluj-Napoca: Babes-Bolyai University Publishing House 1997: 93-98.

225. Popoviciu, T. *On the best approximation of the continuous functions by polynomials* (Romanian). Monografii matematice, Sec. Mat. a Univ. din Cluj, III, 1937.

226. Popoviciu, T. *Les fonctions convexes*. Actualités Scientifiques et Industrielles, 992, XVII, Paris, 1945.

227. Precup, R. Sur l'axiomatique des espaces à convexitè, L' Analyse Num. et la Th. de l' Apprrox. 1980; 9, 1:113-123.

228. Precup, R. *Properties of allure and some applications* (Romanian). Thesis, University of Cluj-Napoca, Faculty of Mathematics, 1985.

229. Prenowitz, W. A contemporary approach to classical geometry. Amer. Math. Monthly 1961; 68:1-67.

230. Prenowitz, W.A., Jantosciak, J. *Join geometries: a Theory of Convex Sets and Linear Geometry*, New York, Springer, 1979.

231. Pumplin, D. The Hahn-Banach theorem for totally convex spaces. Demonstratio Math. 1985; 18: 567-588.

232. Rapcsák T. Geodesic Convexity in Nonlinear Optimization. J. Optim. Theory Appl. 1991; 69:169-183.

233. Rawlins, G. J. E., Wood, D. "Ortho-convexity and its generalizations." In *Computational Morphology*, G. T. Toussaint, ed. Amsterdam: Elsevier, 1988: 137-152.

234. Robinson, C.V. Spherical theorems of Helly type and congruence indices of spherical caps. Amer. J. Math. 1942; 64: 260-272.

235. Rodé, G. Superkonvexe Analysis. Arch. Math, 1980; 34: 452-462.

236. Rosenfeld, A., Pfaltz, J.L. Distance Functions on Digital Pictures. Pattern Recognition, Pergamon Press, 1968; 1: 33-61.

237. Rosenfeld, A. *Picture Processing by Computer.* Academic Press, New York, London, 1969.

238. Rosenfeld, A. Connectivity in Digital Pictures. J. ACM, 1970; 17, 1: 146 – 160.

239. Rosenfeld, A. Arcs and curves in Digital Pictures. Journal of the Association for Computing Machinery, 1973; 20, 1: 81-87.

240. Rosenfeld, A. Adjacency in Digital Pictures. Information and Control, 1974; 26, 1: 119-127.

241. Rosenfeld, A. Digital Straight Line Segments. IEEE Trans. on Computers, 1974; December: 1264-1269.

242. Rosenfeld, A., Kak, A. *Digital Picture Processing*, Academic Press INC., San Diego, 1976.

243. Rosenfeld, A., Davis, L. Image Segmentation and Image Models. Proceedings of the IEEE; 1979, 67, 5: 764-772.

244. Rosenfeld, A., Kak, A. *Digital Picture Processing*, Vol.1-2, Academic Press, San Diego, 1982.

245. Rosenfeld, A. The diameter of a fuzzy set. Fuzzy Sets and Systems 1984; 13:241-246.

246. Rosenfeld, A., Haber, S. The perimeter of a fuzzy set. Pattern Recognition 1985; 18, 2: 125-130.

247. Rosenfeld, A. Measuring the sizes of concavities. Pattern Recognition Letters, 1985; 3: 71-75.

248. Rosenfeld, A. A note on average distance in digital sets. Pattern Recognition Letters, 1987; 5: 281-283.

249. Rosenfeld, A. Fuzzy rectangles. Pattern Recognition Letters 1990; 11: 677-679.

250. Rubinov, A.M., Glover, B.M. "Quasiconvexity via two step functions". In *Generalized Convexity, Generalized Monotonicity: Recent Results,* Crouzeix, J.P., Martinez-Legaz, J.E., Volle, M. eds. Dordrecht/Boston/London: Kluwer Academic Publishers, 1998: 159-183.

251. Rubinov, A.M., Glover, B., Jeyakumar, V. A general approach to dual characterization of solvability of inequality systems with applications, J. Convex Anal. 1995; 2: 309-344.

252. Schaible, S., Ziemba, W.T. (eds.) *Generalized Convexity in Optimization and Economics*. New York, Academic Press 1981.

253. Schneider, R. A measure of convexity for compact sets. Pacific J. Math. 1975; 58, 2: 617-625.

254. Scrader, J. Eine Verallgemeinerung der Fenchelkonjugation und Untersuchung ihrer Invariaten: verallgemeinerte konvexe Funktionen, Dualitäts - und Saltelpunktsätze. Thesis, University of Bonn 1975.

255. Shreve, J.L., Van Den Bos, J., Pedlow, K., The Use Pharmacoeconomic Research by MCOs: An Actuary's View, Drug Benefit Trends 2000; 12(2): 45-54.

256. Siegel, C.L. Integralfreie Variationsrechnung. Nachr. Akad. Wiss. Göttingen, Math.-Phys. Kl. 1957: 81-86; *Ges. Abh. III,* Berlin, Springer, 1966: 264-269.

257. Simon, J.C., Haralick, R.M. (eds.) *Digital Image Processing,* D. Reidel Publishing Company, 1981.

258. Simonnard, M. *Programmation linéaire*. Paris, Dunod 1962.

259. Singer, I. Surrogate conjugate functionals and surrogate convexity. Applicable Anal. 1983; 16:291-327.

260. Singer, I. "Generalized convexity, functional hulls and applications to conjugate duality in optimization. "In *Selected Topics in Operational Research and Mathematical Economics*. G. Hammer and D. Pallaschke, eds., Lecture Notes in Econ. and Math. Syst. 226, Berlin: Springer Verlag, 1984: 49-79.

261. Singer I., Some relations between dualities, polarities, coupling functionals and conjugations. J. Math. Anal. Appl. 1986; 115:1-22.

262. Singer, I. Infimal generators and dualities between complete lattices. Ann. Math. Pure Appl. 1987; 148(4):289-358.

263. Singer, I. Some further duality theorems for optimization problems with reverse convex constraint sets. J. Math. Anal. Appl. 1992; 171:205-209.

264. Singer, I., *Abstract Convex Analysis*. Canadian Mathematical Society Series of Monographs and Advanced Texts. New York: A. Wiley - Interscience Publication, 1997.

265. Sklansky, J. Recognition of convex blobs. Pattern Recognition, 1970; 2: 3-10.

266. Sklansky, J. Measuring concavity on a regular mosaic. IEEE Trans. Comput., 1972; C21: 1355 –1364.

267. Sklansky, J., Cordella, L.P., Levialdi, S. Parallel detection of concavities in cellular blobs. IEEE Trans. Comput. 1976; 25: 187-196.

268. Sokirjanskaia, E.N. Zamecenija k stat'e M. Al'tmana "Bilineinoe programmirovanie". Optimizacija 1975; 16(33): 91-98.

269. Soltan, V. P. *Introduction to the Axiomatic Theory of Convexity* (in Russian), Chişinău: Ştiinţa, 1984.

270. Stamley, W. J., Marr, J. M. Unions of two convex sets. Can. J. Math. 1963; 15:152-156.

271. Stern, H.I. Polygonal entropy: a convexity measure. Pattern Recognition Letters, 1989; 10: 229-235.

272. Stoer, J., Witzgall, C. *Convexity and optimization in finite dimensions/* I. Berlin – Heidelberg – New York, Springer Verlag 1970.

273. Stolzenberg, G. Polynomially convex sets. Bull. Am. Math. Soc. 1962; 68: 382-387.

274. Stolzenberg, G. Polynomially and rationally convex sets. Acta. Math. 1963; 109: 259-289.

275. Stone, M. Postulates for a barycentric calculus. Ann. of Math. 1949; 29: 25-30.

276. Taha, H. A. *Integer programming. Theory, Application, and Computations*. New – York Dan Francisco London, Academic Press 1975.

277. Takahashi, W. A convexity in metric space and non expansive mappings, I. Kodai Math. Sem. Rep. 1970; 22: 142-149.

278. Tietze, H. Über Konvexheit im kleinen und im grossen und über gewisse den Punkten einer Menge Zugeordnete Dimensionszahlen. Math. Z. 1929; 28:697-707.

279. Toader, Gh. Some generalizations of the convexity. Proc. Colloq. on Approximation and Optimization (University of Cluj-Napoca, 1984), University of Cluj-Napoca Publishing House, Cluj-Napoca 1985: 329-338.

280. Toader, Gh. The representation of n-convex sequences. L'Analyse Num. et la Th. de l'Approx., 1981; 10, 1: 113-118.

281. Valentine, F.A. Some properties of C-convex sets, Canad. J. Math. 1950; 2:481-488.

282. Valentine, F. A. Arcwise convex sets. Proc. Amer. Math. Soc. 1951; 2:150-165.

283. Valentine, F. A. A characterisation of simply connected closed arcwise convex sets. Proc. Amer. Math. Soc. 1951; 2:778-780.

284. Valentine, F.A. Three point arcwise convexity. Proc. Amer. Math. Soc. 1955; 6:671-674.

285. Valentine, F. A. A three point convexity property. Pacific J. Math. 1957; 7:1227-1235.

286. Valentine, F. A. *Convex sets*. New York: McGraw-Hill Book Co., Inc., 1964.

287. Valentine, F. A. Local convexity and L_n sets. Proc. Amer. Math. Soc. 1965; 16, 6: 1305-10.

288. Valentine, F. A. Local convexity and starshaped sets. Israel J. Math. 1965; 3(1): 39-42.

289. Vandal, A. Bilinear programming. Ekonomska Analiza 1970; 4: 21-41.

290. Voiculescu, D. Spaces with convexity (I) (Romanian). St. Cerc. Mat. 1967; 19, 2: 295-301.

291. Voiculescu, D. Spaces with convexity (II) (Romanian). St. Cerc. Mat. 1967; 19, 2: 303-311.

292. Webster, R. *Convexity*. Oxford University Press, 1994;

293. Wieczorek, A. Spot functions and peripherals: Krein-Milman type theorems in abstract setting. J. Math. Anal. Appl. 1989; 138: 293-310.

294. Wieczorek, A. The Kakutani property and the fixed point property of topological spaces with abstract convexity. J. Math. Anal. Appl. 1992; 168: 483-499.

295. Yosida, K. *Functional Anlysis*. Springer Verlag, Berlin - Gottingen - Heidelberg, 1965.

296. Yu, P. L. A class of Solutions for Group Decision Problems. Management Science 1974; 19: 936-46.

297. Yu, P. L. Cone Convexity, Cone Extreme Points and Nondominated Solutions in Decision Problems with Multiobjectives. J. Optim. Theory Appl. 1974; 14(3): 319-377.

298. Yu, P. L. *Multiple Criteria Decision Making*. New York and London, Plenum Press 1985.

299. Youness, E. A. E-Convex sets, E-Convex Functions, and E-Convex Programming, J. Optim. Theory Appl. 1999; 102(2): 439-450.

300. Zelinskii, J. B. Various generalisations of the concept of convexity, In *Metric Questions in the Theory of Functions* (in Russian). 158; Kiev: Naukova Dumka, 1980: 38-42.

301. * *Pharmaceutical Research and Manufacturers of America: Methodological and Conduct Principles for Pharmacoeconomic Research*, Drug Benefit Trends 1996; 34 *(2): 28-32.

Author Index

Subject index

Figures index

Tables index

Applied Optimization

36. G. Di Pillo and F. Giannessi (eds.): *Nonlinear Optimization and Related Topics*. 2000 ISBN 0-7923-6109-1
37. V. Tsurkov: *Hierarchical Optimization and Mathematical Physics*. 2000 ISBN 0-7923-6175-X
38. C. Zopounidis and M. Doumpos: *Intelligent Decision Aiding Systems Based on Multiple Criteria for Financial Engineering*. 2000 ISBN 0-7923-6273-X
39. X. Yang, A.I. Mees, M. Fisher and L.Jennings (eds.): *Progress in Optimization*. Contributions from Australasia. 2000 ISBN 0-7923-6286-1
40. D. Butnariu and A.N. Iusem: *Totally Convex Functions for Fixed Points Computation and Infinite Dimensional Optimization*. 2000 ISBN 0-7923-6287-X
41. J. Mockus: *A Set of Examples of Global and Discrete Optimization*. Applications of Bayesian Heuristic Approach. 2000 ISBN 0-7923-6359-0
42. H. Neunzert and A.H. Siddiqi: *Topics in Industrial Mathematics*. Case Studies and Related Mathematical Methods. 2000 ISBN 0-7923-6417-1
43. K. Kogan and E. Khmelnitsky: *Scheduling: Control-Based Theory and Polynomial-Time Algorithms*. 2000 ISBN 0-7923-6486-4
44. E. Triantaphyllou: *Multi-Criteria Decision Making Methods*. A Comparative Study. 2000 ISBN 0-7923-6607-7
45. S.H. Zanakis, G. Doukidis and C. Zopounidis (eds.): *Decision Making: Recent Developments and Worldwide Applications*. 2000 ISBN 0-7923-6621-2
46. G.E. Stavroulakis: *Inverse and Crack Identification Problems in Engineering Mechanics*. 2000 ISBN 0-7923-6690-5
47. A. Rubinov and B. Glover (eds.): *Optimization and Related Topics*. 2001 ISBN 0-7923-6732-4
48. M. Pursula and J. Niittymäki (eds.): *Mathematical Methods on Optimization in Transportation Systems*. 2000 ISBN 0-7923-6774-X
49. E. Cascetta: *Transportation Systems Engineering: Theory and Methods*. 2001 ISBN 0-7923-6792-8
50. M.C. Ferris, O.L. Mangasarian and J.-S. Pang (eds.): *Complementarity: Applications, Algorithms and Extensions*. 2001 ISBN 0-7923-6816-9
51. V. Tsurkov: *Large-scale Optimization – Problems and Methods*. 2001 ISBN 0-7923-6817-7
52. X. Yang, K.L. Teo and L. Caccetta (eds.): *Optimization Methods and Applications*. 2001 ISBN 0-7923-6866-5
53. S.M. Stefanov: *Separable Programming Theory and Methods*. 2001 ISBN 0-7923-6882-7

54. S.P. Uryasev and P.M. Pardalos (eds.): *Stochastic Optimization: Algorithms and Applications.* 2001 ISBN 0-7923-6951-3
55. J. Gil-Aluja (ed.): *Handbook of Management under Uncertainty.* 2001 ISBN 0-7923-7025-2
56. B.-N. Vo, A. Cantoni and K.L. Teo: *Filter Design with Time Domain Mask Constraints: Theory and Applications.* 2001 ISBN 0-7923-7138-0
57. S. Zlobec: *Stable Parametric Programming.* 2001 ISBN 0-7923-7139-9
58. M.G. Nicholls, S. Clarke and B. Lehaney (eds.): *Mixed-Mode Modelling: Mixing Methodologies for Organisational Intervention.* 2001 ISBN 0-7923-7151-8
59. F. Giannessi, P.M. Pardalos and T. Rapcsák (eds.): *Optimization Theory. Recent Developments from Mátraháza.* 2001 ISBN 1-4020-0009-X
60. K.M. Hangos, R. Lakner and M. Gerzson: *Intelligent Control Systems.* An Introduction with Examples. 2001 ISBN 1-4020-0134-7
61. D. Gstach: *Estimating Output-Specific Efficiencies.* 2002 ISBN 1-4020-0483-4
62. J. Geunes, P.M. Pardalos and H.E. Romeijn (eds.): *Supply Chain Management: Models, Applications, and Research Directions.* 2002 ISBN 1-4020-0487-7
63. M. Gendreau and P. Marcotte (eds.): *Transportation and Network Analysis: Current Trends.* Miscellanea in Honor of Michael Florian. 2002 ISBN 1-4020-0488-5
64. M. Patriksson and M. Labbé (eds.): *Transportation Planning.* State of the Art. 2002 ISBN 1-4020-0546-6
65. E. de Klerk: *Aspects of Semidefinite Programming.* Interior Point Algorithms and Selected Applications. 2002 ISBN 1-4020-0547-4
66. R. Murphey and P.M. Pardalos (eds.): *Cooperative Control and Optimization.* 2002 ISBN 1-4020-0549-0
67. R. Corrêa, I. Dutra, M. Fiallos and F. Gomes (eds.): *Models for Parallel and Distributed Computation.* Theory, Algorithmic Techniques and Applications. 2002 ISBN 1-4020-0623-3
68. G. Cristescu and L. Lupşa: *Non-Connected Convexities and Applications.* 2002 ISBN 1-4020-0624-1

KLUWER ACADEMIC PUBLISHERS – DORDRECHT / BOSTON / LONDON

MIX
Papier aus verantwortungsvollen Quellen
Paper from responsible sources
FSC® C105338

If you have any concerns about our products,
you can contact us on
ProductSafety@springernature.com

In case Publisher is established outside the EU,
the EU authorized representative is:
Springer Nature Customer Service Center GmbH
Europaplatz 3, 69115 Heidelberg, Germany

Printed by Libri Plureos GmbH
in Hamburg, Germany